D1629979
THE LESLIE PRICE
MEMORIAL LIBRARY
FIELD HOUSE
BENE
DICTUS
ES
TUA
STATUTA
ME
O DOMINE DOCE
BRADFIELD COLLEGE

THE FINAL CALL

BY THE SAME AUTHOR

Non-fiction

Aircrash Detective
Double Cross
Fire
Sex Slavery
That Thin Red Line

Fiction

Blockbuster
Crash Course
Cuban Confetti
The Ruling Passion
The Price of Silence
Tsunami

THE FINAL CALL

AIR DISASTERS . . .
WHEN WILL THEY EVER LEARN?

BY

- STEPHEN BARLAY -

Author of *Aircrash Detective*

SINCLAIR-STEVENSON

First published in Great Britain by
Sinclair-Stevenson Limited
7/8 Kendrick Mews
London SW7 3HG, England

The details in this book have been checked and found to be correct at the time of writing. However, certain problems may well by now have been ironed out, due to improvements in the aviation industry.

British Library Cataloguing in Publication Data
Barlay, Stephen
The final call: air disasters: when will they ever learn?
1. Aircraft. Accidents. Investigation
I. Title
363.12465

ISBN 1-85619-001-3

Photoset by Rowland Phototypesetting Limited,
Bury St. Edmunds, Suffolk
Printed and bound in Great Britain by
Richard Clay Limited, Bungay, Suffolk

CONTENTS

To Mari and Laci

ACKNOWLEDGMENTS

People on four continents to whom I am greatly indebted for help, advice and guidance are far too numerous to list here individually – besides, it could be unfair to those who had to remain anonymous for a variety of reasons. This is to express, however, my gratitude to at least some of the organizations which rendered invaluable assistance to this work by giving access to their highly sensitive records, and permitting time-consuming interviews with more than 300 individuals on their staff:

Air Accident Investigation Branch
Air Canada
Air Safety Group
American Air Line Pilots' Association
Boeing
British Air Line Pilots' Association
British Airways
British Caledonian Flight Training Centre
CAADRAP
Canadian Air Line Pilots' Association
Canadian Aviation Safety Board (reconstituted as Canadian Transportation Accident Investigation and Safety Board)
Civil Aviation Authority
Cranfield Institute of Technology
Federal Aviation Agency
Flight Safety Foundation
Flugunfalluntersuchungsstelle, Luftfahrt-Bundesamt
Frere Cholmeley
Institute of Aviation Medicine
Institute of Pathology and Tropical Medicine

International Air Transport Association
International Civil Aviation Organization
International Society of Air Safety Investigators
Japan Air Lines
Japanese Aircraft Accident Investigation Commission
Joint Airmiss Section
Joint Airworthiness Registration
Lufthansa
National Transportation Safety Board
QUALITAIR
Royal Air Force Directorate of Flight Safety
SCI-SAFE
Speiser, Krause & Madole
Singapore Airlines
Swissair
U.S. Department of Transportation
Willis, Faber & Dumas

My special thanks are due to Drucella Andersen for organizing and coordinating a vast number of meetings with NTSB personnel; Dick Stafford of the FAA Office of Public Affairs; Mrs. K. A. Corrie of ISASI; John Silver of BA; A. W. L. Nayler, ever helpful librarian of the Royal Aeronautical Society, and Robin Barlay for his help with the arduous task of compiling the Index.

The illustrations on pages 180 and 181 are reproduced with the generous permission of Malév and Lufthansa.

The photographs in the picture section are reproduced by kind permission of Associated Press, with the exception of both pictures on page 2 (Canadian ASB), the bottom picture on page 3 (by courtesy of Lufthansa. © Photograph, East African Standard Newspapers), top picture and bottom left on page 4 (by courtesy of Eric Newton), bottom right picture on page 6 (AIB), top left picture on page 8 (Canadian ASB).

GLOSSARY

of more frequently used abbreviations

AAIB	Air Accident Investigation Branch (U.K.)
AD	Airworthiness Directive
AGARD	Advisory Group for Aerospace Research and Development (NATO)
ALPA	Air Line Pilots Association (U.S.)
APU	Auxiliary Power Unit
ATA	Air Transport Association
ATC	Air Traffic Control
BALPA	British Air Line Pilots Association
CAA	Civil Aviation Authority (U.K.)
CASB	Canadian Aviation Safety Board
CAT	Clear Air Turbulence
CHIRP	Confidential Human Factors Incident Reports
CVR	Cockpit Voice Recorder
DFDR	Digital Flight Data Recorder
EROPS	Extended Range (twin-engine) Operations
FAA	Federal Aviation Administration (U.S.)
FSF	Flight Safety Foundation
GPWS	Ground Proximity Warning System
IATA	International Air Transport Association
ICAO	International Civil Aviation Organization
IFALPA	International Federation of Air Line Pilots' Associations
IFR	Instrument Flight Rules
ILS	Instrument Landing System

ISASI	International Society of Air Safety Investigators
JAA	Joint Aviation Authorities of Europe
McDD	McDonnell-Douglas
MEL	Minimum Equipment List
NASA	National Aeronautics and Space Administration (U.S.)
NTSB	National Transportation Safety Board (U.S.)
PAPI	Precision Approach Path Indicator
QFE	Barometric pressure setting of altimetre so that on landing aircraft altitude will reveal zero
QNH	Barometric pressure setting of altimetre so that on landing aircraft altitude will read equal to elevation of airport above mean sea level
RT	Radio Telephony
SB	Service Bulletin
TCAS	Traffic Collision Avoidance Systems
VASI	Visual Approach Slope Indicator
VFR	Visual Flight Rules

There are no new types of aircrashes – only people with short memories. Every accident has its own forerunners, and every one happens either because somebody did not know where to draw the vital dividing line between the unforeseen and the unforeseeable or because well-meaning people deemed the risk acceptable.

The aim of accident/incident investigation is to identify the causes of mishaps, and make the final call to prevent any repetition. Unfortunately, it is human nature to ignore the inconvenient, and forget lessons. Collective knowledge, embodied in government, ought to be a safeguard against that. It is not.

If politics is the art of the possible, and flying is the art of the seemingly impossible, then air safety must be the art of the economically viable. At a time of crowded skies and sharpening competition, it is a daunting task not to let the art of the acceptable deteriorate into the dodgers' art of what you can get away with.

Stephen Barlay,
March 1990

– 1 –

THEY HAD NO RIGHT TO LIVE

Take that 747 cruising peacefully at 41,000 feet above California. Ten hours out of Taipei, tedium was the mood of its 274 passengers and crew. Riding one of the best-tried workhorses of the skies, they felt safe, didn't they?

'Had the Lord meant us to fly, he'd have given us wings,' shrugged the top-notch engineer, 'but all in all, we ain't doing too badly.' He spoke with the pride of the executive who had helped to create generation after generation of safe and popular aircraft which turned flying into a joy, a simple necessity or even a non-event for most of us. We catch the 8.20 shuttle up country as routinely as we hop on a bus going down town; flying to some godforsaken corner of the globe is less uncomfortably titillating than hitching a ride to Dallas with Wells Fargo or booking a bunk on the old Orient Express might once have been; we may even subscribe to the view that fares not fears deter the non-fliers. And yet, and yet, when we trawl the pits of our stomachs where the knots grow, many of us haul in a premonition of doom, ill-founded, inexplicable, but disturbing nevertheless, whenever the nose lifts and the earth begins to tilt.

We are wrong, of course. Volumes of facts and dazzling figures tell us that air transport is amazingly safe. It kills some 1,100 a year (almost 1,700 in 1989), but we *know* that 3,000,000 people fly safely to their destination every day; and we *know*, statistically speaking, that a baby born aboard an airliner, and staying airborne incessantly, would not be involved in a fatal accident for eighty-two years.

In the Wings Club lecture, 1986, Joseph F. Sutter, Boeing's Executive Vice President, named 'an obsession with safety' one of the great virtues of aviation, an obsession that has served us well over the years. That is why we have faith in air transport (well, in most aircraft, most airports and most airlines in most countries' airspace, anyway) – and entrust our lives to the pilots and regulators, as well as the 5,000,000 spare parts and 140 miles of wire that must fly in close formation to make up a Jumbo. But we also know that any one of those might fail in a small way,* that the 'amazing level of safety' may carry its own damnation, the temptation of complacency, and that whenever the system goes *pop*, statistics – showing no improvement in the 1980s – will not guarantee our right to survive.

Let us pull just three cautionary tales out of thin air, so to speak, and board that China Airlines 747 we left cruising so peacefully on February 19, 1985.

The long journey had been uneventful until it reached some unmarked point in the sky, 300 nautical miles away from San Francisco, where the sun was shining atop a layer of clouds. As it was to be an eleven-hour nonstop flight, the three-man crew of the Jumbo had been augmented by an additional captain and engineer. They had another hour to go. Although they had traversed several time zones that would affect the crew's circadian rhythm (at home the pilots would be fast asleep just then), nobody in the cockpit felt tired. Taking advantage of the relief crew, Captain Min–Yuan Ho had had some five hours' rest though the quality of his two hours' actual sleep in the bunk could never be fully assessed.

Fortunately, or unfortunately, the crew had not much to do beyond watching the instruments, updating the information required by the computers, and just letting the auto-pilot get on with flying the aircraft. Magnificent automation thus relegated the

* The vulnerability of even a huge, welltried aircraft was illustrated in 1988: what turned out to be the most expensive cup of coffee in history was spilled on the central pedestal in the cockpit of a 747, causing two Far East flight cancellations plus repairs at the cost of U.S. $300,000.

pilots to the role of monitors, a tedious function to which people are least suited.

At about 10.11 Pacific Standard Time, the 747 suffered an inflight upset. The crew attributed it to some clear air turbulence.

Airspeed began to fluctuate. When it had dropped, the auto-pilot tried to restore the programmed speed of Mach 0.85 by moving the throttles forward. The flight engineer noted that the No. 4 engine failed to respond properly. He duly interfered manually with the automated system. Still he had no response. Because of the buffeting, the captain switched on the 'fasten seatbelts' sign. He allowed the auto-pilot to continue flying the aircraft, and concentrated on watching the engineer who would soon report that engine No. 4 had 'flamed out'. At that point, the relief engineer in his bunk heard some 'tapping sound' through the aircraft structure, and opened the curtains to inquire what was wrong. He was told to come forward and give his colleague a hand.

The checklist for routine flying on three engines was then produced. Requesting a lower altitude from Air Traffic Control, the captain commanded that attempts to relight the recalcitrant engine should be made although they were still well above the flight level prescribed for this procedure. All their attempts failed. The aircraft decelerated menacingly, and then began to roll. Captain Min–Yuan Ho grew totally preoccupied with the airspeed indicator – to the apparent exclusion of the instrument horizon that showed considerable banking. His main concern was to gather more speed and avoid stalling. He put the nose down using solely the auto-pilot control wheel.

Disastrous news came from the engineer: the remaining three engines had also failed.

The auto-pilot obeyed the captain, and pointed the aircraft towards the ground. The relief captain emerged from his bunk, but was thrown to the floor as the Jumbo dived. Mighty G-forces immobilized him. The engineer could not lift his arms any more, his head was pinned to the central control pedestal.

By now the aircraft was out of control, diving, rolling left and right. Inside the featureless grey goo of the clouds, with no horizon to be seen, the captain became spatially disoriented, and lost track of the Jumbo's altitude.

With 440 thousand pounds of metal and flesh hurtling out of

the sky, the altimeters went wild and made small change out of the precious capital of height and safety. The captain could do nothing about the rolling of the aircraft. He needed the first officer's assistance to pull back the yoke, in the hope against hope that they could stop the dive, raise the nose, force the aircraft to decelerate without disintegrating.

The pilots did not hear the aural overspeed warning, did not feel the stall-warning stickshaker at any stage of this ordeal.

The 747 broke out of the clouds at 11,000 feet, and the captain, aided at last by visual references, began to see which way up he was. He slowly regained control, and stabilized the machine. By the time he succeeded, the 747 had dived 32,000 feet. The engines now restarted smoothly, and a shaky crew reported to Oakland ATC that they had experienced 'a flame out, ah, we . . . emergency . . . we are niner thousand feet . . .'

Climbing again to safer heights, and trying to accomplish some *ad hoc* damage assessment, the crew redeclared emergency at 10.38, diverted to San Francisco, and landed safely. A flight attendant and one passenger only suffered serious injuries. The cabin erupted in cheers for the crew. A passenger was quoted in *The Times*: 'I really thought that was it.' Somebody added: 'It sounded like the engines stopped. It was dead silent for a few seconds before' the aircraft dived, levelled out, and dived again. Those who had no seatbelts on flew against the ceiling, 'popping up like popcorn'.

The aircraft was a mess. Massive aerodynamic forces had bent the wings upwards, cracked, broken and twisted many key parts of the structure and flying controls.

The National Transportation Safety Board was alerted, and a Go Team to investigate was assigned, as usual, right away. There was no time to waste. Hundreds of 747s in service could be doomed.

The NTSB investigators dug deep into the incident – and the cheers for the crew began to die away.

Although the No. 4 engine had two previous 'occurrence write-ups' in the log for special attention by maintenance, now there was nothing wrong with it. The investigation concluded that it had not 'flamed out' but must have been in a 'hung' condition due to the way it had been operated. The flight engineer's recollection of events was not accurate. Had he followed the flight manual and properly prescribed procedures, he could have dealt with the problem.

The other engines did not fail either.

Incorrect diagnosis about the engines need not have precipitated an emergency. The captain could still have averted potential disaster. But how? And what exactly did happen? The Cockpit Voice Recorder (CVR) and the Digital Flight Data Recorder (DFDR), aided and double-checked by other information, produced some endless columns of data. But, even to Dennis Grossi, the NTSB recorder specialist, the 'sequence of events was not quite apparent'.

Perching in a typical cubby-hole allocated to NTSB specialists (the luxury of floor space even for one's feet among piles of documents and stacks of equipment does not seem to be a prerequisite of good work), he recalled with fine understatement: 'We knew it must have been a very dynamic situation on the flight deck, and from the data it appeared that the aircraft had actually *rolled over*, but the pilots had no recollection of that. In fact, their own account of their actions could not be coordinated with what we had from the read-out.' The available facts were applied to the latest technique of *computer animation*. This recreates, second by second, the crucial period of a flight in real time,* and presents a line drawing of a 747 as the star of a terrifying video. Grossi took interested parties through the reconstruction film. They could see when and how the auto-pilot did the flying, and then what the pilot tried to do.

The film stopped with the 747 rolling over. 'Beyond that, we didn't need to go. Our films are not for some sensational PR exercise or entertainment.'

Most revealing of the hair-raising antics of China Airlines Flight 006 is an 'aside' in the eventual NTSB accident investigation report: neither pilot 'had done any aerobatic work since completing their air force training . . . ' Until that February day, that is.

So what happened? 'More than enough time was available to the flightcrew to prevent the upset.' After the flight engineer's erroneous conclusions about the engines, the captain failed to disengage the auto-pilot as taught, and to assume without any delay the more demanding role of flying the aircraft manually.

* See: Chapter 14 on Recorders.

He allowed himself to be totally preoccupied first with the flight engineer's efforts, then solely with the airspeed indicator when other instruments could have saved him from spatial disorientation. He also failed to achieve proper crew coordination and to assign specific tasks to his fully capable first officer. His performance was not impaired by boredom, monotonous environmental conditions, fatigue, over-reliance on automated systems and the effects of lengthy and dull monitoring duties, but these factors might have contributed to his apparent reluctance to re-enter the 'control loop' and take active command.

The memory of cheers for the flightcrew faded away. Within a year, the chief of the airline had to resign in the wake of some further accidents and severe criticism regarding management, maintenance and safety record. At the end of the day, only the aircraft came up trumps: without the structural soundness and aerodynamic capabilities of the 747, an extra 274 fatalities aboard lucky Flight 006 would have turned the bad record of 1985 into a horrific one.

'It's a great testimony to the strength of modern aircraft that we don't have a lot more accidents,' said Geoffrey Wilkinson, an outstanding former head of Britain's Air Accident Investigation Branch. 'But the risk is that we may develop a degree of over-reliance on it. A modern aircraft will tolerate quite a lot of mistakes, and some people may even take diabolical liberties with it. The way that 747 was handled, it had no god-given right not to break up. And, on paper, those people on board had no right to survive.'

Their chances were certainly no better than those aboard the SAS DC-9 that virtually crash-landed at Trondheim in Norway. In February 1987, through a misunderstanding between the two-pilots, the aircraft dropped fifteen metres on to the runway. The captain had no idea how disastrous the effect might be, but he accelerated in the last second to minimize the damage. In fact, the engine mounting bolts had been virtually extracted out of the fuselage, and both engines were hanging at an angle, about to fall off. A wing had cracked, fuel was streaming out of the tanks, and the Auxiliary Power Unit (APU) had been driven deep into the fin. Nevertheless, the aircraft became airborne once again, climbed, flew for sixteen minutes, and landed safely. According to an engineer from McDonnell Douglas, the manufacturers, the

DC-9 behaved like a dumb bumblebee that was unable to fly but flew on regardless, as if only to defy all the laws of aerodynamics.

Engines these days are so reliable that they hardly ever fail in a big way – hardly ever unless something turns them into potential time-bombs. When engines fail, they threaten to shatter those fragile statistics of safety. 1983, for instance, was a good year. Apart from the 269 people who perished when the trigger-happy Soviet Air Force shot down a Korean Air Lines 747, fewer than 700 scheduled airline passengers were killed that year, and a third of even those were also victims of suspected sabotage and acts of war. On May 5, 1983, the people who boarded Flight 855 from Miami could not have known that their fate might have upset the fine record of the year and increased the number of fatalities by almost twenty-five per cent.

The Eastern Airlines crew were at the airport well before the scheduled departure time of 8.55 in the morning. They completed their preflight checks and duties. The routine 'walk-around' inspection of their 'ship N334EA', a Lockheed L-1011 Tristar, produced no sign of any oil leaks from any of the engines. With just one minute's delay they were off. There were 162 passengers and ten crew on board. Some thirty-seven minutes later they were due to land at Nassau in the Bahamas.

The flight engineer was preparing the landing card data when, at about 9.12, the captain noticed a warning light: the No. 2 engine oil pressure was low. Perhaps the engineer should have spotted the problem before the warning light came on, but the news was of 'no great shakes', and he had never had any oil pressure problem with the L-1011. This time, the pressure kept fluctuating well below the required minimum, so the captain commanded that the No. 2 engine should be shut down, and chose to return to Miami. (They were closer now to Nassau, but the weather in the Bahamas and a light aircraft ahead of them might cause some delays.) They got clearance, and made a U-turn.

The engineer was now busy with the shutdown procedure for about four minutes, during which the No. 3 engine oil pressure gauge illuminated. Then the same happened to the No. 1 engine. The oil quantity gauges for all three engines read zero. Routine reasoning by the crew concluded that there must be some faulty indicator problem because simultaneous oil exhaustion would have a 'one in a million chance'. They radioed the Miami maintenance base for advice: what sort of common electrical fault could affect those instruments all at once? A few minutes later, having received the answer, they checked the appropriate circuit breaker, but found no discrepancies. Then an even more serious warning came from another source: the 'No. 3 engine conked out.' It could no longer have been an instrument problem.

The flight was cleared to descend to any altitude and make a straight-in approach. Meanwhile, several attempts to restart No. 2 engine failed. As a precaution, the Coast Guard were alerted, and cabin crew were told to prepare for ditching. And not too soon, either. At 12,000 feet above the ocean, No. 1, the only remaining engine, also failed. The APU provided the necessary power and hydraulic pressure for the crew to control the flight without engines, but, from now on, they had only one way to fly – downwards, in silence.

The cabin crew followed the ditching checklist. When the flight engineer announced that ditching was imminent, passengers came close to general panic. Many of them would not respond to instructions. They could not don their life vests and brace for impact without assistance. Others had difficulties with even retrieving their life jackets from under their seats, or could not open the plastic wrappers, or had to struggle to slip on and tighten the vests. The routine preflight demonstration by the cabin crew was long forgotten or had never been observed. Only forty-six people claimed to have glanced at the safety briefing cards. Now flight attendants stood on seats to give another impromptu demo and be seen. The crucial rule not to inflate the life jackets inside the aircraft was ignored by many who might not have heard the warning because although the engines were silent, several passengers were screaming throughout the emergency.

Six . . . five . . . down to 4,000 feet with a hush . . . and still some twenty two miles from Miami . . . At last the captain's determination and ceaseless efforts to restart at least one engine were rewarded. Descent was arrested at 3,000 feet. Although, unnecessarily, the passengers were left in the brace position for the remaining nine minutes of the flight, the aircraft could climb again. It levelled at 3,900 feet, and limped home to land without a hitch on just the one of its three engines.

Guided by the facts known about the mishap suffered by ship

N334EA, the investigation focused on the engines. A massive shock was in store: losing all the lubricating oil, the engines had run dry; location bearings were in a molten condition from heat; and engine No. 2 could be restarted only because it was shut down early on before it would be damaged as disastrously as the other two. 'What made it all even worse was that the near-catastrophe was not caused by some costly and complicated engineering problem. Fortunately, it is not a common occurrence that – to use a domestic analogy – a tenpenny washer should be allowed to threaten to bring down an entire household, bricks, mortar and inhabitants alike,' said a fuming accident specialist of the American pilots' union whose comments in private would have produced sheaves of blank pages if all the expletives were deleted.

Because all that happened was that three 'washers' called O-ring seals were missing.

Each RB211 engine must have a master chip detector that will reveal if certain components are not functioning properly. It is an essential part of the engines' oil system. O-ring seals, warning gauges and other protective measures guard it against leakage and loss of oil. The engine manufacturers used to require its inspection every 250 flying hours. Due to cautious worldwide monitoring of inflight engine shut-downs, the Federal Aviation Agency and Rolls Royce made a recommendation to reduce inspection intervals to a nominal twenty-five hours. Eastern Air Lines did the job every twenty-two hours though, if an aircraft was away from major maintenance bases, it could go for some forty hours without it. Complex supervision was in operation at every stage of maintenance work, and master chip detector checks were one of the merest routines: highly qualified Eastern mechanics who had worked on this particular aircraft must have done it more than a hundred times each – and in the two years before the incident, a hundred thousand detectors had been changed.

Like every job, the work was governed by a set drill. The foreman's logbook and Work Card 7204 were the bible of this particular task, and each step taken would have to be signed off on completion. Special Training Procedures (STPs) were posted from time to time. They were required reading for all mechanics, but there were no follow-up procedures in operation to ensure that everybody did read them. The two men 'who replaced the master chip detector assemblies on N334EA said they had never read the (relevant) STP 49-81'.

Usually, the mechanics got the detectors from the foreman's office *with the O-ring seals already in place*. This time they found no detectors in his cabinet so, for the first time ever, they had to draw them from the stock room. Nobody noticed that, this time, the O-ring seals had not yet been fitted, and nobody looked for them. Although both mechanics held high-level ratings, 'their failure to complete correctly the relatively simple task of installing master chip detectors and their failure to follow approved Work Card 7204 procedures are indicative of unprofessional work habits'.

How come that such habits had gone undetected? The investigation revealed some confusion, slip-ups and loopholes in the maintenance management, training and inspection procedures (they were revised within twelve days of the incident) that prevailed despite the fact that *this was the twelfth occasion that detectors or O-ring seals had been left off, and the fifth unscheduled landing due to master chip detector installation problems*. The FAA maintenance inspectors were also criticized: they had been aware of such problems in the past but did not consider them a major issue that would require some high priority special surveillance.

The NTSB made several important recommendations for maintenance programmes and their supervision as well as for emergency and ditching procedures. With goodwill all round, a major hazard was probably eliminated. But some of the background of this case, mentioned in not much more than asides in the final report, continues to give concern to many people.

In Washington, Philip Silverman, a partner in the law firm Speiser, Krause & Madole said: 'Nobody should get the impression that nobody is minding the store. There're ample safeguards, regulations and preventive measures, but the question is: how much can slip through? There's always the human element. Okay, in that Eastern case, two workers left out those O-rings. They were re-trained, not fired, then returned to the line. But is that a safety problem? Is it a shop-floor or management problem? If there were quality inspectors, somebody should have caught it before it could kill.'

In Frankfurt, Ferdinand Füller, an engineering executive who is also a technical delegate to the Lufthansa Flight Safety Department, spoke about the O-ring seal as if life and death depended on it. Perhaps because they do: it can cause a haemorrhage in the system and make an aircraft bleed to death. 'An O-ring costs less than two dollars. We use thousands of them. Rubber ones for fuel, plastic types for the hydraulic fluids. They're all black, and mostly the same size. They're just right for Murphy's Law: if anything can go wrong, some day, somehow, it will go wrong. If they get mixed up and somebody fits the wrong one, it may cause a small leakage or a big one. If it's a sliding part, it may get stuck in the flight control or any other system. In a non-return check-valve it may cause very serious problems. We used to store them in rising numerical order to make it easy to find the right one. Now they may be anywhere at all because only the computer keeps track of them. So now we had to introduce new safeguards: when an O-ring is needed, Luftansa fitters must follow the book, never accept it without its original, undisturbed wrapping, and must not install it without checking the part number – even if it seems to fit perfectly.' BA went even further, introducing a costly modified unit with a ball-spring that actually prevents the installation of chip detectors without seals.

Airlines are not the only ones to know that trifling even with such petty items is not to be taken lightly. After the *Challenger* disaster in 1986, the booster rocket was redesigned for the next American space shuttle. In 1988, on the successful completion of the tests, the building of *Discovery* was well under way when a batch of O-rings for the booster was found to be faulty. The damage looked suspicious enough to trigger off an immediate investigation by NASA and the FBI, knowing full well that a potential saboteur might target 'such an insignificant item' – a mighty midget that could strangle a giant destined for space travel.

Benefiting from other people's mishaps is, of course, a crucial part of any learning process. When, after the Miami case, Eastern tightened its precautions, it reviewed and adopted several safety measures that were already routine – but only at other airlines. For example, seven months after the incident, the use of a so-called *grasshopper* pin was introduced because it cannot be installed unless the chip detector is locked properly into the RB211.

Apart from government agencies, manufacturers and international aviation organizations, it is usually the duty of the airline's own safety department in the first place to keep itself informed about industrywide experiences, to study them, and to spread the gospel within the company. The influx of data from all over the world is so huge these days that it requires full-time attention. Unless this is a specific task for a whole range of specialists, invaluable information can get snowed under.

At the time of Flight 855 losing all three engines, Eastern's Flight Safety Department consisted of a manager and an engineer. Among their numerous duties, they had to monitor flight operations and maintenance functions in order to identify problems and trends through the computer analysis of masses of information pouring in every day. Before 1983, this used to be the job of the department's own analyst. Then the position was eliminated. The facilities remained available, the personnel did not, even though, after Miami, the amount of data and the workload increased. Nobody could say for certain that the former, larger safety department would have spotted what precautions taken by other airlines – now deemed to be essential – should have been copied long before, but numerous, particularly European, specialists are convinced that cuts in the name of economizing can be introduced in this field only at the passengers' peril. And yet this is exactly what happens in several countries that follow the latest American trend.

Much of the work of a flight safety department could be regarded as an incessant review of 'safe routines'. It was not lightly that Captain Alan Terrell, director of Qantas flight operations, named *complacency* as today's number one safety issue.

'We are the first generation of pilots who may go through a whole career without having a genuine emergency; many pilots have completed fifteen years of flying without having suffered an engine failure. Although this is undoubtedly a blessing, it does beg the question, how will they behave if they are eventually put to the test?' wrote Jack Diamond who gave an excellent technical account of one of the most hair-raising events on record.*

Speedbird 9, was to be just yet another 'milk run' – a five-hour night flight by a British Airways 747 from Kuala Lumpur to Perth on June 24, 1982. The weather was fine, the take-off was fine, the

* The LOG, BALPA, April, 1986

forecast was fine, and so was the savoury Malay satay being served to the crew and 247 passengers alike. Everything was fine except for the crew toilet because it happened to be occupied when Captain Eric Moody wanted to use it 37,000 feet up, south of Jakarta. That was why he had to go down to the first-class section. The aircraft was flying on auto-pilot, the shop was minded by a good crew, he knew, so he could afford to delay his return, hoping to snatch a few minutes' relaxation. But his chat with purser Sarah Delane–Lea was interrupted: his presence in the cockpit was requested urgently. The acrid smell of smoke rising from the floor vents made him climb the stairs faster. The sight that greeted him on the flight deck was less than reassuring: the windscreen seemed to be red – ablaze with a fine display of unseasonal fireworks. His reaction, immortalized on CVR, was not suitable for peak family viewing.

'St Elmo's fire, that was my first thought,' he told me. But his second thought argued back: 'That's crazy . . . it can't be . . . not at 37,000 feet, not in clear skies, not without something like an electrical storm showing up on the radar!' Defined as a corposant (holy body from the Latin *corpus sanctum*), St Elmo's fire has long been known to sailors who may witness this electrical brush discharge appearing as a mysterious glow around the mast-head. It is also a familiar sight to fliers but, as Moody recalled, it would occur in cirrus cloud or even on the ground, around the engine, in very moist conditions, but not up here, like now, surely. And why was that sharp, awful smoke emitting from the air vents everywhere?

According to the radar, the weather was still fine. The flight was still smooth, but the radio began to play up. Was the aircraft enveloped in some static discharge? How come they could still talk to another aircraft but not Jakarta? The display grew wilder, as if they had strayed into a Star Wars action sequence. Invaluable minutes were ticking away, and the baffled crew had no experience, no drill, no hear-say, nothing to guide them.

Behind them, all was well in the flying ballroom of a cabin (if you see one bare, before seats are installed, it seems just too bizarre that it would ever leave the ground). In first class, people had finished their meals and were preparing to sleep. From the wings forwards, there was nothing unusual to attract any attention. But in the back rows, people saw fireworks light up the sky. The engines . . . the engines must be on fire. The cabin crew reacted with a brave dose of humour and well-controlled gibberish that would avert panic – they hoped.

From the flight deck, the engine intakes seemed aglow. More cause for bafflement. Then blow number one from senior engineer Barry Townley–Freeman: 'Engine failure No. 4.' It was a low-key call without a trace of alarm. According to an aviation psychologist it might have been 'crucial in setting the mood for Moody and the rest of them'. Engine shut-down and fire drill were initiated – a routine that would cause 'no great shakes'. An ominous sound of familiarity for the present reader? There was more to come.

Within the next few minutes, the engineer kept calling out: No. 2 was a goner, then No. 3, and finally, No. 1.

Eric Moody stared at his instruments with incredulity: four can't fail; four did fail; four engines fail only in training exercises when keen instructors try to torpedo experienced pilots' self-assurance during periodic refresher sessions in the simulator. But this was no exercise. All the usual assumptions for four-engine generator failure were denied by the confusing information that bombarded the crew – and Moody knew that none of it was suitable for standard reasoning because he had practised barely a few months earlier how to deal and cope with a four-engine failure. How come that all the instruments were still working and the auto-pilot was still functioning normally? If there were engine fires, why did the red warning lights fail to come on? He got only *ambers* that indicated overheating. And yet the airspeed was dropping. What was going on? We ought to turn back towards land. But what's our safety height? Where are we going to hit something?

'In an emergency,' Captain Moody told me, 'the hardest is to do nothing. I learned it the hard way as I sat on my hands preventing them from doing things. I used the auto-pilot as an extra pair of hands, and let it put us on a gently descending slope. I remembered my old British Airways training pilot who used to seem a bit too pedantic, and a real stickler when it came to routine, because he claimed that going by the book is statistically safer than thinking. But now we needed time to think because all our combined considerable training and experience could not tell us what the hell was going on. In the circumstances, with our firm faith in the machine and routine, we could accuse nothing and no-one but ourselves: what did *we* cock up?'

So a few more vital seconds were lost. The engines had to be shut down. An emergency had to be declared. At 20.44 local time, senior first officer Roger Greaves tried to alert Jakarta with the message he thought he would never need to transmit: 'Mayday, Mayday, Speedbird 9. We've lost all four engines. We're leaving 370' (their cruising altitude). In the movies, pilots shout it, scream it, choke on it. Greaves just said it.

They changed course to return to Jakarta, and retreated to their last resort, the emergency drills according to written checklists that would ensure that not even the most trivial item could be overlooked in the heat of the moment.

Attempt after attempt to relight the engines failed. Descending unstoppably, a warning horn sounded at 26,000 feet: cabin pressure had dropped. As the captain and the engineer began to don their oxygen masks, the first officer got into difficulties. He pulled his mask out of stowage – and it fell to pieces. Yet another dilemma for the captain: if he continued to keep descending slowly, it would buy them extra time, but it would expose Greaves to all the pain of anoxia. Moody chose to put the nose down to increase the rate of descent, and began to plan and prepare for ditching. Once again the checklists helped him. Memories of a training film floated back. He also recalled childhood visits to the seaside where he had once heard that the old flying boats never flew at night when it would be impossible to see the swells and estimate the height above the waves. That was useful to remember but no comfort. He then made an unforgettable announcement to the passengers:

'Good evening, ladies and gentlemen. This is your captain speaking. We have a small problem. All four engines have stopped. We are all doing our damndest to get them going again. I trust you are not in too much distress.'

He knew the cabin crew were doing *their* damndest.

While the struggle to restart the engines continued, Moody had at last a moment to himself: as he could find no more to do, nothing else to think of to solve the mysteries – for the first time, he was scared.

When just 13,000 feet of air separated them from the sea, No. 4, the engine that had failed first, started up. Perhaps the fact that it had been shut down without any delay might have saved it from further damage, whatever the cause.

The rest of the apocalyptic journey belongs to aviation history. It was the highest degree of airmanship with which the crew saved the aircraft, nursed it upwards (only to run into more St Elmo's fire), nursed it downwards (the Jakarta glide path, an essential aid, was inoperative), and executed a blind yet smooth landing even though all they could see was a diffuse glare because the windscreen had gone completely opaque. (Moody had to rely on instrument call-outs by his colleagues. Forward visibility from the cockpit was nil – on the ground they would need a towing vehicle because they could not see enough even to drive along taxiways.)

After forty minutes in hell – thirteen minutes with no engines – they were safe at last. As Captain Moody rose from his seat, he saw that his hands were dirty as if he had done some weeding. The engineer suggested that it looked like volcanic ash. Ash at 37,000 feet? What a daft idea! But that was only the least significant of the mysteries puzzling the crew. And now a new fear grabbed them by the throat: how could they, how could anybody, go on flying if they could not trust themselves, their routines or their machines, if they could not *understand* the risks and *explain* what happened?

Once the passengers had left the airport, Captain Moody and his crew sat back and tried to relax in the first-class cabin of *Speedbird 9*. Telex messages from London were pouring in. What happened? How could it happen? And why? The pilots' unease grew more acute: 'We must have gone wrong somewhere. The buggers back home got us by the balls.' Then they had an unexpected visitor, a pilot of UTA, the airline that flies a great deal in the area. He listened to their tale, and shrugged: to him there was nothing mysterious about the drama; he himself had experienced something rather similar only two or three days earlier.

It was caused by volcanic eruption.

Did he report it? No, why should he? Everybody knew that old Mount Galunggung was at it again; this must have been about the ninth occurrence in 1982, and it would explain why their hands were dirty . . . Yes, it must have been volcanic ash. But how could volcanic ash incapacitate an aircraft? Oh, well, over the years, the active chain of eighty-two volcanos, stretching from Krakatoa to New Zealand, had produced many a drama up and down the airways and sealanes.

But then, why weren't we told? Moody kept thinking, why

wasn't I warned, why wasn't I briefed how to handle anything like that? Was I the only one who didn't know?

Unbeknown to him, at almost exactly the same time, a flight crew was being briefed in Sydney. They received the weather reports that augured a fine night for flying. But when they glanced at the usual satellite pictures, something unusual caught their attention. It was a peculiar, white, conical shape that said nothing to anyone. Even the professional meteorologist thought it was 'some blemish' in the photograph. It would emerge only afterwards that the 'blemish' was the first known satellite picture of a volcanic eruption – the one that hit *Speedbird 9* on the night of June 24.

The following day, Moody and his crew inspected their battered aircraft in daylight. It seemed incredible that *ash* could have caused all that damage. The leading edges, engine nacelles, turbine blades and the nose cone seemed to have been sandblasted. Not only was the paint gone, but there were deep abrasions. The windscreen and landing light covers would not have qualified for bathroom windows: they kept out even the light. The engines had suffered the worst damage. The turbine blades, with volcanic ash fused to them, looked as if they had just come out of a furnace. No. 4 engine, the first to fail and be shut down, had suffered the least damage. That was why it had the best chance to come back on line and save the aircraft.

Fresh shock waves hit the crew in the shape of an engineer from Rolls Royce: he had already been in Jakarta for a few days to investigate the 'ash erosion of the engine of a Caravelle that seemed to have been sandblasted in the area only a week before'. Even if the main mystery was solved by such infuriating revelations of known but uncommunicated advance knowledge, the crew had still to endure the final act: a deputation of air safety specialists from British Airways descended upon Jakarta, 'and they put us through a real grilling,' said Moody. 'The interrogation lasted for seven hours. We were forced to remember every detail, every thought, every bit of action until we felt guilty of everything, and ready to confess anything. Only then did we get a clean bill of health. But they had to know it all. And so did we. Without that we couldn't have regained our confidence in ourselves and the aircraft, without that we could not have continued flying free of fear and doubts.'

The crisis management of *Speedbird 9* became a textbook case. But why had the industry not been warned in good time about the hazard? The Indonesian government was certainly aware of renewed volcanic activity in the area. Others, too, including some aviation companies, must have known about it. It was a historical fact that eruptions could have very serious consequences. The vast explosion that had blown Krakatoa out of the sea in 1883 killing 35,000 people and causing the biggest tidal wave on record was not forgotten. Mount Galunggung had its own reputation as a killer. Periodic seismic monitoring recorded nothing significant in 1981. But in April, 1982, there was an eruption, and 30,000 people were evacuated. There was an observation station on the slopes of the volcano: it had no radio contact with the world.

A devastating series followed the April eruption. They lit up, then blackened the sky, and the hot volcanic ash knew no upwards limit as it savaged aircraft flying along the busy route between Australia and Malaysia. In the vicinity, everybody was concerned, many suspected that the big one was yet to come. And so it did on June 24, in the form of several massive eruptions from seven p.m. to eleven p.m., peaking shortly before eight o'clock. Twenty-seven people were killed, twenty-two villages were destroyed, while, mercifully, 35,000 people had time to evacuate. But nobody thought of closing the airways or warning *Speedbird 9* which had taken off just nine minutes after seven, and headed into the critical area.

After the incident, the route was closed, then reopened, only to be closed again. The world was alerted; the satellite pictures were studied and the identification of eruptions was standardized. The observation station on what was left of Mount Galunggung received a VHF radio; airline flight manuals were amended. ICAO set up a network to watch out for volcanic activities and flash warnings to all users of the relevant airways. On August 3, barely fifteen minutes after yet another Indonesian eruption, Australia notified and alerted Indonesia! Those who tell us that air safety is a forbiddingly expensive and complex affair, ought to familiarize themselves with a footnote from Captain Moody: 'By the 3rd of November, three hundred and fifty eruptions had been chalked up in 1982. Forty of them were big'uns, though ours was still the worst. In all, they vomited up fifteen million cubic metres of lava, and bombarded the sky with hot volcanic ashes. One of their targets was a 747 of Singapore Airlines that followed our path in reverse nineteen days behind us. When the pilot noticed that his engines were overheating and his instruments had begun to behave peculiarly, he shut down the engines without any delay, made a 180° turn back to Jakarta, descended to a safe height without risking any further fatal damage to his engines, re-lit two of them, and landed safely, simply because he *knew* for sure what his problem was. How did he know? He read a concise, extremely useful report about our incident in *Flight International*. That's how expensive this particular piece of accident prevention was.'*

People aboard *Speedbird 9* owed their salvation to a combination of luck, durable machinery and, above all, exceptional airmanship. Both the flight and cabin crew won a whole range of awards for skill and bravery, and their greatest reward was that their near-disaster taught the industry a lesson that would continue to save lives. It was befitting that Captain Moody should also receive the Hugh Gordon–Burge Memorial Award, named after a man who used to run BEA's air safety department and firmly believed that lives lost in an air disaster would become an unacceptable waste if nothing was learned from their fate.

Shortly before his tragically premature death, Captain Hugh Gordon–Burge, a former member of Britain's Air Accident Investigation Branch, told me: 'Once I was called to Ankara where we had a fatal accident. When I got there, a local colleague received me at the airport, and suggested dinner. I said I'd rather see the accident site. No need, he countered, open and shut case of pilot error: the bloody fool climbed out too steeply, and stalled. But I wanted to know *why* this bloody fool had climbed too steeply. Because I happened to know that "bloody fool" and I knew he was not the fool who'd fly by the seat of his pants. Eventually we found out that his pitch indicator was faulty. That was a fault that could have killed others, too, if it wasn't recognized and cured with the minimum possible delay.'

So does the aviation industry retain its healthy obsession with safety? Does it still welcome the inclination to do that little extra

* Captain Moody has never stopped flying because he claims that what you can understand cannot frighten you. He still dines out on his hairy experience: his lighthearted account of the spine-chiller makes a popular after dinner speech with his fees being donated to RADAR, a charity for the disabled.

above the legally required minima, that little more on which the current amazingly high level of safety is built? Nobody would argue with Olof Fritsch, ICAO's vastly experienced safety chief, that 'prevention is the cheapest way to achieve safety, because if you think that prevention is expensive, you've got to try having an accident.' When, however, prevention fails, learning from accidents is our last line of defence. But anyone who believes that *it's easy to be wise after the event* must be living in a fools' paradise.

– 2 –

BLOOD RESPONSE

If we accept that there are no *new* accidents – and there is no data to support the opposite – how come that we still have accidents at all? Is it only the interplay of an unavoidable services of coincidences? Is it individual carelessness? Do we have to accept that as long as it is people who make and operate aircraft, computers and other equipment, mistakes will be made?

In the last few years, watchers of the industry have identified several potentially dangerous trends. Some of them are new, others may be 'old' villains which have so far escaped suspicion. Mentioned most frequently are: sharper competition, excessive or misinterpreted deregulation, the withering of the safety cushion behind us, changes of style in management, the hitherto unrecognized responsibility of managers and managerial policy, more crowded skies and airports, too much faith in technology at the expense of slips in selection and training, the deeper meaning of 'human error', the worldwide proliferation of airlines, and the extended lifespan of geriatric aircraft.

It is, however, hotly disputed by many of the most influential authorities that these are hazards at all. They want proof, i.e. accidents on a statistically significant scale, to make them sit up and take notice of the 'theories' that they tend to regard as scare-mongering. But keen guardians of air safety, particularly accident investigators, want to sound the horns of warning in good time, and broadcast the final call for us to board the bandwagon of prevention *before* that statistically significant proof hits us with demands for new marvels of what they classify as *tombstone engineering*.

Accident investigators are the goalkeepers of the safety game. Their work is our last line of defence . . . This author who used to subscribe to those views now offers to eat his words. Investigation, however admirable, is no better than the Maginot line used to be: it can be circumnavigated much too easily. Our final line of defence is what happens *after* a cause or causes of accidents and incidents have already been found. Ignore, shilly-shally, do nothing – and more crashes will follow.

'It always takes an accident – or two – for the airlines, manufacturers and regulatory agencies to respond to a problem that seemed an acceptable risk before,' said Don Madole, a leading aviation lawyer, counsel to ISASI, the International Society of Air Safety Investigators, as we sat in Washington, surrounded by massive volumes of law reports, aviation antiques like a beautifully sculpted wooden propeller, and the video-happy investigator's paraphernalia in the haughtily modest, patrician building of Speiser Krause & Madole. His colleague Phil Silverman added: 'It's hard, of course, to decide what is an acceptable risk. Among other things, economy is always an essential factor of assessment. Nobody wants to weigh down an aircraft by an extra bolt or bulkhead if the chances of failure are no worse than something like two million to one. Actually, the whole world operates that way. I may stop at a parking meter without putting in any money because I only want to run into a shop and pick up a cigar. The odds are on my side not to get caught. But if I do, I'll be furious that a lousy quarter could have saved me twenty bucks. And perhaps next time I won't take a chance on it.'

'That's what I call a re-active response,' said Dave Kelley, Chief of the Operational Factors Division of the NTSB. An ex military pilot and air traffic controller, he shook his head almost in disbelief: 'It . . . it is just unfortunate that so very frequently corrective action comes only after catastrophies, especially when in many of these cases, the accidents are foreshadowed by clear evidence that the problem existed long before, like the cargo door of the DC10.

'And that's not all. As if it wasn't bad enough that the response tends to be re-active rather than pre-active, there tends to be a dangerous, even unforgivable time gap between accidents and the corrective reaction. When after an investigation we have a hearing, maybe even a congressional hearing, we often discover that there's a long paper trail laid out all the way back to some old recommendation which, even if it was formally accepted, suffered delays, postponements, inadequate funding, you name it. Oh, yes, they were going to implement it, and oh, yes, they're still planning to, but you see, there were money problems, further evaluations, legislative backlog . . . excuses and more excuses – and eight years or more have gone by without anything getting done. We pushed and pushed the regulators, who, in turn, submitted budgets for the legislators to mandate the use of equipment that would help avoiding collisions, didn't we? So when was anything positive decided about it? Only *after* the Cerritos horror, needless to say'.

We were not talking about some banana dictator's fleet of one DC-3, but about the United States, the world's leading user and creator of aviation, chief setter of our safety standards.

The huge grey Emmenthaler in Washington's Independence Avenue that houses the NTSB also contains their 'old foe' and sparring partner, the natural target of their comments, criticisms, and recommendations, the Federal Aviation Agency. So, one would think, the NTSB's missives do not need to travel far. But the route they have to take sometimes seems to be too circuitous to reach, alert and stir the regulators into prompt action.

Jim Burnett, who was hauled from a judge's bench into the NTSB, and whose jovial appearance may easily mislead the casual observer, is a fierce critic of the aviation establishment. Despite all five Members of the NTSB being appointed by the U.S. President, they are dedicated to the independence of their Board which reports directly to the highest level of government. (The Members' main privilege seems to be offices with elbow-room in a building where space is at a premium.)

Burnett said: 'I arrived in Washington on January 4, 1982, became Acting Chairman on the ninth, and four days later, an Air Florida flight crashed into the Fourteenth Street bridge right over there,' he pointed towards the plate windows, and the boyish glitter behind his glasses took on a certain darkness. The precise direction of his pointing finger was not the window but a photograph on the wall next to it showing the crash site. 'I went over there, learning my job the hard way, tried to establish a command center, watched bodies being fished out of the frozen river, and just knew that I'd never want to go to another scene like that thinking that I hadn't done my utmost to prevent it. That would be too hard to live with. If you've ever viewed aviation from an accident site, it's impossible not to keep pushing for safety, even if it means occasional conflict with the FAA or anyone else, and even if it may cause political controversy. Luckily, we now have a good relationship with the FAA – though it's not trouble free, of course. That improvement is one of the things I'd like my Chairmanship to be remembered for. The other is that it was achieved and maintained without any impeachment of our independence. The NTSB comes to an accident site with no *baggage*, no fears, no favours, no prejudices. And that must remain so.' Brave words spoken with great conviction, but a critical undertone is audible.

'The NTSB has no regulatory power. Their recommendations tend to be pretty good, and both sides claim that some seventy per cent are accepted. That's probably true,' said Captain Don McClure, who chairs the accident investigation board of ALPA, the American Air Line Pilots Association. 'But I'm talking about the bureaucracy, the time scale of implementation. I know, of course, that new ideas cost money and take time to put into practice. That's why airlines tend to do nothing unless all of them are mandated to spend and act. For they're afraid of losing some of their competitive edge.

'Take the classic hazard connected with that old accident in Manchester, U.K. Oops, sorry, it only *seems* to be an *old* case, for the report hasn't been published yet, has it?' (That was in the summer of 1988, when the long-delayed report on the Manchester tragedy, yet another full year away, was the butt of many quips in the investigators' international circles where the apparently final version of the British draft report was still circulated inviting comments. Several specialists seemed to be convinced that such excessive caution was a new disease afflicting the AAIB, and that in Britain it might not be enough to dot the i-s these days, but each dot must also be certified well in advance to avoid any controversy. Critics of the meticulous or, rather, cumbersome British system acknowledge that it is devised 'to be fair all round' but suspect that it mostly serves 'to protect ass all round'. The British Airtours 737 burned out in August 1985. A few days later the cause of the disaster was already known but in fairness, as discussed in Chapter 17, much research remained to be done by the investigators – simply because no one else would do it! In 1989, Britain introduced new regulations against 'vexatious' objections by airlines and manufacturers who would go to extreme lengths to make themselves appear in the best possible light.)

'The delay in publishing the report did not delay the transmission of the British investigators' urgent, useful recommendations,' said McClure. 'The case suddenly brought fire deaths in aircraft into the focus of worldwide attention – as if it was news to us that people still perish in survivable accidents when they're burnt to death only because smoke incapacitates them.

'We always knew that smoke would rise towards the ceiling, and advised people to escape, on all fours if necessary, even at the possible risk of getting trampled on – but we left the emergency lights near the top where smoke would obliterate them! No wonder that many survivors complained: "The lights have failed." After Manchester, various markers – what pilots call *new runway lights* – along the aisles had to be retrofitted by law to all airplanes. A good example of preventive action? Not quite. I'd call it blood response. The idea was around for *decades*! It took yet another disaster to push through the legislation.'

NTSB investigator Barry Trotter thinks it would be wrong to blame the legislators alone for slowness: 'Very often the airlines and manufacturers also drag their feet until blood response becomes inescapable and gives them the impetus to *do* something. Take the crash I had to look into in Bolivia.

'Approaching La Paz, an Eastern Air Lines flight hit a glacier at 19,600 feet. The spot was adjacent to a busy regular airway where pilots were allowed to maintain a minimum height of 18,000 feet. It was obvious that they wouldn't need to stray much to fly into that mountain at that altitude.

'The crash occurred in a very hostile, inaccessible environment. A huge amount of snow covers deep and dangerous layers of ice everywhere. We flew over the site, and saw a sprinkling of objects that looked as if they could be parts of an airplane broken into very small pieces. As we had no way of getting there ourselves, we used some native climbers to bring down a few pieces of wreckage only to let us establish for sure that, yes, what we had seen were the remains of the Eastern flight. Ten months later we organized a second assault on the glacier to recover the flight recorder, but it was unsuccessful. So, to this day, that may be the only large airplane wreckage over land that has never been recovered.

'Rumours persist that the locals are selling bits of aircraft as souvenirs to tourists. It could be true: at about 17,000 feet up the mountainside, there're a couple of small mining villages, and the locals can somehow run 2,000 feet up to the crash site and back without oxygen. But that's not the point of the story. What *matters* is that, although the full details of the accident remained a mystery, the local authorities and the carriers using that airway now introduced a new safety measure: the minimum altitude was raised! So pilots will have to maintain a minimum of 25,000 feet until they're right over the airport. Yes, that will make sure that they keep well clear of the glacier. But my questions have never been answered: Why only *now*? Why didn't anyone think of this simple precaution before?

'I've recently come across a major European air carrier that uses many old aircraft, few of which are equipped with GPWS (Ground Proximity Warning System). In many European countries, the use of GPWS is still not mandatory despite their frequent incidents on approach in bad weather, and only some miraculous luck has prevented them from having a high number of major accidents . . . so far. We used to have this problem over here, and it's a fact that the much-delayed introduction of ground proximity warning is now saving many lives. So, have we learned all that was to be learned from experience? Far from it. Our commuter airplanes are exempt from this regulation. They may fit GPWS, but it's costly, and it's up to them. Can anyone tell me why? Only because they can't kill hundreds of people in one accident? Do we have to wait for several tragedies to awaken everybody concerned?

'Oh, yeah, we can get awfully smart *after* the event.'

But can we? Accident investigation may be 'the costliest form of prevention' but there is no guarantee that the omens will be recognized every time, and that the accidents that can be avoided will be prevented.

In 1980, John Boulding, then Chief Air Safety Investigator of BA, warned the U.K. Flight Safety Committee that BA had an incident: the baggage door on a twin turboprop Hawker Siddeley

748 blew open after take-off and hit the tail. Boulding said he did not like that door because it opened outwards, which in itself might not be a risk if the locking mechanism, operation technique, and 'door open' warning systems were perfect – but were they? Boulding's comment was only noted and filed by the CAA and the manufacturers.

The shrewd reader can guess, of course, what is coming. Yes, there was a fatal accident. And it was avoidable.

At 17.28 on June 26, 1981, a BAe HS 748, registered as G-ASPL and owned by Dan-Air, took off from Gatwick for Castle Donington, the East Midlands airport. It was to be a Post Office mail flight with no passengers. In addition to the two-pilots, there was a PA (Postal Assistant) on board whose main duty was to act as a fire monitor in the cabin. PAs, used only on mail flights, were temps, many of them holidaying university students. They needed four days' ground training (including instruction in the operation of aircraft doors), and a minimum of two flights (eight sectors) under supervision. Already enrolled to study Aeronautical Engineering, the PA aboard G-ASPL was known to be 'keen, conscientious and capable'. Soon after six o'clock, he served coffee to the pilots, and reported, as the cockpit voice recording reveals, that: 'The indicators on the rear port (baggage) door are showing red.' Then, the aircraft was descending. The co-pilot mentioned a 'strong updraft' to which the commander replied: 'No, the rear passenger door . . . port passenger door's showing *unlocked*.'

The warning light might not have caused great concern because the system was *known* to be mischievous. But thirty seconds after 18.09, the CVR recorded an ominous noise typical of rapid cabin depressurization that would be consistent with the opening of a door, followed by sounds of severe vibration. Problems for the pilot then mounted so fast that, only eleven seconds later, he put out a Mayday call to air traffic control: *we'd like to come straight in we've er had a violent depressurization er it looks as though we've lost our back door and having a severe control problem.*

Emergency services were alerted, the aircraft disappeared from the controller's radar, and all flights in the area were asked to look out for it, but, three minutes later, the first reports of a crash came in. Several witnesses saw the last stages of the flight. The pilot was obviously fighting for his life. He tried to cope with severe rolling and pitching, regained level flight from a dive, but then tremendous forces made his wings fold upwards, and G-ASPL began to disintegrate.

There were no survivors. The wreckage was spread over several fields near the village of Nailstone. Almost three miles back along the flight path, well before the beginning of the break-up, several small articles, including a cardboard box from inside the cabin, had been scattered. The examination of wreckage, confirmed by CVR and the flight recorder data, revealed the tragic sequence.

The pilot was correct: the baggage door had opened and was torn off. It was found adjacent to but detached from the fuselage. The shape of indentations on its outer skin were consistent with impact damage caused by hitting the *leading edge of the tailplane*. Inside the creased skin, rubber deposits were found – they would have come from striking that leading edge. Had the door just flown off, it would have landed well away from the fuselage, and disaster could have been averted. But apparently, the door attached itself firmly to the distorted tailplane, and created a new set of disastrous flying characteristics. Wind tunnel tests also supported the material clues that, in such circumstances, no pilot could have saved the aircraft.

The investigation then delved into the history of G-ASPL, the door design, and its locking mechanism. Eventually, the official report would state: 'Although the lodging of a door on a tailplane may seem an unlikely occurrence, it is interesting to note that it had happened twice before to BAe 748 aircraft.' *Interesting* was a fine example of British understatement by accident inspector Shaddick of the AIB. Particularly because there was more to come. The log of the crashed aircraft revealed 'a continuing history of door defects' and warning light problems that needed several repairs. The manufacturers' records showed that since 1962, baggage doors of 748s had come open *thirty-five times*. Full details were available of thirty-one such incidents. On thirteen occasions, the doors actually separated from the fuselage, and five of those struck the tailplane.

As a result, a whole range of engineering and operational modifications were introduced over the years to make the locking mechanism more secure, the warning lights more reliable, and the locking technique foolproof. The urgency of finding a complete solution must have increased enormously after 1974, when the locking problem of the DC10 had caused mass slaughter in France and an international uproar. Yet clearly, it remained possible, easy in fact, for any PA who happened to be aboard G-ASPL to shut the baggage door and assume, incorrectly, that it was fully locked, while the warning system could still not be relied on to alert the pilots.

Within a week after the Nailstone crash, John Chaplin, now retired Group Director Safety Services of the CAA, the British Civil Aviation Authority, convened a meeting of the investigators, operators and manufacturers. 'Although we knew a door did come off, we were nowhere near envisaging *how* it could have come off,' he said in his office which had a panoramic view of London. 'So we tried to cater for all possibilities to prevent a recurrence. In such situations, safety is of primary importance, and we mustn't defer action purely because of the risk of a lawsuit. It sometimes gives our lawyers the kittens. But, fortunately, major manufacturers tend to take the same view. I've never seen any hesitation by Boeing, for instance, when seemingly urgent steps had to be taken.'

Similarly, long before the findings and a list of recommendations could be published, re-active prevention swung into action. Numerous further mandatory and optional modifications were developed, and some sections of the manufacturers' maintenance manual were rewritten. On July 9, 1981, barely a fortnight after the accident, British Aerospace sent out an Alert Service Bulletin by telex to all operators of the 748. As the investigation progressed, several more such telexes followed. The CAA made their contents mandatory. The course and urgency of action were most commendable, but they were still begging the questions: Why only now? Why did it need to be blood response? Was it not obvious to anyone that if doors kept coming off and hitting the tailplane, the damage could one day be fatal?

So, who could have and should have spotted the problem? The incidents were spread out over almost two decades. Many incident reports that reached the manufacturers were sketchy, and some cases came to light only by implication when operators ordered replacements for doors that had been lost. All but three of the thirty-five incidents happened to other than U.K. operators and, even after the establishment of MOR, the Mandatory Occurrence Reporting system, the manufacturers were not obliged to pass such foreign information on to the British airworthiness authorities except when a defect 'affected the continuing airworthiness of the product'. Presumably, the risk with the door was not seen to be such an exception although the sheer volume of incidents 'should have been sufficient to indicate an urgent need for investigation and rectification of their underlying causes' that had to be more significant than errors and simple negligence of well-trained crews. And yet the suspicion must have been lingering for years: from 1976, regular 'defect meetings' were held, fifteen more incidents were considered, and the minutes of those meetings were sent to the CAA – but only two of the fifteen cases 'became known to the Safety Data and Analysis Unit and were recorded in the MOR information bank'. Why?

The accident report included several important recommendations 'to identify, analyse and eliminate recurring defects'. In the six years since then much has been done in the right direction, and it would be most comforting to be able to conclude that perennial troublemakers now belong to the past. Unfortunately, this is not the case. Some day in the not-too-distant future, computers may be enabled to spot hazards that lurk behind seemingly unrelated incidents, but, at the moment, the state of the art is such that computerized occurrence watchers still have to rely on 'old men with long memories' to initiate research and alert the machine.*

The 'state of the art' is a frequent – and frequently valid – explanation for accidents, for the lack of some safety feature, and even for the occasional inadequacy of incident investigation that could not or would not go deep enough to eliminate the need of blood response. And it is this genuine consideration for the given state of the aircrash detectives' art that ensures that no investigation, however full and final, should ever be regarded as closed.

To the mythical 'Icarus accident' the blood response was: don't fly too near the sun, my son. But was it a useful piece of advice? I once devised a 'reopening of the Icarus file'† beginning with a brief historical investigation. But what started out as a cursory

* See Chapter 5.

† In *Aircrash Detective*, published by Hamish Hamilton in 1969.

exercise to illustrate the twentieth-century approach and techniques led to a few rather startling conclusions.

The legendary flight and 'the first structural failure of aviation' have been an enigma for millennia. Our initial review of original accounts raised an additional puzzle: is there any explanation of the oddity that an island and a sea, both named after Icarus, are twenty-five miles apart?

In the absence of wreckage and flight recorder data, our basic 'facts' had to come from the least reliable source: witnesses. Apollodorus, the earliest 'witness' to the legend, reported in about 140 B.C. that Daedalus warned his son 'neither to fly high, lest the glue should melt in the sun and the wings drop off, nor to fly near the sea, lest the pinions should be detached by the damp. But the infatuated Icarus, disregarding his father's injunctions, soared even higher, till, the glue melting, he fell into the sea called after him Icarian, and perished.' Strabo made the same error as witnesses who have related only hear-say ever since. But he added that 'alongside Samos lies the island of Icaria . . . ' and 'Icarus fell here having lost control of their course.' Arrian concurred, repeating the legend that Icarus fell on that island. He also attributed the accident to the sun melting the wax when Icarus had flown too high.

The engineering evidence comes from Ovid who, having collected details from Greek authors and hear-say, describes in *Metamorphoses* how the wings were constructed. Daedalus 'lays feather in order, beginning at the smallest, short next to long, so that you would think they had grown upon a slope.' He then fastened the feathers together with twine and wax. Finally, he 'bent them with a gentle curve' to make them birdlike. The wax he used was quite soft and could be moulded by the thumb. When the wings were ready, the two began the flight from Crete to Sicily.

Daedalus 'balanced his body on two wings and hung poised on the beaten air'. He told Icarus to follow him and he kept turning back to watch and instruct his son 'in the fatal art of flying'. As they flew, flapping their wings, fishermen, shepherds and ploughmen believed them to be gods.

Ovid's 'flight log' reveals some geographical and factual confusion. He reports that Samos was passed 'on the left and Delos and Paros', and 'Lebinthu was on the right and Calymne' when Icarus ascended to a higher altitude. Earlier witness statements are then repeated: the wax was melted by 'the scorching rays of the nearer sun' and the undisciplined or inadequately trained pilot fell into the sea. Daedalus, who at this late stage of the journey turned less and less frequently to watch his son, now conducted an air search, and kept calling out until 'he spied the wings floating on the deep, and cursed his skill'. But was the sighting of the wings evidence that Icarus had fallen into the sea? Elsewhere Ovid says Daedalus 'buried the body in a tomb and the land was called from the name of the buried boy'. So did Icarus fall on the island or was his corpse washed ashore there?

Nowhere is any night flying mentioned, therefore it must be assumed that all this was in a day's flight. But as Sicily is to the west of Crete, why did they fly north-north-east? One possible reason would be the presence of a low pressure system in the Aegean sea area around which the two flew like birds *with the wind* to conserve energy. This theory would explain why they flew north first, intending to turn west later, but in this case, after Icarus lost his wings, his body would fall while the feathers would travel farther with the wind, floating along a curved trajectory, and coming down some thirty miles to the north. *This would mean that either the Icarian sea or the island of Icarus was always wrongly marked on the maps.*

Since many mythological stories have been proven to be based on at least some facts, could one find another explanation?

The Icarian sea is part of the Aegean between the islands of Patmos and Leros, and the coast of Asia Minor. It is about thirty-two miles square. The island of Karus lies about twenty-five miles away to the north-west of Patmos which is at the edge of the Icarian sea. It would seem logical that the island should be inside or near the edge of its sea namesake. So why is it not? The classics can dismiss it as a mystery or some human error. But modern accident investigation can produce a theory that would fit the fragments of evidence into a neat pattern, and negate any slanderous thoughts about 'some silly old ignorant Greek'.

According to the Meteorological Office at Bracknell, it is improbable that the present climatic and wind conditions in the Aegean would be very different from those in mythological times. The wind blows mostly from the north; gales of thirty-four knots or more are infrequent, but unpredictable gusts and sea breezes towards the shore can occur at any time. I sought advice from some professional investigators, and they reasoned that if Daedalus had decided to fly into headwinds and chose an island-hopping route – across the present airways of the area – so that they could have 'refuelling stops' and rest on land whenever necessary but at least every night, the flight towards the north would be understandable and, considering the likely speed of the summer wind, it would be quite feasible that they had travelled some 180 miles in daylight.

To calculate the altitude of the flight, Fred Jones, then head of the Accident Section of the Royal Aircraft Establishment, was an ideal choice: he is a keen bird watcher. Ovid reported that the fishermen below thought the pair to be gods – thus perhaps humanoid in appearance but definitely not birds – which means they could not have been flying higher than about 3,000 feet. We also have evidence from the bird's eye view: from the air, Daedalus recognized the feathers of his son's wings floating on the water, and once again, the 3,000 feet maximum would make sense.

So now we have the intrepid travellers flying north at not higher than 3,000 feet, Daedalus possibly at a lower altitude. The sun must already be sinking slowly. The sea breezes of dusk may persuade Daedalus to make for the shore by having an easy ride on the back of the wind or he may continue to fly into headwinds hoping to reach the next island before nightfall. Either way, the wind direction would set the fatal pattern.

Icarus grows more and more confident, and ventures higher up. He may get into turbulent conditions, and an attempt to restore control may break off his wings – a phenomenon not unknown to pilots of swept-wing aircraft.

But the contemporary witnesses were adamant that it was the wax that gave way, and that the failure was not stress-related. They might have been right, too – except that the actual cause was the opposite of what they imagined. The wax, instead of melting from the heat of the sun, grew brittle and cracked in the evening breeze and the cooler air of the younger pilot's higher altitude. A decrease in temperature of two or three degrees centigrade can be expected with every 1000 feet upwards. Even if he climbed from, say, 500 feet to only 3,500 feet, this alone would account for some six to nine degrees loss in temperature.

Once the wax gave way and structural failure occurred, the body would plummet almost vertically and could land on or near the island named after Icarus, but the feathers would float along a certain trajectory that can be calculated like that of modern air crashes. (In the 1970s, an aircraft in flight lost a part with no disastrous consequences; the pilots noted the height and speed at the moment of the occurrence, radioed the data to an AIB inspector, whose calculation on the telephone enabled searchers to recover the part many miles from the airport even before the damaged aircraft could land.)

As RAE accident specialists were not *entirely* familiar with the flying characteristics of feathers, Fred Jones climbed to roof level in a Farnborough hangar, and dropped feathers of various sizes. It was found that in the still air the average feather would fall at a speed of six inches per second. If Icarus's wings disintegrated at 3,000 feet, the feathers would ditch in the sea in 6,000 seconds. During that time, they would be carried horizontally by the sea wind which could blow at about seventeen knots or thirty feet per second. The descending feather would be carried about thirty-four miles towards the mainland, so that Daedalus could, in fact, 'spy' them 'floating on the deep' right in the middle of what is now known as the Icarian sea. (If the wings fell off in larger sections than a single feather, their trajectory would be steeper and shorter, but they would still fall inside the Icarian sea.)*

While unfortunately, aerodynamicists cannot yet reproduce the secret of the flight of Daedalus, a feasible explanation for the mysterious separation of Icarus's island and sea could thus be concocted by the modern investigator whose techniques have come a long way from the days when, at the beginning of World War I, a British military Court of Inquiry declared that the cause of a particular accident was 'perfectly clear. What goes up must come down.'

But is it right to call these cases *accidents*? A random pick of three dictionaries has produced some subtle differences in the definition of the word.

* An analysis of the Icarus accident could go far beyond the material aspects. Writing in *ISASI forum*, April 1988, Major T. A. Bailey blames deficient pilot training, wilful disregard of orders, and over-confidence for the accident. All these probable causes would tally with the witness accounts, but they would still ignore other possible factors: the structure had no tolerance beyond the stated limits; eyesight, Icarus's altimeter and sole instrument, might have been faulty; boredom and fatigue by nightfall might have played crucial parts.

Concise Oxford Dictionary: '1. Event . . . without apparent cause or unexpected . . . ; 2. Unintentional act, chance, misfortune . . . esp. one causing injury or damage . . . ' Does No. 1 mean that a crash ceases to be an accident once investigators have revealed the cause? Does No. 2 mean that Captain Moody's dive was not an accident because it produced no casualties?

American College Dictionary: '1. An undesirable or unfortunate happening, casualty, mishap; 2. Anything that happens unexpectedly, without design or by chance . . . ' The three key words, 'undesirable,' 'unfortunate' and 'unexpected,' do not mix happily. All crashes are undesirable, but far from all depend on misfortune or on being unexpected because they might be the result of ignorance, negligence, preventable error and outright miscalculation.

Chambers' Twentieth Century Dictionary: 'That which happens: an unforeseen or unexpected event: a chance: a mishap.' Is the key word 'unforeseen' or 'unforeseeable'? What's the difference? If it was only unforeseen, it would be more of a mistake than an accident, because with better understanding of the problem, the hazard should have been recognized, controlled, catered for or eliminated.

A re-definition of the true meaning of the word 'accident' appears to be overdue even in a dictionary sense. It would certainly help those who are concerned with safety and who strive to eliminate mishaps that are no accidents.

Ira Rimson, a Safety Engineer in Springfield, Virginia (ex Navy investigator whose dedication to safety stems from 'seeing many deaths caused by design-induced human errors') had no hesitation in quoting one of the great gurus of investigation and safety philosophy: 'It would be hard to argue with Gerry Bruggink who says that most accidents are caused by "uncritical acceptance of easily verifiable assumptions".'

The most dangerous assumption is, of course, the evaluation of a certain known risk: how likely is it to cause a calamity? Be it human or material, operational or conceptual, what is it worth to design it out of the system? All aviation hazards can be eliminated – but only if we stop flying or, the next best thing, if we accept the economic penalty, and let, say, a Jumbo carry a vastly inflated array of safety features with never more than one passenger. Beyond that come all the compromises we must accept because experience shows that the aviation fraternity plays the game of probabilities exceedingly well . . . most of the time.

Geoffrey Wilkinson relies on experience, not statistics, when he says: 'I've seen hundreds of accidents, and every one of them had at least three cut-out points. If any one of them failed to open the floodgates, there would have been no accident. The three cut-outs may be separated by seconds or months or even years. If the first is crossed, the accident becomes a possibility; after the second, it is a probability; beyond the third, it's an inevitability. By then it's a matter of time. Like that Manchester 737 case. As soon as the conditions were right, it could have happened any time.' So, once the causes exist, only our failure of recognition will turn an impending disaster into the *excuse* of calling it an *accident*.

There was an air of painful resignation about Chuck Miller – another guru of crash investigation, former Director of the NTSB Bureau of Aviation Safety – as he sat on one of the last two chairs, surrounded by pictures stacked on the floor, before moving house from McLean, Virginia. A big man, with a cannonball skull to house and support the weight of all there is to know about accidents (was his *necklace* really just a post-surgical support?), he seemed to repeat for the umpteenth time: 'When I taught accident investigation at USC (the University of Southern California), I regularly challenged my colleagues and students, offering to buy a beer for anyone who could mention a single crash to which I can't quote an identical previous one at least in principle. I'm sorry to say, I've never bought a beer to this day. If you've been in this business long enough, you never see anything new. Apart from some very rare hazards brought about by completely new technology, causal factors are just a repeat performance.'

John Owen, the Accident Investigation course director at Cranfield: 'An accident can be described as a combination of factors which focus in time and space; a logical chain of events – in an unforeseen order.'

And finally, one must add what George Parker, an Associate Professor of USC said at the Tenth International Seminar of ISASI: 'An event that is "accidental" would also have to be unexpected. There is nothing unexpected about a repeat cause factor that has been previously investigated, analysed, and reported.'

So, whenever blood response is called for, i.e. curing the hazard only in the wake of slaughter, we cannot really talk about accidents at all. Mishaps, disasters, calamity, mistakes or negligence, yes, but not accidents.

Some of the cases already reviewed belong to the 'unexpected' category – the in-flight failures, the all-engines failures, the sort of disaster when aircraft just fall out of the sky into story-book escapes or the grave. But what was so unexpected about the initial predicament of the Boeing 727? Why had 234 lives to be wasted?

THE STING IN THE TAIL

The 727 entered service in 1964. It belonged to the 'second generation jets' with swept-back wings, T-shaped tail, and rear-mounted engines that were dubbed, after the first crashes, the 'sting in the tail'. The aircraft was designed for 'up and down' service, to use smaller airfields where the big jets could not go, climb steeply out, almost drop in at a high sink rate, fly barely ten per cent behind the speed of sound or slow down like a propeller-driven machine. Its popularity matched its climb and sink rate: a star overnight, and an ogre just as quickly.

Its troubles began in August 1965. A United Airlines flight to Chicago was about to land when it crashed into Lake Michigan. First a bomb was suspected, then the theory gave way to guesses because the little of the wreckage that could be recovered gave insufficient clues, and the flight data recorder was never found. With too little to go on, the investigators could not issue a report.

The mystery bred fear.

By the end of October, many people had premonitions of a doomsday, when all the 300 727s already in service fell out of the sky in one go. (The imagination was fuelled by the fate of the BAC One-Eleven, whose apparent 'stalling problem' might have been caused by over-testing, and which was yet another type with rear-mounted engines.)

In November 1965, Captain O'Neill prepared to land American Airlines Flight 383 at Cincinnati. He had been a captain for eight years, and was about to complete his seventh conversion training flight on a 727 under the supervision of Captain Tellin. Near the airport the weather deteriorated, and it began to rain. Instead of a visual one, the flight was to make an instrument landing. In the final stage, witnesses saw the aircraft bank rapidly, and crash. Of the sixty-two people on board, four survived miraculously, including an AA pilot, travelling as a passenger on flight 383, who had noted an alarmingly rapid descent.

A massive investigation was mounted without delay. Thomas Saunders, in charge of the CAB (later NTSB) GO team, was aware of the 'sting in the tail' suspicions. Prater Hogue, the Boeing coordinator on the scene, got worried: did the 727 perish because of some hidden characteristic? 'The company's attitude was to let the chips fall as they may,' he once told me. 'If the accident had anything to do with the design, it was better to find out now than after another crash.'

Saunders was hard pressed for an answer: was the crash due to some design fault or a pilot error? Within forty-eight hours, the investigators recommended the formation of a new group to examine the performance characteristics, handling and special capabilities of the 727. While Saunders was on the phone, propounding his ideas to his boss in Washington, a local TV station flashed the news: yet another 727 had crashed at Salt Lake City. Again it happened in the landing stage. *'Ground the killer with that damned sink rate'* became the cry up and down the country.

Salt Lake City was a so-called survivable accident though almost half of those aboard perished in the post-crash fire. But for the first time, the invesigators had a live crew to interrogate. It emerged that, shortly before the impact, the co-pilot had tried to apply power and so arrest the fast sink rate, but the captain brushed his hand away saying 'Not yet!' Was it 'human error'? The captain, in his hospital bed, claimed he gave the engines plenty of time, at least thirty seconds, but they did not respond properly. This was contradicted by the engines which now ran without a hitch on a test bed, adding to the temptation to yield to the *bloody fool descended too fast* theory.

Meanwhile, a detailed reconstruction of the fire and its devastating effect produced a barrage of urgent recommendations by the investigators to William F. McKee, the Administrator of the Federal Aviation Agency, who answered them promptly: FAA experts were 'pleased to report that we have had many of them [the suggestions] under consideration for some time'. At last a solution to problems which had been known admittedly for at least 'some time', would soon be found. The seating of stewardesses would be rearranged so that they were near the exits; the weakness of fuel lines would be improved; the close proximity of fuel lines and electrical leads would cease; a new system of emergency lighting would be introduced just nineteen days after Salt Lake City. Heavy black smoke, produced by the interior furnishings, had also prevented escape and contributed to the killings.

The engines and the aircraft were now gradually exonerated. So how come that, one after another, well-trained pilots behaved like bloody fools? Were they well enough trained for an aircraft which responded to lighter efforts than other types? A month after Salt Lake City, new training procedures came into force as a *prevention*.

The 'ground'em!' campaign gathered more momentum in February 1966, when an All-Nippon Airways 727 crashed into Tokyo Bay, killing 133 people and creating the worst single air disaster to that date. Only two minutes earlier, according to the air traffic controller, it had still been at 10,000 feet. How could it have dived so fast? While the investigation got under way, the FAA changed the landing procedure for the 727. The Japanese established strong suspicions of fatigue damage to the tail engine mounting, and their urgent warnings were flashed to all users, but it turned out to be yet another red herring.

Back in Washington, the Salt Lake City accident report was rushed out in record time. In June 1966, it indicated pilot error as the probable cause. Something else could be read between the lines: the training history of the pilot contained some black marks. Why was this not picked out, corrected or taken seriously enough at the time by those responsible for his training? Was this then the key question? A resounding 'yes' answered it day after day. The improved training technique and better, augmented instrumentation began to make their marks: the repeat crashes discontinued, and insurers acknowledged, belatedly, that the track record of the 727 had become 'much better than average'.

Blood responses come in all shapes and sizes. A documentary trawl for examples has produced some classics.

JUST A KNOCK

It is now a routine accident investigation procedure to examine aircraft for crashworthiness.

In the early 1960s a British registered aircraft crashed in Yugoslavia. It was a survivable accident. The plane did not break up, and it was only after considerable delay that the subsequent fire engulfed the cabin. Yet dozens of people died.*

The aviation pathologist who conducted the medical investigation found that many of the victims had suffered non-lethal bone injuries and pulmonary embolisms which were due to fat and bone marrow being carried by the bloodstream into the lungs. The indication was that the injuries must have occurred well before death. But if the injuries were non-lethal how would they contribute to the death, if at all?

The answer was in the passengers' near-identical condition: many of the dead had fractured shins. In the final moments of severe deceleration, their bodies would have been restrained by seat-belts, but their limbs would have flailed violently forwards. The rest of the reasoning required only the measurement how high above the floor the shin injuries had occurred: the height coincided with the level of a crossbar that was intended to strengthen the back of each seat. That innocuous-looking thin metal strip must have broken those flailing shins, incapacitated dozens of passengers and so prevented their escape.

The response was almost immediate: seat design was modified on that type as well as those still on the drawing board, and, although it did not bring back the dead, the lesson was learned by all – one would have thought.

In 1988, I had just finished a substantial meal, and stretched to relax on a Hungarian Malév flight. As my legs moved forward, perhaps with undue abandon, my shins hit the crossbar in the back of the seat in front. Although the news of the pathologist's vital discovery had been freely available for almost a quarter of a century, the TU-134, also known as the BAC-ONE-ELEVENSKI because of its remarkable similarity to the British model, was still carrying the lethal crossbar. Fleets of several hundred TU-134s are a mainstay of Soviet and East European airlines to this day, and many of them have been sold to third world countries. Perhaps their makers and users are still awaiting some blood-letting nearer

* Hardly any of the type remain now in service. The case has been disidentified at the request of the pathologist to avoid stirring up images that might haunt the victims' relatives.

home before wasting money on such petty modifications just to prevent a knock.

JUST HOT AIR

When a tyre blows out on a bicycle or a car, the effects may range from a nuisance to a disaster. On aircraft, in view of the number of wheels, the risk might seem to be reduced, and sometimes even pooh-poohed. In the late 1960s, when I wrote to an American airline executive, he answered calling my questions 'somewhat tyresome'. The introduction of the big, heavy jets, however, exacerbated the problem, and the 1970s left no room for feeble puns. Tyres blew by the hundreds, and caused serious accidents as well as numerous incidents. Blood response became overdue.

On March 1, 1978, the runway was wet at LAX, the Los Angeles International Airport. That in itself was no problem, and 'final calls' for passengers to board were announced as scheduled. Continental Airlines Flight 603 was also cleared for departure on time. Captain Hersche accelerated his DC10, almost reached V_1 speed and the point of no return where normally he would have to take-off no matter what, when he heard a loud bang and experienced severe vibration. There was no time to analyse and investigate. An instant decision had to be made, and the captain chose to abort. He applied brakes and reverse thrust, followed standard procedures as required, but unfortunately, the aircraft could not be stopped on the remaining 2,000 feet of wet runway. It skidded into rough ground; the landing gear snapped off and ruptured the fuel tanks. Flames shot up, four passengers died.

Examination revealed that it was the No. 2 tyre that had blown out causing the bang and vibration as well as the subsequent failure of the No. 1 tyre. 'The pilot's decision – based on the information he had at the time – was obviously correct; but if he had been aware of the tyre problem he might have continued the take-off'* to dump fuel and land safely. The accident was followed by seemingly endless litigation with huge claims and counter-claims. Continental accused McDonnell Douglas of lying about flaws in the design of the landing gear that was meant to snap

* Dr Alan Roscoe, in *Flight Safety Focus* No. 2/87 of the U.K. Flight Safety Committee.

cleanly in an accident without damaging the tanks. The manufacturers alleged that it was the pilot's poor judgment that had caused the accident. Additionally, legal flak hit the two manufacturers who had supplied the two damaged tyres, and the company that had retreaded one of those tyres. Verdicts awarding dollar millions in damages were handed out only to be reversed on appeal, and, at the time of writing, much of it is still argued venomously. But beyond feeding the lawyers, it was obvious from the case that tyre failures were a potentially lethal menace, and pilots needed immediate information if they were not to misinterpret bangs and other mysteries.

Another of the many painful reminders came in 1980, when occupants of a runway clearance vehicle noticed that a Miami-bound PanAm DC10 burst a tyre during take-off at Heathrow. They radioed a warning to the tower. The pilot of the accelerating DC10 overheard the message, and just about managed to stop before the end of the runway. In the course of the successful evacuation, a passenger suffered a broken leg. This pilot was lucky: he was *told* what exactly had happened. But the investigation brought out numerous disturbing facts.

It emerged that, after the Los Angeles accident, the NTSB had made fourteen Safety Recommendations, but only some of them 'have been implemented, either in part or in whole'. Following the LA case, McDonnell Douglas also issued 'a number of Service Bulletins that provided improved integrity and protection to systems on the landing gear and in the landing gear bay'. In America only one, in Britain all except one of these were made mandatory. It is probable that the overwhelming majority of DC10 users in other countries followed the American example, and so the blood response was wasted.

Colin Allen, who was in charge of the Heathrow investigation told me: 'Once again, only one of a pair of tyres failed, but as usual, that in turn caused overloading on the other, and created a critical condition. That was a highly unsatisfactory state of affairs. We also found that both No. 8, the initial failure, and No. 7, too, had been six years old. Their ageing carcasses suffered some overloading during a heavy landing the day before the accident. Both tyres had been retreaded seven times. Our eighteen recommendations included experiments to establish the extent of degradation of tyres caused by age, usage and retreading. We also recommended that tyre status monitoring equipment should be installed in all large public transport aircraft, so that the pilot should know at once if something went wrong.'

John Boulding, now Secretary of ASG, a remarkable British air safety pressure group of dedicated volunteers, commented: 'Retreads are no extra hazard in themselves but they're more liable to get damaged, may have extra weaknesses if retreading is not done with extra great care, hygiene, etcetera, and when they go, they tend to cause more damage than the new ones.'

If still more warnings were needed, they were well on their way. In September, 1982, two months after the publication of Allen's report, a Spantax DC10 captain experienced some airframe vibration during take-off at Malaga. It gave no cause for concern (actually, a piece of tread had come off the right-hand nosewheel tyre), and the experienced pilot continued his run. But then, something much more dramatic happened. The cockpit floor, the seats, the joystick columns and the panels began to shudder violently.

Captain Manfred Reist, the Swissair Deputy Flight Safety Manager, who analysed the situation in an excellent training film dealing with Stress Management, told me that the Spantax captain must have felt 'the plane's airworthiness was seriously compromised. The stress was too great, and his decision had to be made in a *second*: STOP! REVERSE!' Not knowing what was wrong, the pilot thus chose the seemingly lesser of two evils despite the fact that he was hurtling down the runway beyond the compulsory lift-off speed. Eventually, the investigation 'concluded that the captain had been under an abnormal amount of situational stress at the time, and would therefore have hardly been capable of logical reasoning'. The decision to abort was found to be 'an expedient one' in the given circumstances, even though the aircraft overshot the runway, crashed into a concrete building, fire broke out, and fifty-four people died during the evacuation.

A few months later, the captain of an Eastern 727 was also at a loss when he heard a loud bang and received '*doors*' and ''*right gear*' warning lights at 10,000 feet up, as he was climbing out of Miami. Then both his hydraulic systems packed up. He tried trouble-shooting while dumping fuel, and prepared for an emergency landing. After a fly-by, allowing observers to inspect the condition of the undercarriage, he landed safely. It would soon be discovered that the incident sequence started with some overheating in the wheel bay. That in itself should not have created a dangerous situation, but, due to the resultant high temperature, a badly worn and improperly retreaded tyre exploded and caused massive damage to the hydraulic system. Once again the recommended response (fortunately no blood this time) reflected some problems that sounded alarmingly monotonous: establish training programmes to teach pilots how to assess, recognize and deal with such occurrences; eliminate certain retreading practices; 'final inspections of retreaded tyres should rigorously follow the guidelines . . . '

The PanAm, Spantax and Eastern cases as well as many others happened despite the fact that some airworthiness authorities had already tightened the regulations to make both tyres and wheels safer at the end of the 1970s. Meanwhile, new retread regulations came into force, the manufacturers introduced better tyres, and a new laser light method of holographic non-destructive tyre inspection test became available for quality control. (Ansett, the Australian airline, invented its own ultrasonic tyre tester, and checks every tyre, new or retread.)

Then airframe manufacturers added tyre pressure monitoring systems to the pilots' armoury to identify and prevent the risk of blow-outs. 'These are now standard fit on all new Airbus, Boeing and Douglas aircraft, and Concorde has had a similar system for many years,' wrote a strictly anonymous spokesman of Dunlop's Aircraft Tyres Division, in a letter, dated March 14, 1988, that went on: 'The result of all these actions was nicely summarized by the Engineering Director of a major Far Eastern airline when he said at a recent conference " . . . ten years ago I was worried by tyre failures every week. Today they are a thing of the past".' Not everybody, however, shares this testimony of confidence.

In 1986, a speaker complained at the Flight Safety Foundation seminar in Vancouver: 'Blown tyres cause more rejected take-offs than engine failures, yet we don't train crews to handle tyre bursts.'* In 1988, Captain Don McBride, who chairs the Technical and Air Safety Division of the Canadian Air Line Pilots Association, said to me in Ottawa: 'Most of the stopping criteria cater for engine failures that are all too clear to the pilot anyway. But some eighty per cent of the rejected take-offs are due to tyre

* *Flight International*, 25.10.86.

failures, hydraulic problems, unusual noises, unidentified bangs – and boy, oh boy, when you have to stop a big heavy airplane in an emergency approaching V_1, when red lights come on, and you don't know what the hell is going on . . . well, we're back to the old, old problem, and with a sick airplane, I'd rather go off at the end of the runway than up and *then* down.'

Thus the risk remains perennial because the question of retreads has still not received a universally valid answer. The Dunlop executive wrote: 'Aircraft tyres are retreaded "on condition". The number of retreads per tyre varies with the aircraft and the operator. In Europe the airlines generally impose their own limitation at five retreads. In other areas some airlines have no limitations. The Dunlop limit is arbitrarily set at ten but very few tyres achieve this number and those that do so are generally nose-wheel tyres which have a comparatively easy life.

'The trend today is towards limiting tyre life by total cycles. One cycle being equal to one take-off and landing . . . '

There was also a move by the FAA towards limiting retreads to six only. That proposal has been entirely dormant for almost three years.

The Lufthansa system is an interesting example. On some aircraft, six, eight or even ten retreads are allowed as long as the carcass is good. If pressure drops eighty per cent, the tyre can be reflated, but no more than twice. If it goes below the limit, it must be changed, and the tyre on the same axle must also be replaced simultaneously because the overloading might have created a hidden hazard. The system depends on cautious monitoring of even minute incidents. This was illustrated by freighter Yankee Echo, a Jumbo that kept bursting a right-hand tyre almost weekly. Ferdinand Füller described how that rogue elephant was then watched in its natural habitat at Frankfurt. Doing a regular run, it was always loaded at a spot that was a real sun-trap with an air conditioning outlet blowing hot air straight on to the tyre on the right. The fully laden 747 would then have to taxi some three and a half miles to the far end of the runway – and there it was, yet another blow-out. A slight alteration of the loading area, restricted usage of air conditioning, and stricter selection of tyre types eliminated the waste and the risk.

Another, at last receding, problem with tyres also testifies to the point that even blood (particularly if it is somebody else's) often fails to provoke immediate response. Manufacturers have recommended nitrogen for tyre inflation for more than twenty-five years because it is safer than air. The airlines resisted the idea simply because of the extra initial cost involved. Some made the move reluctantly, others had to be pushed. A BOAC VC10, for instance, had a fire in a tyre on the ground at Rome in the early 1960s. It could not have happened if nitrogen was used. That settled the issue for BOAC (and British Airways, their successors) as well as for many major airlines. Others continued to 'economize' waiting for their own in-flight fires to convince them – and the regulatory authorities let them vacillate.

On March 31, 1986, a 727 of Mexicana crashed in Mexico. NTSB sleuth Brian Richardson, who flew down to help the local investigators, told me: 'The airplane had a brake malfunction and, on take-off, the brake got real hot. After the wheels had been retracted, and the 727 was climbing through 29,000 feet, there was an explosion. The main gear tyres blew up inside the wheel-well like a bomb. With 167 fatalities, the crash was the worst in Mexican history to date. The phenomenon was known, of course, long before but Mexicana was still using air in the service of the main landing gear tyres. It was a wholly preventable accident, and it had worldwide implications that demanded corrective action at last. Late that year, the FAA issued an AD (Airworthiness Directive) which mandated the use of nitrogen in main gear tyres of transport category airplanes.' (Current British regulations specify the use of inert gas with a maximum of five per cent oxygen in any tyre; nitrogen has always been mandatory for Concorde; in some countries, the *air-cum-savings* formula is still more attractive than the *nitrogen-cum-safety* proposition.)

JUST A DOOR

To many readers, this simple heading will be a reminder of only one accident: the Turkish Airlines DC10 that killed 346 people near Paris in 1974.* Although several books have been written

* The accumulation of DC10 cases in this chapter is not meant to imply that the frequently suspected, much maligned but rightly criticized aircraft is any less safe these days than other big jets. It came about only because the accidents reviewed happened to provide some glowing examples that highlight particular problems.

about the slaughter in the beautiful forest of Ermenonville, and although we retain indelible memories of the subsequent legal battles that were splashed across front pages for ten years as the manufacturers vomited counter-accusations, spat out lies, devised cover-ups, and generally tried to wriggle out of their responsibility, it is worth recalling at least one aspect of that foreseeable but unforeseen tragedy as perhaps the ultimate example of missed opportunities for timely response to the original warnings that did not spill actual blood, and were therefore almost ignored.

In 1970, McDonnell Douglas tested Ship No. 1 of the DC10 model for pressurization, and the cargo door blew out. Consequently, the latching mechanism was modified.

In May, 1972, after three operators had reported failures of the cargo door latching, McDD issued an SB (Service Bulletin) to *suggest* some further safety measures.

On June 12, 1972, American Airlines Flight 96, a DC10, made an intermediate stop at Detroit en route from Los Angeles to New York. The ramp service agent had considerable difficulties with closing the cargo door, and had to use his knee to force it shut. Five minutes after departure, climbing through 12,000 feet above Windsor, Ontario, Canada, the pilots heard a 'thud', and, at the same time, dust and dirt flew into their faces. Crucial parts of the flight control systems went dead. Rudder pedals and thrust levers assumed a life of their own, and the aircraft began to yaw. Back in the cabin, the air went 'foggy' and the floor in the aft lounge area collapsed into the cargo compartment. It was only with quite exceptional airmanship that Captain Bryce McCormick and First Officer Peter Whitney saved the crippled aircraft.

The investigation discovered almost right away that the cargo door had only apparently been locked. When it blew open, and was torn away, the floor collapsed due to sudden depressurization, and severed the control cables that run the length of the aircraft underneath. The report would not be completed until some nine months later, but because of the urgency, the NTSB issued immediate recommendations for a really secure locking mechanism, and for 'the installation of relief vents . . . to minimize the loading on the cabin floor' if sudden depressurization occurred in the cargo area. 'Sudden loss of pressure in this cargo compartment for any reason should not jeopardize the safety of the flight.'

FAA Administrator John Shaffer, a Nixon appointee, responded within twenty-four hours stating that all operators were checking their cargo doors in accordance with the manufacturers' Service Bulletins, and that the problems were being studied. He concluded that 'while a preliminary investigation indicates that it may not be feasible to provide complete venting between cabin and cargo compartments, your recommendations will be considered with respect . . . '

Service Bulletins do not make any work compulsory. They concern anything from recalcitrant soap dispensers to suggesting 'handle with care' when 'bomb disposal' might be indicated. At the time of the Windsor emergency, American Airlines, for instance, had not yet done anything about the fortnight-old SB – nor were they *required* ever to comply with it. It was up to the FAA to respond to the NTSB recommendation with an AD (Airworthiness Directive) that would make the work universally mandatory. It was already drafted, but Administrator Shaffer chose to make a 'gentlemen's agreement' with the manufacturers who would issue a mere SB and a repair kit, and so protect themselves from any embarrassment an AD might cause. That suited their 'vigorous' marketing policy rather well at a time when they were pushing the DC10 with all their might.*

A senior executive of another manufacturer (who wished to remain anonymous) told me: 'I just couldn't believe what was going on. It was plain stupid. They argued, at least privately, that all their DC10s were flying inside the United States, and so they could control them with SBs alone. But they should have known that you can't beat off the competition by withholding ADs – not for long, and not without undue risk. If one company has an accident, the rest of the industry mustn't take advantage of it. But in the sixties, some McDD salesmen would stoop to all sorts of tricks. They'd even list SBs and ADs concerning other people's aircraft to make potential buyers think twice about them, and so promote their own DC8s. I always tell our sales and PR people

* In many cases, though not in this one, the dividing line between an AD and an SB can be much too fine. To this day, some people believe that, for instance, the Lufthansa 747 crash at Nairobi in 1974 might have been prevented if Boeing and the FAA had responded with an AD rather than an SB.

that I'd cut their balls off if they ever use that kind of data against our competitors.'

Frank Fickeisen, Flight Control Systems Manager of Boeing, recalled in 1988: 'After Windsor, we re-examined our own cable runs. They go from the flight deck to the tail, and individual elements could be liable to damage by all sorts of failures, fractured discs, exploding tyres, etcetera, but we found the total system safe even without modification.' (Boeing and others would, in fact, introduce venting – but only after Paris, the next big crash. In 1989, the latest 747-400 version was still designed without upper deck floor venting that would be introduced only under severe European pressure from 1992 onwards.)

In 1972, not everybody involved with the design of the DC10 was happy with the SBs. Eventually, determined digging by lawyers and journalists would discover the '*Applegate Memorandum*'. Written by the product engineering director of Convair* barely a fortnight after Windsor, it severely criticized McDD's proposed 'bandaid' measures: ground tests had already 'demonstrated an inherent susceptibility to catastrophic failure if exposed to explosive decompression', so it was in 1970 that immediate corrective steps should have been taken. The latching mechanism was still not good enough, said Applegate, who warned that in the years ahead 'cargo doors will come open and I would expect this to usually result in the loss of the airplane'. Unfortunately, the memo was never forwarded to the FAA or even McDD.

Applegate's prophecy came true nineteen months later. A Turkish DC10 took off from Paris, lost its cargo door at 12,000 feet, suffered catastrophic depressurization and loss of all controls, spewed out six passengers and their seats, and crashed in the forest, killing everybody on board.

The subsequent mud-slinging revealed that the 'gentlemen's agreement' had been broken in more ways than one. At the time that the post-Windsor warnings and SBs were issued, the aircraft that would crash at Ermenonville was still on the production line, known as Ship 29. McDD was in such a hurry to fulfil a hardwon order to Turkish Airlines, that it failed to subject it to its own required modifications. (The records showed that the job had

* A division of General Dynamics that built the fuselage, floors and doors of the DC10 for McDD.

been done, but that was not true. The three highly-qualified inspectors, whose jealously guarded stamps appeared on the job sheets, fully denied that they had ever worked on Ship 29. There could be only one explanation: some faker must have stamped the blank spaces *after* the crash.)

The NTSB made its own blood response: having recognized that losing track of their own recommendations was a loophole, Chuck Miller initiated a follow-up procedure. Others reacted in their own way. Boeing, for instance, introduced *key* or *significant* SBs to give extra weight to certain changes.

When in 1975 the FAA issued the AD mandating, at last, that all wide-bodied aircraft should have a vent system and a strengthened cabin floor, it was estimated that the modifications to all American Jumbos would cost about 60,000,000 dollars. The lawyers* who had led the revolutionary fight for proper compensations and the unmasking of the guilty parties, found it painfully ironic that the payments to the victims' dependents in this single case came to only 2,000,000 dollars more than prevention would have cost.

In a world so utterly disgusted with the whole affair, the FAA fared lightly. It made no statement to explain its own role, because Administrator Shaffer had already cleared out his desk and departed carrying the can, leaving the Agency apparently blameless. But the system began to change. 'Yes, Shaffer was at fault, and he went because at the time the Administrator was considered to be personally responsible for air safety. The responsibility structure is different now,' Matthew Scocozza told me in 1988 when he was Assistant Secretary of the U.S. Department of Transportation. 'I've been in transportation ever since graduating from law school in '73. In my early days, all the top people in charge of air safety were political appointees, like the man to whom it was a reward for campaigning hard for President Carter. Now more professionals get the jobs. The Administrator is Number One dealing with safety, but it is the Secretary of Transportation who bears the ultimate responsibility. It's almost like in Japan. They have a Director General for civil aviation but . . . remember when they had the big Tokyo crash? It was the Minister who resigned! That's the way we're moving. It's healthier. The Secretary is a

* Speiser, Krause & Madole.

cabinet officer who has direct access to the President, handles policy, oversees everything from a general standpoint. That relieves the Administrator from political debts, and frees him to make even controversial decisions.'

But there remain some grounds for serious concern. Critics feel that the Secretary now tends to get too closely involved and act almost like an Administrator. Matthew Scocozza said 'aviation is not a small industry any more. It carries 500,000,000 people a year, and airlines spend tens of billions on new equipment.' In these circumstances, with a politician in ultimate charge yet again, safety may suffer if the interests of an industry begin to carry just too much political weight and get confused with the good of the public and governments.

– 3 –

BOGEYMEN OF THIN AIR

If in doubt, plead 'Act of God'. Failing that, blame the pilot. Combine the two, and you have the perfect let-out: the weather.

When the introduction of the jets brought hopes of 'flying above the weather', *jetstream* became 'the dirtiest word' in the air. This high-velocity wind, circling the earth along a narrow path some 30,000 feet up, offered extra speed and fuel-saving free rides to aircraft, but it was soon to be classified as an associate of murderous downdraughts.

Then there was *turbulence*. Bobbie Allen, Director of the CAB (later NTSB) Safety Bureau in the 1960s, once despaired: 'If I were trying to write a best-selling novel, I'd put "turbulence" in the title – there seems to be so much interest. We don't know if we have a turbulence problem.' Turbulence headlines sold newspapers, no doubt, but was it just a media event when the angry air was named as the probable cause of crash after crash? Najeeb Halaby, who then headed the FAA confessed: 'I can discern neither a pattern nor a panacea, so I don't sleep so well nights. This is a terrible admission to make, but I wouldn't know what to do about it if we did prove turbulence was the cause. I guess we think that the system of corrections we've launched will be the answer.'

He guessed right – for the time being. But then came something even more mysterious and sinister *clear air turbulence*, the sneaky CAT. Weather radars could spot thunderstorm cells, and warn about 'probable turbulence ahead', but CAT had nothing to do with stormy weather, and would strike out of clear blue skies.

Needless to say, all these phenomena had always been up there, but we encountered them only when flying became more adventurous in terms of speed, height and the streamlined structure of the jets. Once recognized and understood, new flying techniques could deal with them. So, apart from obvious mistakes or negligent oversight, all was well – well, almost – until something else was found lurking not 'up there' but 'down here', near the ground. It was not even a case of now you see it now you don't because you never saw it. Like Jack the Ripper, it was there, it had to be there, because countless deaths testified to the presence of a killer. It was selective in choosing its targets, and unless it just vanished in thin air every time it had to have an ephemeral life, for it might slam one aircraft on the ground without so much as touching the two others that sandwiched the victim with barely a minute's separation.

They called it a *vicious windshear*. They called it a *ferocious downdraught*. Many aviators were inclined to pooh-pooh it for they had dealt with both successfully throughout their working lives. But goalkeepers of safety were worried.

The first opportunity to obtain real data came in 1973. On December 17, Flight 933, a DC10 of Iberia Lineas Aereas de España, flew into a mucky afternoon at Logan International Airport, Boston, Massachusetts. The clouds were hanging low, visibility was reduced by moderate rain and drifting fog. Including the crew, there were 167 people on board. They were cleared to land. The pilots were aware of the weather conditions. Runway lights were in sight when the captain disconnected the auto-pilot though not the autothrottle system. The aircraft was aligned with the runway, and the engineer called 'minimum decision height'. That meant they were too low too soon but no problem to the captain who overrode the system, advanced the throttles, and pulled gently on the yoke. 'Still low', came the warning, and the captain applied more power. Despite his efforts, the aircraft continued to descend. Rapid calls of diminishing altitude erupted: fifty, forty, thirty, twenty, ten . . . and the aircraft struck the approach light pier, then the embankment of the harbour, bounced up and down, and burst into flames as it began skidding off and alongside the runway. Fortunately, there were no fatalities. The only three serious injuries were suffered during the evacuation.

According to reliable witnesses, Flight 933 had come in 'desperately low' and 'too low to recover'. Verbal exchanges among the crew were recalled from memory (there was no Cockpit Voice Recorder on board, and none was required), but, this time, the investigators had an invaluable aid at their disposal: a DFDR (Digital Flight Data Recorder). Although the casing was battered, the recording was intact, and it helped to reconstruct the flight profile, the rate of descent – and the wind. (The data for heading, airspeed and altitude told the investigators how the aircraft would have behaved in 'no-wind' conditions, and this plot could be compared with the actual flight path. The force of the wind had to be responsible for the difference.) So for the first time, the NTSB could determine 'with confidence' that the impact was caused by a significant low-altitude windshear. The facts were substantiated by simulation that recreated the weather conditions – and yet again, they reduced the crew's ability to complete a safe landing.

The final report analysed the problem, and called for urgent research. Among its seven recommendations it suggested that *all pilots should be trained specifically for flying out of low-level windshear*, and that it was essential *to develop equipment to measure and report windshear*. There was some response, but the Boston crash did not become a *sit-up point* for the industry, perhaps because the windshear in this case was not the kind that would normally drive an aircraft into the ground, and because there were other factors involved. (The pilot was just shifting from auto-pilot to manual control, and so probably lost valuable moments to identify the cause of the sudden, rapid, involuntary descent.)

The list of recommendations arising out of windshear-related incidents kept growing. In January 1974, a PanAm 707 crashed short of the runway at Pago Pago, killing 96 people. Heavy rain and windshear were blamed. And then, in 1975, another daytime disaster, at JFK, New York. Once again, the aircraft crashed into approach lights during the landing phase: Eastern Air Lines Flight 66 was destroyed by fire on the ground, and 113 of the 124 people on board were killed. This was the sit-up point.

Severe weather conditions did their best to conceal the real cause of the JFK accident. The handling of the aircraft offered a chance to blame the pilots: their 'delayed recognition and correction of the high descent rate were probably associated with their reliance upon visual cues rather than on flight instrument references'. But now both CVR (Cockpit Voice Recorder) and DFDR data were available, and the mysterious bogeyman could be pinned down in the accident report: 'The adverse winds might have been too severe for a successful approach and landing even if' the crew had 'relied upon and responded rapidly to the indications of the flight instruments'. But how could they? They did not know what the nature of real threat was or what exactly to do against it. During the previous year, Eastern Air Lines staff-bulletins had dished out some suggestions how pilots could detect low-level windshear, but 'did not provide specific flying techniques to overcome the effects' of such dangerous conditions.

So what are those too severe, adverse winds? First there is a *downburst*. Hitting the ground, it fans out like an inverted mushroom, and creates horizontal winds in all directions. On approaching a microburst during take-off the pilot flies into a headwind of up to eighty knots that rapidly increases his indicated airspeed; following his instinct or Flight Director, he pulls up to try to maintain climb out airspeed, but barely a mile away, the aircraft is slammed down by the vertical wind-shaft forcing it to descend. Losing indicated airspeed, the nose of the aircraft is now lowered in order to try to maintain climb out speed of at least $V_2 + 10$. To complete the dramatic scenario, all within a minute or so, the wind changes to the opposite direction: the aircraft is in a nose-down attitude, and the sudden strong tailwind gives a further reduction in indicated airspeed even to the point of stalling. As this happens, the stickshaker operates cueing the pilot to push the control column further forward in a desperate attempt to increase airspeed – which drives the aircraft into the ground. During approach, the problem is no less serious. (See page 56.)

The anatomy and nature of such downdraughts were discovered and explored by T. T. Fujita, Professor of Meteorology at Chicago University. He studied the JFK and similar accidents, and coined the name *microburst* for the elusive menace, a name that sounded precipitous, dramatic and sinister enough with a touch of science fiction flavour to make everybody sit up. If *turbulence* made

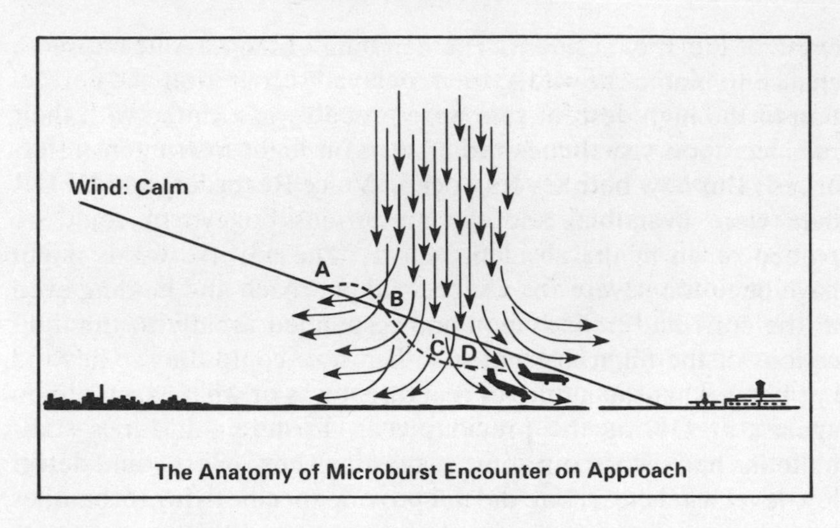

The Anatomy of Microburst Encounter on Approach

best-sellers, *microburst* was destined to become a blockbuster. But were the investigators impressed?

'Hey, wait a minute, I said to myself, what's so very new about it all? The new findings sounded too familiar even in the light of our own fourteen recommendations at the end of our JFK report which was now two years old,' Bud Laynor told me. An engineer with industry and Navy background, who had been trained 'on the job' as a crash investigator seventeen years before, he dug into the NTSB archives, and emerged with some shaming discoveries. He unearthed, for instance, a CAB report concerning an American Airlines DC3 accident, which occurred on July 28, 1943.* Sixteen passengers, including an infant, and all the crew were fatally injured. One of the survivors, a U.S. Air Force pilot described his initial fright: 'Immediately prior to the crash there was quite a sudden change of pressure, and of course a light feeling on the seat, which indicated a sudden loss of altitude to me.' He also noted some pressure on his back, revealing to him the application of full power. Like others at the time, he was used to flying by the seat of his pants, and his interpretation was perfectly correct. As for American's company policy regarding

* John Owen, a former Inspector of the AIB, now a director of the Cranfield accident investigation course, recalls that in Britain, a Hermes take-off accident in the early 1950s was the first to be attributed to 'a sudden shift in wind direction'.

thunderstorm areas? Contemporary staff-bulletins warned against but did not forbid pilots flying through storms: 'Avoid them . . . as a safety factor and for passenger comfort.'

In the absence of then simply unimaginable recorded evidence, the investigators had to rely on plain reasoning alone, but what their report described most accurately as the cause of the 1943 crash was microburst by any other name. It was followed by recommendations which were to be repeated in essence in 1947, and in the 1950s, again and again, and in the 1960s, and in technical papers by Delta pilot Bill Melvin warning against the low-level windshear problem: 'Sweptwing jets? They only accentuated the risk because the aerodynamic characteristics and engine response times of these airplanes made the situation more critical in some windshear situations.' And still more repetitions of unheeded warnings in the wake of more accidents: Flying Tigers DC8 at Okinawa in 1970 – four killed; non-fatal cases in May and December 1972; thirty-eight killed at St Louis in 1973. In all there were some seventeen cases actually associated with windshear before the Iberia crash at Boston in 1973.*

Many cases, attributed to thunderstorms which might have had microbursts embedded in them, could have been re-interpreted after 1975. Could have been, indeed, but were not, not yet, anyway. For, despite the new, fuller understanding of downbursts brought about by the JFK crash, the scare was still not penetrating deep enough to command the attention it deserved. Professor Fujita's analysis was regarded as controversial 'because only a handful of meteorologists, at the time, could visualize that a downdraught descends as low as 300 feet above the ground before spreading out violently'.

Experienced pilots had a sensation of *déjà vu*. For example, Captain Harry Harrison, manager of the simulator division of the British Caledonian Flight Training Centre at Crawley, Sussex, recalled that 'during the war, we lost lots of aircraft to something we didn't even know to have existed – jetstream. In 1951 I flew a fighter plane above clouds, and got lost, ending up miles away from where I thought I was. It baffled me. Did I make a stupid

* William G. Laynor: Summary of Windshear Accidents and Views about Prevention, SAE Technical Papers 681, 1986; also a Study of Windshear, by U.S. ALPA, 1986.

navigation error? It was only several years later that I fully understood: jetstream did it to me. More recently it was the microburst that puzzled us the same way until in the light of better understanding, many of our so-called pilot error cases would have to be reassessed.'

The American ALPA suspected a potential repetition of the sad drum altimeter saga: 'For some twenty years, that instrument was highly susceptible to misreading, yet the NTSB kept accusing pilots of having failed to look,' said Harold Marthinsen, Director of ALPA's investigation department. 'Now another batch of mistakenly blamed pilots should have been exonerated. The DC8 that crashed at Okinawa in 1970 was a classic case. It was a cargo flight, so there were not many fatalities. Having run into a heavy rain shower, it hit the water short of the runway. After the Fujita paper we felt reasonably certain that most probably the Okinawa crash had been caused by a microburst. But it remained chalked up as crew failure. We went to the NTSB and asked for a revision of several similar cases, and a change of the probable cause where obvious. It was refused. The Board wanted a full petition for each review, which is a massive task.'

It is easy to sympathise with ALPA's chagrin: every innocent pilot should be exonerated irrespective whether it was actual blame or damaging implication that tainted his name. But the investigating authorities' position is equally clear: although their investigations are never 'closed', and although they are open to accusations of bureaucratic cover-up, they cannot review their reports on just anybody's, even ALPA's, say-so. Like a court of law, they need new evidence and strong arguments for a new 'trial'. And yet, an urgent reassessment of some pre-1975 accidents, that seemed now potentially attributable to microburst, might also have generated a more determined response from the regulatory authorities and the industry as a whole.

'Instead of leaving it to the judgment of individual companies – some more conscientious than the others – immediate better training techniques to deal with microburst could have been introduced universally, and they might have saved lives,' said Bud Laynor. 'And training would have been only the first, perhaps the easiest step to take. At the same time, the development of hardware should have been pursued with complete devotion because, in tandem with training, that is the ultimate solution. For you can't *see* the wind, you cannot even be sure that heavy rain would always be associated with microburst, so you cannot avoid it without proper information. In the mid-western part of the United States, in high plane areas like Denver, Colorado, the *home* of the microburst, you often experience the phenomenon *without* any rain.

'About twelve years ago, the FAA introduced a Low-level Windshear Alert system that gave us some patchy protection. We were still a long way from developing airborne warning devices and ground-based microwave doppler radar that would let you *see* what's ahead. The alert system failed to warn and protect an Eastern Airlines 727 pilot at Atlanta in 1979, but luckily, his company training did include techniques to combat windshear, and that helped him to turn a potential accident into the virtual non-event of a missed-approach.'

Others were not so lucky.

Aircraft fly constantly over the neighbourhood of New Orleans International Airport. It is most unlikely that, on the showery summer day of July 9, 1982, those who lived a good half a mile beyond the end of Runway 10 would have bothered to look up when hearing the roar of just another departure. Yet it signified that eight of them had some fifty seconds to live.

PanAm Flight 759 took off into gusty, swirling winds at 4.08 in the afternoon, climbed to about a hundred feet, then despite its nose-up attitude, began to descend unstoppably, and crashed through trees into a residential area. Apart from the eight on the ground, 145 people on board were killed by the impact, explosion and fire. Six houses were destroyed, five more were badly damaged.

Bud Laynor, who had participated in the investigation of windshear-related accidents since the early 1970s, commented: 'Microburst was the cause. It would have been difficult for the crew to recognize it and react to it in time. The airport's alert system could not give them adequate warning. Two windshear alert advisories had been issued – one in response to the pilot's request – but those were not alarming enough to delay take-off. In the light of the information available to him, his decision to proceed was reasonable.' A departing DC-9 had encountered windshear some seven minutes earlier and reported it to the controller, but that was not passed on to Flight 759. The windshear that caused the catastrophe was not detected until after the take-off. The airline's own training programme included written instructions for dealing with severe windshear during initial climb, but the records did not show if this particular pilot had done any actual 'hands on' simulator exercises.

The accident report added yet another fourteen to the twenty-four recommendations already sent to the FAA in the previous twelve years. The NTSB noted some 'modest progress' in wind-shear alert and preparedness, but apparently, the real sit-up point had still not been reached and, during halfhearted alterations, business remained as usual. The alert system was repeatedly shown to be of limited use; controllers failed to relay in proper terms the little information that was available; pilots took inappropriate decisions; and only luck saved lives in microburst incidents at Denver and Detroit.

A special problem was that many non-American pilots and airlines appreciated the risk even less, and believed that there was not much to worry about if they did not fly to the *home* of the microburst. 'Denver might have been unique, of course, in having an average of 330 microbursts a year,' said Captain Harrison, 'but similarly catastrophic windshear could also develop above deserts and in any country with a hot climate. And even the high plateaux and extreme climate are not the sole prerequisites for dangerous wind conditions to occur. When a steady, say, ten-knot wind is blowing, it isn't much affected by trees, buildings, etcetera. As its speed increases, friction with the ground, hills and buildings can cause quite severe turbulence, big fluctuations, gusting horizontal windshear, and eddies that we have been forced to live and cope with for many years. But the vertical windshear or microburst – trust the Americans to come up with a good, dramatic word for it! – is something else. It doesn't need a storm. The rising and falling currents associated with the development of cumulonimbus clouds can create vertical wind hurtling down at eighty knots on what appears to be a perfectly clear and calm day. And most pilots were unprepared for that even in the middle of the eighties.'

The day was far from being clear or calm on August 2, 1985, when Flight 191, an L-1011 Tristar of Delta Air Lines left Fort Lauderdale, Florida, for Los Angeles with an en route stop at Dallas/Fort Worth International Airport in Texas. It had 152 passengers and a crew of eleven on board. The pilots were told to expect isolated thunderstorms on the way. Passing New Orleans, they in fact changed their route to avoid bad weather even though that necessitated a few minutes' hold over Arkansas. Thirty-three seconds past 17.43, they were cleared to descend towards Fort Worth, but the captain felt uneasy about the heading offered to him: he could see a 'pretty good size' weather cell that way, and said: 'I'd rather not go through it, I'd rather go around it one way or another.' So he got a new heading, and told the first officer: 'You're in good shape. I'm glad we didn't have to go through that mess. I thought sure he [the Approach Controller] was going to send us through it.'

At 17.52, reacting to a warning to all aircraft about rainshowers in the area, the first officer said: 'We're gonna get our airplane washed.' After a report of gusting winds, the CVR recorded an unidentified crew member saying: 'Stuff is moving in.' At 18.04, the routine landing checks began. A few seconds later, the first officer noticed lightning 'coming out of that one'. 'Where?' asked the captain. 'Right ahead of us.' The descent continued along the final approach course. '1,000 feet,' was called, and the captain cautioned the first officer to watch his indicated airspeed. Rain drumming on the roof could be heard in the cockpit. Two seconds later the captain warned again: 'You're gonna lose it all of a sudden, there it is.' And another five seconds later: 'Push it up, push it way up.'

On the voice recording, against the monotonous background of radio traffic with other aircraft, 'way up, way up, way up' is heard. Then the sound of engines spooling up. 'That's it,' says the captain, and then: 'Hang on to the # #.'*

At forty-four seconds past 18.05, the Ground Proximity Warning speaks up in a metallic voice: 'Whoop whoop pull up,' and the captain responds with the command 'TOGA!' to get the Take Off/Go Around switch activated.

The tower is heard telling an aircraft to turn on to a taxiway.

18.05:46 seconds: 'Whoop whoop pull up.'
18.05:47 – Captain: 'Push it way up.'
18.05:48 – 'Whoop whoop pull up.'
18.05:49 – 'Whoop whoop pull up.'

* In the transcript a # stands for a 'nonpertinent word'.

Over the repetitive, emotion-free robot voice come noises similar to landing or impact. A warning horn wails for 1.6 seconds.

18.05:53 – A pilot says '#'

18.05:55 – 'Oh #,' and noises of second impact.

18.05:58 – The recording ends.

Phases of the crash were witnessed and clearly marked out on the ground. People on State Highway 114 saw the aircraft break out of a wall of rain, touch down momentarily a good mile from the north end of the runway, become airborne again but only to hit a car killing the driver on the westbound lane. Finally, it snapped a light-pole on the highway, cut a furrow in the ground towards an airport water tank which it struck with its cockpit and left wing. The fuselage then spun round and disappeared in an exploding fireball from which, somehow, the tail emerged skidding backwards. Apart from twenty-four badly injured passengers and three cabin attendants (mostly in the aft 'smoking' zone), everybody on board was killed. Miraculously, two passengers walked away shocked but unscathed. Had Flight 191 crashed to the left instead of the right of the runway, several taxiing aircraft could have been hit.

The investigators' conclusions were a devastating comment on the state of the art.

There was a terrific microburst with 'six distinct reversals of the vertical wind components' along its southern side. The pilot who was flying the aircraft successfully transited the first part of the microburst but that was not enough to avert the disaster. 'The captain's decision to continue beneath the thunderstorm did not comply with Delta's weather avoidance procedures; however, the avoidance procedures did not address specifically thunderstorm avoidance in the airport terminal area.' By 1985, some 600 deaths and scores of serious incidents had been found attributable to microbursts. Now Delta's Dallas/Fort Worth accident was determined to have been due to the following *probable causes:*

' . . . the flightcrew's decision to initiate and continue the approach into a cumulonimbus cloud which they observed to contain visible lightning; the lack of specific guidelines, procedures and training for avoiding and escaping from low-altitude windshear; and the lack of definitive, real-time windshear hazard information . . . '

Faulty decision, lack of adequate training, lack of microburst detection and warning. This time, at last, familiarity bred some action rather than contempt. The industry and the regulatory authorities began to make their moves without waiting for the completion of the NTSB report and yet another salvo of not at all surprising recommendations.*

How come that experienced pilots kept making faulty decisions?

How come that the sight of an 'opaque rain shaft' with lightning did not induce the pilots of Flight 191 to offend the Mammon of schedules, and *waste* time on a missed approach, let alone a diversion? Were they suicidal or careless? Of course not. To understand how their minds worked it was enough to refer back to something like a NASA technical memorandum† written a full decade earlier: a large part of the decision-making process 'involves the pilot's judgment of probabilities; he is attempting to make wise decisions, often in the face of uncertainty. In addition, he must consider cost and safety tradeoffs . . . ' He must not be left to his own devices alone.

Delta's training techniques and instructions to-pilots received much criticism, but virtually all airlines would have deserved the same. The mere availability of in-house magazine articles and video tapes on windshear were more like casual entertainment than a compulsory curriculum. And it was obviously insufficient to offer relevant training in the flight simulator only if 'sufficient time' for it could be spared. Windshear training was not required by the FAA – and the rest of the world was only too happy to follow the negative example.

Taking a breather in the middle of a punishing schedule of globe-trotting, John Purvis, the head of Boeing's own air safety investigations, told me in Seattle: 'Long before the Dallas/Fort Worth accident we knew it wasn't enough to *tell* pilots what to do about microburst. They'd have to experience it in the simulator. They'd have to be trained to go against the grain, against instinct, against their traditional training. Eastern, American and a handful of others showed real respect for the problem, a few airlines dished out halfhearted guidance for inadequate

* At the end of the accident report, there is a history of recommendations and FAA responses to specific windshear accidents. The coldly worded list of wasted opportunities glows with by now fossilised predictions of doom.

† A Method for the Study of Human Factors in Aircraft Operation, 1975.

techniques, while the rest introduced no training at all for windshear.

'Instead of making a serious effort to reach all-round agreement on the right way of eliminating the risk, a lot of people chose the easy way out by blaming the FAA, but it wasn't as simple as that. In 1984, nine months before Dallas, the industry made an unofficial proposal to the FAA, but it had to go to the Department of Transport – and it just stuck there! It sat real close to the fella who's the head of it right now. Not that anybody would have argued that it was a genuine issue of concern, oh no, but DoT politics held back the FAA from issuing even a small-dollar contract to develop consensus about a training programme for dealing with low-level windshear. Why? Because both government and industry were worried about the possible liabilities: what if something was recommended – and it went wrong? Before doing anything, they wanted to put in all sorts of clauses that had nothing to do with air safety but everything to do with legal cover for everybody.'

'It's very sad that it needed the Dallas accident to change all that,' added Earl Weener, Boeing's Chief Engineer of Airworthiness and Product Assurance. 'But then at last a contract was issued to establish how to recognize windshear, how to train for and deal with microburst, and how to avoid it – because not all windshears can be successfully flown through. The idea was for a core group consisting of Boeing, Lockheed, McDonnell Douglas, UAL and Aviation Weather Associates to come up with a proposed solution, then consult organizations such as the NTSB, ALPA, ATA, NASA, Allied Pilots, and other interested parties for inputs. Eventually, the final report and covering letter (that went to FAA Administrator Engen on February 25, 1987) amounted to a unique document. It demonstrated an industry-wide unity in addressing the problem, and it's looked upon as a sort of hallmark of cooperation. You'll probably have to wait a long time to see all those parties agree to that extent on anything again. And at the end of all that, it cost less than two million dollars! Peanuts – compared to what could have been saved by speedier action.'

Meanwhile, everybody tried to do something. As soon as the FAA circulated the above document for consultation, Delta issued 'a temporary revision' of guidelines to its pilots. In a testimony before Congress, the NTSB called for a 'priority program' to buy doppler radar for airports to detect windshear. So where did the kiss of universal awakening lead?

'To the admission that special training was the first key to success,' said Harry Harrison of the B. Cal. centre.* 'I guarantee that if any seasoned pilot comes to us and encounters a microburst on any one of our simulators without warning and training, he won't "survive". When our microburst programme is not wanted, it must be cancelled from the control computer. Unfortunately, this was sometimes forgotten, with the result that the next crew using the simulator would encounter a microburst and inevitably *crash*. It happened to experienced crews and even training captains. Microburst training is now given as standard to all pilots converting here, including our many third world pilots. They all need it, too. Their first reaction is sometimes akin to panic – and not without good reason. Without practice, they could be lost.'

They are not to be blamed. When a professional simulator training captain gave me a demonstration of the correct technique to be used, this is what happened: 'Shall we go to Zurich? Okay.' At the touch of a button, Zurich airport filled the screen. QNH was set, preparations for take-off were completed. By then it did not just look like Zurich: we *were* over there. The realistic feel of the machine was incredible. We had hazy visibility of less than three miles. No sign of any cumulonimbus build-up. Yes, microburst could be lurking in an occluded cold front. Radar picked up some clouds but showed them at a safe distance. The aircraft was heavy – 250 tons. Take-off at V_2 would be at 180 knots plus ten. We were hurtling down the runway. The 'co-pilot' called 'V_1 . . . rotate . . . V_2' and warned 'wind increasing . . . thirty . . . forty knots . . . Gear up . . . ' At that moment I asked something, but suddenly an alarm was wailing: 'Pull up . . . pull up.' A horn sounded. The captain increased the pitch. At twelve degrees the stick-shaker came into play. The captain pulled on the controls to pitch up to fifteen degrees, but he was a fraction too slow – and a crash was inevitable. I ducked as the ground came up. 'That's one we didn't get away with,' said the captain with a touch

* Now owned by British Airways and Rediffusion, the makers of the simulator.

of hurt pride, 'but at least in the simulator you live to try it once again.'

Yet it was not his fault. Harrison, acting as his engineer, had noticed a big build-up of headwind and should have warned the pilot at once to put on all the power he had in reserve (about five per cent only, yet enough to pull him out), but he was distracted by my question for a second, and by then it was too late.

A landing, just as hairy, was then demonstrated. Windshear hit us out of the blue, and we 'survived' only through some first-rate piloting.

'In a nutshell and in lay terms, the trick is to be prepared, know exactly what you have to do, disconnect the auto-pilot because it could kill you, disregard instinct and even the flight director, maintain pitch and use maximum power,' said Harry Harrison. 'But to this day, special training is still not compulsory everywhere for every pilot. On any British *performance A* aircraft, both captain and co-pilot must be fully qualified for the job, and have it entered in their licences. Not so in the U.S., where only the captain need have it. The co-pilot must be trained, of course, but need not have qualified for it. The reason is cost. An FAA rep told me that the risk was too low to justify the extra bucks. It's crazy. What if the captain has a heart attack?

'Apart from training, a major problem is that *being prepared* means different things to different airlines. We demand, for instance, that an approach should be stabilized at 500 feet or else overshoot and go round. That gives extra protection against microburst. Many airlines enforce no such precaution.

'Some hilly airfields are more exposed to low-level windshear than others. Hills can't be moved, so we have to live with that, but planners of future airports must cater for the risk. In built-up areas, like London's new Docklands airport, there will be more turbulence problems. Strong crosswinds can create difficulties at exposed single-runway airports like Jersey, Manchester, Newcastle or Cardiff.'

Through a survey of twelve major airlines, Boeing studied a sample of crew caused accidents.* One of its findings was that operators who buy training from other airlines will try to cut costs by requesting a less extensive programme than the seller uses for its own crews. They often buy training from the lowest bidder, and that results in serious lack of standardization. Because recurrent training may be bought from several airlines, a pilot may end up equipped with a set of widely differing operation manuals. As a result, airlines that maintain less than sixty per cent coverage of the fleet with company-owned simulators tend to have worse than average accident rates.

The study found that 'the criticality of recurrent training . . . is universally recognized' and that it should place 'emphasis on accident-related operational topics such as windshear . . . ' Not all of the twelve operators in the survey trained captains and first officers to equal standard; only half of them type-rated all pilots to the same level. A few gave recurrent simulator training four times a year. One that tried to cut costs by halving the frequency of the programme, soon reverted to the four times per annum formula because of 'a noticeable degradation of piloting skills'.

Since 1985, scores of incidents with happy endings have been notched up, and many lives have been saved by good flying techniques. But there is a risk of over-reliance on training alone. As Harry Harrison admitted: 'Some microbursts can be of such severity that they're virtually unflyable.' And in 1987, Jim Burnett, then NTSB Chairman, urged all pilots to 'take thunderstorm warnings deeply to heart, even if it plays havoc with your schedule. The alternative is far more destructive.'† Those warnings, however, have remained patchy and unreliable.

So Jim Burnett still regarded the need to provide better detection and prompt information of primary importance. In 1987, testifying yet again before a House of Representatives Subcommittee concerned with hazardous weather and aviation safety, he emphasized that the oft-recommended 'priority programme' for introducing truly effective equipment had only just begun. And in 1988, Dave Kelley, Chief of the NTSB Operational Factors

* L. G. Lautman and P. L. Gallimore: Control of the Crew Caused Accident, *Airliner*, April, 1987.

† Speech to Allied Pilots Association at Arlington, Texas.

Division warned that the 'radar equipment we have at major airports is still nowhere near the type we want and need to actually see and measure microbursts. I know that there're funding problems, and better devices are to come on line in the nineties, but time is pressing'.

John 'Jack' Ryan, the FAA Air Traffic Operations Director, sounded more confident: 'In the United States we're moving in the right direction. We'll have ground-based and airborne sensors to detect windshear. The new data-link will alert all pilots directly and cut out the communication detour of alerting first a controller who, while dealing with heavy traffic, then has to find the time to alert, in turn, pilots in the affected area.' (The Honeywell windshear detector for the MD-80 series will offer pilots the choice between guidance and fully auto-pilot controlled escape manoeuvre.)

But sceptics, baked to taste in the Washington heatwave, expected further postponements of compulsory windshear warning due to the heavy pressures on the FAA from all sides to introduce legislation for costly improvements in interior fire resistance in aircraft, better CVRs, DFDRs, traffic flow and anti-collision aids, to mention but a few of the worthy causes:

'Everybody has his own priority,' said Anthony Broderick, Associate Administrator of the FAA. 'Just now we're running into a capacity problem right up to 1992, when we're going to use all available excess capacity for the installation of collision avoidance and windshear warning systems.'

Thus Americans show at least some determination to bury the microburst and collision risk in one go. Japan and some other countries are ready to follow suit, but Europeans still object to having to fit the windshear devices. This is because of the considerable cost as well as the general attitude to TCAS (Traffic Collision Avoidance Systems) that will become mandatory in the U.S.A. for domestic airlines by 1992, and for all internationals a year later: 'In Britain the Civil Aviation Authority is worried that if pilots used TCAS in the crowded airways near Gatwick or Heathrow airports they could constantly be getting signals telling them to take immediate avoiding action which would only put them into even worse danger.'* The Americans disagree, though

* Harvey Elliott, *The Times*, 28.1.88.

they also have *somewhat* busy airports. And like it or not, Europe and the rest of the world, including airlines that still refuse fully to appreciate the windshear problem, will have to go along with the system if they want to continue flying to places like the 'homes' of the microburst.

'It's never easy to adapt to new techniques and new ways of thinking,' said Roy Lomas, Head of the Air Safety Branch of British Airways which was probably the first non-American airline to introduce specific microburst training. 'In this case, the problem was that it took a lot of convincing to make pilots pay serious attention to something they would never actually see, and to train experienced men to go against their instinct when confronted with a microburst.'

In 1987, when a Northwest Airlines DC9 crashed at Detroit,* the cockpit voice recording included an exchange between the tower and the pilots who, anxious to 'get out of here before it starts raining', were not much startled by what they reported to the ground: 'You just had a microburst out here on two seven I don't know whether you saw it or not but the dust just exploded down there.'

'Okay thanks which way was it going there?'

'Every which way.'

And another voice in the cockpit: 'That's what a microburst is, you idiot . . . Which way was it going – the last I saw it was headed eastbound . . . [Sound of laugh] At Mach three.'

This section of the recording turned out to be a red herring for the eventual investigation. The tragic accident had nothing to do with windshear. But the pilots were concerned about severe crosswinds, and hurried to get away from the ever-present bogeyman that might still have the last laugh if delays in the introduction of major counter-measures are allowed to continue well into the 1990s. For the conflict is a sad illustration of the perennial game of safety football – *The Angels* v. *The Can Carriers*.

* See *Chapter 14*.

– 4 –

ON THE SIDE OF THE ANGELS

Annex 13 and the *Manual of Aircraft Accident Investigation* are the Bibles of the painstaking autopsy that follows every aircrash. The *Manual* is now somewhat out of date due to technological progress since its last edition in 1970(!), but *Annex 13 to the International Convention* has needed no more than a handful of amendments in almost five decades to keep its guiding principles in line with the ever-improving techniques of investigation.

For the regulatory (rather than substantial and interpretive) purpose of investigations, the *Annex* defines '*accidents*' as occurrences in aircraft operations that cause fatal or serious injuries to people; structural failure of or substantial damage to aircraft requiring major repairs; and the disappearance or total inaccessibility of an airplane.

'*Incidents*' are classified as occurrences other than accidents that can affect the safe operation of aircraft.

The *fundamental* objective of every investigation is set as 'the prevention of accidents and incidents. It is not the purpose of this activity to apportion blame or liability.'

All member states claim that they subscribe unequivocally to this shining ideal, though they admit 'minor' practical variations which they attribute to simple pragmatism in deference to their different laws. The Accident Prevention Manual, an excellent, more recent ICAO document for the guidance of members, comments with the following cautious understatement:

'Although *Annex 13* clearly states that the purpose of an accident investigation is accident prevention rather than the apportioning of blame or liability, in reality the distinction is not quite so clear.'

So is it not time to acknowledge that many states, including some of the giants of aviation, pay no more than lip service to their signatures, and wrap their practices in the shroud of hypocrisy?

Even in countries where the investigators' final report is inadmissible evidence, it influences heavily and inevitably the judgment of courts, authorities and insurers. The investigator himself, with or without permissible reference to his report, is very often the sole and best expert source of information for subsequent blatantly blame-finding exercises. Gerard M. Bruggink said: 'The threat of litigation and punishment associated with accident investigation . . . makes it more difficult for the parties involved to be straightforward in their testimony.' Thus 'the factors that underline flawed performance tend to remain obscure and escape correction.'*

In many countries, the investigation is an outright judicial process, while elsewhere it is a job for the police who can and will utilize the safety investigators' findings as they wish. And what may be even worse, as John Owen points out in one of his Cranfield lectures, in some countries, 'due to limited resources of manpower, government departments or agencies who have executive responsibility for air safety, airworthiness or flight standards, are sometimes required to conduct investigations themselves.' Thus they may have to judge their own or their colleagues' performance with natural self-forgiveness or automatically punitive measures in their minds.

Ian Awford, a leading member of the London law firm Barlow, Lyde & Gilbert, summed it up in an article†: 'The ideal situation would be to divorce completely the functions and product of the accident investigator and his report from any subsequent battle on liability with the courts hearing evidence produced by the parties' themselves. This, however, would be impossible in most cases, partly because of the investigator's key role and access to

* To Kill A Myth, lecture presented at the ISASI seminar in Atlanta, Georgia, 6.10.1987.

† ISASI *forum*, No. 3, 1984

evidence, partly because the 'expense, and sheer magnitude of effort, resources and technical skill needed to mount an independent investigation may be beyond the capability' of some parties (e.g. an individual versus an aircraft manufacturer), and this 'substantial imbalance' would create 'potential injustice'.

An appalling tale comes not from some godforsaken corner of the globe but Canada. According to an entirely reliable source who wished to remain anonymous, an investigator was pressurized by a Justice Department official to release the wreckage of an aircraft. He refused to comply, and even shredded some documents. The pressure increased, allegedly his 'phone was tapped, and he received some veiled threats but, being close to retirement, he was determined to stand up for *Annex 13* and his convictions. Consequently, he was eased out of his job. The good news is that it is most unlikely that, these days, anything like that could happen again: the Canadian system was fully reorganized, and better safeguards were introduced.

'Once investigations are misused for the legal blame-seeking process, everybody wants confidentiality for the wrong purpose,' said Olof Fritsch, Chief of the ICAO Accident Investigation and Prevention Section. 'The judicial system mustn't take a free ride on the investigator's back. Otherwise, testimony in good faith by crew or controllers, and admissions of honest mistakes by anybody involved are bound to dry up. We already see it happening. Surviving pilots have begun to claim loss of memory which may or may not be true. Yes, it's a Catch-22 situation: in an ideal world they ought to be encouraged by full immunity to be frank, but once immunity is granted, everybody will want it.'

Only some armed forces' investigations enjoy a relatively high degree of purity in the spirit of *Annex 13*. In the U.S. Air Force and Navy, if a pilot survives a crash, he will have to face two completely separate processes. In principle, he may walk up to the strictly technical safety investigator and confess freely all the mistakes he has made; then turn round, face a court of potentially punitive inquiry, defend himself, give excuses or invoke his Fifth Amendment right to refuse testimony that may incriminate him. 'Unfortunately, even that system isn't foolproof,' said Ira Rimson. 'In the Navy, for instance, where *mishap* was the word for accidents, we encouraged total candour for the sake of safety, but obviously, we could not withhold the facts from the second, the quasi-legal inquiry. Only the confidential reporting systems, run by NASA in the U.S. and the Institute of Aviation Medicine in the U.K., can guarantee total immunity provided that the confession doesn't reveal wilful misconduct. But then, not many pilots are suicidal enough to make wilful mistakes.'

Wing Commander Jerry Witts of the RAF Inspectorate of Flight Safety told me: 'When appropriate, we hand over the full technical investigations to the specialists of the civilian AAIB, while we run in-house inquiries the result of which cannot be used or made available to disciplinary actions that may be initiated, quite separately, by the Commander in Chief. Usually, the hardest thing is to stop people blaming themselves for mishaps. Our extra difficulty is that we train to fly to the limits of the aircraft performance *envelope*, to fly visually and dodge radar, yet protect the taxpayers' money and safety on the ground.'

Unfortunately, such desirable confidentiality can never be achieved in civilian investigations. Despite their endless, well-meaning protests to the contrary, accident investigators do contribute to the apportioning of blame. It cannot be helped. Take, for instance, two cases already reviewed. Although investigators were looking only for causes and future preventive measures, their findings indicated indisputable blameworthiness of mechanics, supervisors, habits, management, airlines and the FAA in the famous missing O-ring near-disaster, and manufacturers, operators, individuals, and the reporting system in the postal flight disaster when the door flew off. In fact, the ambiguity of the exercise can be traced back to the investigators' Bible. While the *Old Testament*, the *Annex*, emphasizes the purity of the purpose, the *New Testament*, the *Manual*, charges the faithful with the duty to point out individuals' and organizations' acts of omissions, mistakes and negligence uncovered by the investigation.

A British investigation functionary insists that AAIB reports never blame or acquit anybody, and only list the causal factors, but Bill Tench, a former Chief Inspector, admitted at the Toulouse Air Safety Symposium in 1988, that 'some implications of blame may be inevitable'. As Bruggink put it, if there were no flaws, there would have been no accident: 'Shortcomings cannot be identified without also identifying' those who are responsible for them. 'Whether we like it or not, that process has the inescapable connotation of blame. It is our unwillingness to accept that premise that prompted me to suggest that our thinking about the accident phenomenon may need reorientation.'

A similar rethink may also be overdue regarding ICAO member states' attitude to accident investigation. Irrespective of their pretence of uniformity embodied at least in general terms in being signatories to the same set of international conventions, their laws and approach differ widely, and ICAO cannot and will not police them to change their ways.

WHOSE JOB IS IT ANYWAY?

It is the responsibility, both right and duty, of the state in whose territory an accident occurred to conduct the investigation. It receives help from the states of manufacture and registry, as well as from a wide range of experts such as operators and manufacturers. But the variations in attitudes – sometimes openly acknowledged to differ from ICAO standards and recommended practices – are apparent from the start.

Peter Martin, a prominent British aviation lawyer, senior partner in Frere Cholmeley, highlighted the problem: 'The most fundamental difference in investigations exists between countries that have so-called civil and common law jurisdiction. In France, Italy and other civil law countries where it's the public prosecutor's duty to investigate as a police matter any violent death or severe injury and damage to property, the difference between aircrash investigation as a discipline and police investigation are not sufficiently clear. In France, it is the Examining Magistrate who is in charge. He is assisted by police and professional investigators, but the "separation of powers" creates a serious potential conflict which they manage to overcome only when relations between the Ministries of Transport and Interior happen to be good. In many other countries, where there is less discipline in solving such conflicts, there can be great pressure from the public prosecutor resulting in too much haste to find a scapegoat rather than the cause, and a ministry can interfere to suppress embarrassing revelations.

'Italy and Spain are classic examples where the job is much too open to political intervention.* Investigations in Spain belong

* Was it purely with safety and no blame in mind that in 1988, for instance, an Italian magistrate who conducted a crash investigation recommended that nine senior officials should be tried on charges of responsibility?

to the military, with nothing whatsoever to do even with the Ministry of Justice. It's a left-over of the Franco era. If, like the Soviet Union, they don't want to publish and submit a report as merely *recommended* by the *Annex*, ICAO can do nothing about it.

'Currently, there're new moves afoot in Europe to turn investigations into an international effort, but the states guard their sovereignty sometimes against their better interests.'

Recognizing that not all is well in the states of Europe, the EC Transport Commission made some sensible wide-ranging proposals to the European Parliament, but the mills of Brussels grind slowly and sometimes no more than to the nearest convenient halt. Rationalization and coordination of air safety, affecting common problems from collision avoidance to security, and including the creation of a European Accident Investigation Board, may remain not much more than pipe-dreams when even France, a country of well advanced aviation, tends to provoke serious criticism. After the DC10 catastrophe at Ermenonville, for instance, it became clear to the lawyers that 'there would be no public inquiry into the cause of the crash', and that 'the investigation was in the hands of the French government, which was not noted for publishing any findings that might embarrass the aviation industry'.*

Perhaps nowhere are these issues of self-defeating pride more poignant than in the smaller and so-called third world countries where the regulatory and investigative authorities are the same. (The most frequent excuse is a lack of funds combined with shortage of specialists.) And Awford's article revealed that courts sometimes act 'directly on the content of the accident report' to determine liability, and 'there have been instances where reports have not even seen the light of day or even worse where false conclusions have been published in the media'.

In several countries, the available funds and personnel are so meagre relative to the workload that, even before they have any conception of the possible complexity of a case, they set a limit they can afford to the length of the investigation. Boeing investigator John Purvis found that 'in the third world, thirty people may be present at an accident site, but our man may frequently

* *Lawsuit* by Stuart M. Speiser, Horizon Press, New York, 1980.

turn out to be the only one who has ever been to a crash before. We can help, but finally they must write their own report – or appoint someone to do it for them.'

'Getting outside help? Sure, a fine principle. Used to work well,' said Geoffrey Wilkinson, 'but now there's a problem. It's called false pride. India, for example, prefers to go it alone. Zambia was more sensible at the time of the Lusaka crash: they asked for, welcomed and accepted help.'

'Many countries are plainly incapable to investigate at all,' said Peter Martin. 'They don't know even political independence in a true sense – how would they understand the spirit of *Annex 13*? In their haste, any answer can be *the* answer.' With sadness, a renowned pathologist and medical investigator added: 'Not only in the third world but also in some parts of southern Europe local investigators are most interested in proving that their countries were not at fault.'

Colin Allen, now an air safety consultant, painted an equally bleak picture: 'A little knowledge is always dangerous. Emerging nations send their future investigators to courses, and hope that's all that will be necessary. Occasionally, because they have been on a course, these new investigators may think they know it all. Face-saving and pride make them refuse any further outside help. They are then liable to fall into the obvious traps such as wasting time on red herrings and ignoring sound advice from foreign experts and specialists who may be compelled just to hang about for weeks on end until, frustrated by local politics, they have no choice but to pack their bags and go home.' And it certainly takes a different kind of pride, wisdom, and self-assurance to recognize the facts of life and say what Dr Harald Widmer, a lawyer then entrusted with overseeing Swiss accident investigation, once told me: 'We cannot do our job in the highly sophisticated style achieved by the Americans or the British, but we're lucky that we can draw freely on the experience and resources of manufacturers and Swissair.' This certainly did not belittle Dr Widmer's standing or dent his country's reputation.

Perhaps it would help to improve the global standards of investigation if certain interested parties had the *right* to be present and participate without affecting or reducing in any way the responsibility of 'the state of occurrence'. When, for instance, a

BAC-One-Elevenski crashed at Rieka, Yugoslavia, the British AIB was allowed to assist. The investigation, with its enhanced professionalism, led to what Geoffrey Wilkinson regarded as an important rare event: the modification of *Annex 13*, which now at least recommends to grant permission for limited participation to the country that has 'a special interest' in a case 'by virtue of fatalities to its citizens'.

Manufacturers are one of the obvious interested parties that ought to have similar privileges. 'I have six professional investigators in my department, plus company-wide access to any Boeing specialist,' said John Purvis. 'We always offer our support, and usually it is accepted because even the NTSB lacks the full financial and technical back-up we have. If, say, something is broken and found, we ask the government investigator concerned to take his pick but let us have the other half, a sort of mirror image of the fracture, for testing. We then submit our analysis which they can judge for themselves and use or reject.'

Usually is the key word: it is not a right to help. In July 1989, a Korean DC10 crashed near Tripoli airport. Information was withheld from the world, Libyan authorities did not request any help from U.S. specialists, the manufacturers' investigators were held at Tripoli for several hours and then refused entry into Libya, and the flight recorder was analysed by France in such secrecy that McDonnell Douglas knew only what had been reported in the press.

The pilots' representatives, also, ought to gain an official status in crash investigations. In America, the high all-round technical skills offered by the pilots' unions help ensure that their investigators are always allowed to participate in the work of the NTSB. The Air Line Pilots Association has an Accident Investigation Department headed by Director Harold Marthinsen, and specialist committees on national as well as regional and even company levels. Not so in Japan, for example, where according to a document compiled by the Flight Crew Union Federation, the government and particularly police investigators seem to adhere stubbornly to some sort of *keep out* principle.

In Ottawa, Captain Don McBride, head of the Canadian ALPA's Technical and Air Safety Division, complained: 'We used to have a sort of tacit right, but then, under a new Act, there was some resistance to our involvement. It's beyond me why they were often reluctant to bring us in at an early stage. Some government investigators lack the operational experience and understanding, and the result is that ALPA must remain critical of the professionalism of some reports. This could have been avoided if they allowed us, as they do now, to participate in virtually all accident investigations. We have twenty-two trained investigators to help. They're needed, too. Since lawyers have invented "legalese" as a preventive self-defence, some operating manuals can confuse even our members. This could create special problems with blame when human factor causes get mistaken for the media's beloved pilot error.' And in Montreal, Gary Wagner added: 'Of course we recognize the potential conflict of interest. That's why ALPA will assign two men: one to help the investigators, and one to help protect the pilots. We find these functions perfectly compatible.'

There could also be a good case for the operator's right to take part. Captain Heino Caesar, director of Lufthansa's considerable safety organization, has no doubt about it: 'We ought to be there by right – not only by special permission as if it was a favour. The operator has a vast amount of in-house experience to offer, sometimes more than the manufacturer. Besides, it's not enough to get some early warnings as the investigation progresses. We ought to be present, wherever the crash occurs, as the facts emerge day by day, maybe over the years, because otherwise we don't get informed as fully and speedily as we ought to. Mostly, the earliest glimpse of the facts we get is at the draft report stage! That's not good enough.'

But who is the real operator? 'Occasionally this can be confusing,' said Principal Accident Inspector Christian–Heinz Schuberdt of the German Flugunfalluntersuchungsstelle at Braunschweig. 'Take the recent taxiway incident, still under investigation, at Frankfurt. It was a DC8, owned by Gabon, operated by BalAir, a Swiss company, with French pilots, but the flight had an LH number because the cargo on board was booked with Lufthansa, making it a Lufthansa flight for all intents and purposes. So who was the real operator?'

It is unfortunate that much of the argument *against* wider participation comes on the basis of sometimes true, sometimes imaginary confidentiality. It took ICAO sixteen years to obtain a flow of vital reports of incidents for the benefit of all members because the states considered the information 'too sensitive'. (States that remain in or are emerging from the communist limbo still keep information to themselves. Bulgaria for example has declared its readiness to submit reports and reveal its findings as long as they concern foreign aircraft: 'In all other cases the question is resolved according to the concrete circumstances.') Confidentiality can be used as a euphemism for the opportunity to conduct a cover-up. In some states, CVR and DFDR details are withheld even from interested parties in the name of confidentiality; elsewhere tit-bits of the transcripts, including unnecessary details of private conversation, are revealed prematurely to implicate possibly the pilot or the operator in order to whitewash another party such as a country's own air traffic control.

Annex 13 (#5.12) is clear about the consensus of intent concerning flight and voice recordings, but, due to differences in laws and the philosophy, numerous states differ openly or otherwise from the ICAO standards and recommendations, as shown in a Supplement to the *Annex*. Australia tries to safeguard DFDR and CVR recordings but finds it necessary to make them 'available to any court which issues a subpoena requiring them' as evidence. Austrian confidentiality 'expires the very moment the Final report has been released'. Canadian legislation 'precludes any possibility to guarantee . . . protection from disclosure'. In Greece, where the courts are in charge, they have the right to determine what is and what is not to be treated as confidential. Japan finds it 'difficult to comply with the provision of 5.12'. (An exceedingly modest statement in view of prevailing laws and practices.) New Zealand cannot guarantee compliance but will take 'all practical steps' to 'minimize the extent and occurrence' of disclosures. South African findings have no protection from a subpoena served by the courts. In Sweden, the Freedom of Information Act precludes confidentiality. Switzerland simply 'cannot comply' with 5.12.

The protection of recordings is 'contrary to German legal regulations'. Manfred Küppers, Chief Inspector of German Accident Investigations, told me: 'My office enjoys full independence from the regulatory authorities, and reports to the Minister of Transport directly. But in our country, it is the public prosecutor who is officially in charge of inquiries into fatal accidents and serious injuries, so his word carries greater weight than anybody else's. It means he has automatic right to the full voice recording. Normally, he'll respect our investigation for the sake of future safety, and we try to protect confidentiality. Sometimes he may not even know that we have a voice recording, and we may not volunteer the information. But if he asks for it, he must have it. It's hard on the crew, and hard on us because the law doesn't give us the right of secrecy. With the press it's different: we actually try to reveal to them as much of the proven facts as possible because otherwise they will only invent their own stories.'*

The law firm Frere Cholmeley was involved with a case concerning the midair collision of two light aircraft near Hanover. It was not untypical of the German authorities' attitude that, as John Balfour put it, 'the investigators were quite open with the lawyers, allowed us access to evidence, witness statements, and emerging conclusions. They listened to everything we had to say – informally. But the final report, a rather short one, took no account of the representations, and there was no formal route to query its findings.'

In the United Kingdom, where the independence and confidentiality of investigations are worn on everybody's sleeve with pride, it is often forgotten that foreign investigators' proposed findings cannot be kept confidential, as required by the *Annex*, unless the information is 'obtained in confidence'. Legally, inquiries into every unnatural death are in the domain of the coroner and police, but fatal aircrashes must be investigated simultaneously by the AAIB which reports directly to the Minister of Transport. 'Theoretically, there's room for conflict,' said Peter Martin, 'but in practice, in the good old English way of compromise, both the coroner and police will wait for the AAIB's conclusions, and use the technical information available from the report. So everybody retains his exclusive role, and the performance is held together by compromise without conflict.

'Until 1961, when the Cairns Report led to the separation of investigations from central government, there were opportunities for "intellectual coruption", i.e. to protect the government from

* The German accident investigation bureau is known internationally to be a good one, but many specialists criticize its being too small for its workload. It has no spare capacity to run a full and valuable incident reporting system, and its relevant statistical data are admittedly 'just the tip of the iceberg'.

responsibility for, say, aircraft certification. Incidentally, isn't it incredible that such separation in marine investigations was introduced only in 1987, following the *Herald of Free Enterprise* ferry disaster?'

The AAIB's findings must be published, but the law allows it to conduct its business in private. The final reports are not admissible evidence in court, but the investigator can be called to state facts with a reference to his report which, if favourable to one party or another, will exercise some inevitable influence.

In specially important cases public inquiries can be initiated, but British investigators prefer to talk to other experts without the need to explain it all to the press and public as they proceed. (It was only half in jest when John Owen, ex-AIB, said: 'My advice to new investigators is that they should provide all possible assistance to reporters – short of help.') Lawyers can ask 'specific questions regarding evidence,' said a British investigator, 'but fortunately, unlike in some Scottish inquiries, they must not conduct a fishing expedition by browsing through all records associated with the subject.' Before a report is issued, every interested party is invited to comment on the proposed draft. This used to give them the chance to defend themselves in advance no matter what delay it might have caused, and helped everybody to avoid even a hint of controversy in public. So regulations and bureaucratic outlook used to team up to keep everything under wraps as long as possible though it was not always clear whose interest that prolonged secrecy served. In the wake of the Manchester experience, it will now be possible to impose a fine on any party that uses the *right to comment* for obvious time-wasting. It must be emphasized that safety was never compromised by the long wait for comments because findings and recommendations are always revealed to the authorities and the entire industry on a world-wide scale without any delay. Lately, the AAIB has begun publishing interim progress bulletins, and the school of 'publish and be damned' is gaining popularity to eliminate the lengthy delays of confidential consultations.

Annex 13 aims to protect the confidentiality of any evidence unearthed in order to safeguard the availability of information for future investigations, but the courts can demand to see any evidence or records.

Donald Cooper, the current British Chief Investigator of Accidents, expressed the hope that if some U.S.-style Freedom of Information law is ever introduced in the U.K., account would be taken of accident investigation interests to avoid the loss of all-round voluntary cooperation. For example, foreign manufacturers are normally very forthcoming, he says, but if they thought that their evidence and analysis, freely volunteered, could be readily obtained by lawyers or the press, such cooperation could be prejudiced. To get the same information formally through the courts would take the accident investigator a long time. If there was complete freedom of information in this area, organizations might start holding evidence 'off the file', and introduce 'double book-keeping'. American investigators, for instance, ask sometimes for information to be passed to them only verbally, because once any written evidence is in their possession, they may be unable to guarantee its confidentiality.

The United States Freedom of Information Act eliminates secrecy beyond some strictly defined periods of delay that serve as limited safeguards. Undoubtedly, the Act can create problems, and some people who feel threatened may clam up, but it would be a total fallacy to claim that the Act has demolished the effective investigation of accidents. It is all a matter of degrees, of course. Numerous specialists, including NTSB investigators, told me how much it would help them to borrow documents to study freely in total confidence, but confidentiality and secrecy are different concepts, and there can be no doubt about the success of what the late Ed Slattery of the NTSB once fought for so hard – the open process. The *sunshine* Board meetings where any harmful theorizing can be rebutted by anyone, the daily status briefings of press and interested parties at accident sites, and the public hearings have inspired a general faith in the system that has greatly reduced the chance of any cover-up, no matter how high and mighty the potentially guilty may be.

The frequent leaks that do occur are most unfortunate and damaging to NTSB investigations (some parties to the case may try to whitewash themselves, and manipulate public opinion), but such irresponsible acts are insufficient evidence against conducting official business with openness. What experienced NTSB investigators like Ron Schleede and Steve Corrie would like to see are prolonged better protection of CVR transcripts, and stronger legal restrictions against premature leaks.

The weakness of the open house policy is what Captain Don McClure, chairman of the American ALPA's accident investigation board, describes as the *aha* effect: 'U.S. investigations have become a media event. It's a disservice to the profession and public safety. We get piecemeal information and, especially in the early days after a crash, this triggers off speculation and false conclusions by the press and everybody else. It's the *aha* effect – one day it's *aha*, this is the cause, next day it's *aha*, it's something else. Ed Slattery used to run a tight show. Now there're so many parties leaking things with impunity that the official briefing by no less than a member of the NTSB Board is awaited with less interest than the whispers.' It is no good blaming the newshounds alone – sheer inventions that can be rebutted with ridicule are usually less damaging than half-truths that come straight from an unauthorized horse's mouth.

McClure was far from being alone in expressing concern. Even Matthew Scocozza of the Department of Transportation agreed with some of his views but concluded a most forthcoming statement on an optimistic note: 'I used to work on Capitol Hill before the NTSB achieved its full independence, so I'm familiar with the way things have developed. After the Carter presidency, during the Reagan transition period, I wrote documents for the Administration calling for the abolition of the NTSB because I felt it was a sensationalist organization with an identity problem, and not very well run. Remember the DC10 crash in Chicago? A man is racing down the runway holding up a bolt – I still remember the headlines: *"I found it!"* The investigators were so anxious to come up with a quick answer that they almost forgot their own procedures. The famous bolt wasn't the real issue. There was a maintenance problem – that's what led to the engine pylon failure. That kind of sensation-grabbing only holds up the real investigation. Since then, much more discipline has been introduced, and now they're doing an excellent job, using publicity as only one of their weapons. Calling for their abolition? I am ready to eat my words.'

Among the critics, Chuck Miller has no such words to eat because he has always been pro-NTSB. He believes that 'the big investigations are still done in a fine way. But all too often, they just try to get the report out in time instead of going behind the cause all the way. Particularly so in general aviation cases. They put down "pilot error" too readily. And in the smaller cases, not all parties get heard before a brief report is filed. Perhaps fully certified private crash investigators in the pay of some parties ought to be admitted. Otherwise, one of these days, a huge lawsuit may hit the NTSB. Because if they're seen to have asked only some of the relevant questions, or fail to properly examine or protect evidence, they may not be able to claim the old "state of the art" defence against allegations of negligence.'

Miller admits, of course, that much of this is a problem of capacity. The NTSB has only 320 staff in Washington and ten field officers. Their superb capabilities just about keep their heads above water when they have to handle some 3,000 air crashes as well as some 500 rail, highway, marine and hazardous material mishaps every year. The organization is headed by a Board of five members who are not investigating specialists.

'That's because actual investigation is not our duty as Board members,' said Acting Chairman James Kolstad who was raised on a farm where intense interest in flying and the family airplane dominated conversation over the dinner table. 'If we were dealing with only one mode of transportation, maybe we'd need specialists for Board members. But our job is a process of synthesis: constant reading and assessment, with Board members approaching the problems from different angles. We need good judgment, ability to absorb and present facts, and a feel for politics. Not all that long ago, all members used to be political appointees, with not even any transportation background, but that proved unsatisfactory. Now we have people with different experiences – Burnett was a judge, Nall was a lawyer, Dickinson was in transport planning and has a PhD in civil engineering, and Lauber, the technically most qualified among us, has a PhD in experimental psychology and spent his whole career in researching the problems and prevention of so-called pilot errors.'

And newly appointed Lee Dickinson added: 'When we read a report, it is each member's responsibility to make certain that all avenues have been explored, that the analysis is thorough and based on accurate data, that the conclusions are well argued and wholly defensible, and that the recommendations are practical. The Safety Board is also responsible for determining the "probable cause" which is voted on by the five Board members. In a few cases, one or more members may disagree and voice an objection. If the other members are not persuaded to change their votes, the disagreeing member may file a "dissenting opinion".

'All of the Safety Board's work is conducted in the public domain, including on-scene accident investigations, probable cause determination, and recommendations. It is the law that no more than two Board members may discuss ongoing investigations and related safety issues in private. Therefore all deliberations are conducted in open *sunshine* meetings.'*

'These days, we question our investigators more thoroughly than ever before,' said Jim Burnett. 'It keeps them on their toes, and improves the product. It also prepares us for the battles we have to fight on behalf of our staff whom we must relieve of the political and bureaucratic duties that could consume their valuable investigating time. And these battles can be quite considerable when it comes to seeing through our safety recommendations – the main product of a good investigation. Some people claim that the NTSB doesn't live in the real world. But that is not true. We set realistic targets, try not to make unreasonable demands and lose credibility, but we lead, not follow, when we talk about safety, and we always push to the limit without fear of controversy'.

While the leading investigating bodies of international renown like the NTSB and the AAIB grew bit by bit, topsy-turvy, Canada had a unique opportunity, when reorganizing the Aviation Safety Board, to start virtually from scratch, focus on complete independence, learn from both Britain and the U.S.A., and create a potentially top-notch agency. They took a 'global' approach to investigation, employed their own legal adviser, medical and air traffic control specialists, set up a unit to form and monitor recommendations, and another to deal solely with a confidential reporting system, not to mention a first-rate laboratory with a hangar and workshops that can handle anything from engineering problems to recorder read-outs and computerized flight reconstructions. They have three-dimensional photographic and video equipment, and can do non-destructive testing, while, as Ron Hayman, director of the lab said, 'the NTSB needs to rely a lot

* In practice, the members' assistants will convey certain thoughts to other assistants, so that everybody can prepare his argument for the meeting in public. If more than two members want to discuss something face to face, it must be a *sunshine* meeting to avoid even a shadow of potential conspiracy.

on manufacturers and the FAA, and the Brits still lack most technical facilities'. (AAIB recorder read-outs are done in a shack within a hangar, and most engineering analysis work is farmed out to the RAE.)

Alas, that great opportunity was wasted for years. The organization was top-heavy with too many managers and up to ten Board members, and investigators could not even write their own reports because that must be done by *language specialists*. 'All that left too much room for politics and in-fighting, bureaucratic caution, political considerations when accident reports ought to be free from all that,' said Captain McBride of CALPA. An accident at Gander (see p.193) brought most of the problems into the open. That case and the much-disputed, conflicting reports caused furious controversy in political as well as aviation circles. A House of Commons committee was set up to reorganize the CASB, and the pilots' union testified 'that the CASB was wrongly split' between Board members and staff who 'thus tended to work at cross purposes rather than in unity. The Director of Investigations did appear to be a law unto himself,' and CALPA demanded access and answerability at all levels.* The way to such key improvements opened up with the reconstitution of the organization that became known as the Canadian Transportation Accident Investigation and Safety Board, headed by just five 'collectively knowledgeable' Board members.

Viewed from the open plateaux of the North American system, Japanese investigations are probably at the murky extreme end of the spectrum. It is the police (and even worse, the local police) who have the final word. The staffing of AAIC, the Investigation Commission at the Ministry of Transport, shows every sign of an incestuous relationship with the regulatory authority though Mr Fujiwara, himself an Airworthiness Certification inspector turned investigator, claimed with great conviction that these days the AAIC can and will criticize anybody freely, and that their 'recommendations will be accepted usually without any objection or delay'.

In a country where competing air lines find enough passengers to fill several Jumbos a day even on short local routes such as

* Letter from CALPA 2.2.1990.

Tokyo to Osaka, where just one of those airlines, JAL, is the world's biggest international carrier, where corruption scandals seem to be on-stream at all levels at all times, where politicians buy support with favours and cash gifts, where employees are expected to give impressive presents to their bosses, it could seriously damage the quality of crash investigations that the process is ruled by secrecy providing a hot-bed for rumours which gather momentum and credibility with the frequency of strenuous denials.* The law that created the AAIC stipulated that investigations should conform to ICAO standards, and should not compete with criminal procedures. Although according to Article 10, members of the Commission are not allowed to 'divulge to others secrets which may have come to their knowledge in the course of their duties' and although 'this also applies after they retire from their posts', they can release, when deemed necessary, information about work in progress. The Commission has the legal right to question witnesses and interested parties whose false statements, refusal, obstruction and evasion are liable to a fine of 30,000 yen. (The sum is only some £140, but the liability lends police type powers to the investigators.)

'Until 1985, the Commission was extremely silent and conservative,' said Eiichiro Sekigawa, a senior, much respected aviation commentator. 'Since then they've grown more open, but the system doesn't give them true independence. They report to the Minister, their boss, who is also the boss of the regulatory agencies. If they criticize those agencies, it also implies criticism of the Minister who is responsible for them. It's just not on. In Japan, you don't criticize your boss.'

The Flight Crew Union published a document that took a similarly severe line alleging that AAIC investigations are patchy, not conducted in accordance with ICAO standards and recommendations, lack in-house expertise, and as Hideo Fujita of the union put it, 'they're all too often undecided about the cause while being much too ready to simply blame the pilot. They claim to have independence since 1975, but nothing has changed. The findings are not fully reliable. Mr Narabayashi, a former test pilot, who was ready to criticize the government, was just eased out of his (air accident inspector) job.'

* As seen in the of the crash of JAL Flight 123.

Leaks are more devastating in Japan than anywhere else, because the tradition is that those who are accused are expected to show humility and not to stand up defending themselves in public. When the full CVR recording of Flight 123 was published and led to wild speculations barely a week after that huge crash, Yunosoke Tsukamoto of IFALPA was driven to issuing a statement that 'instead of heaping accusations' on all concerned, 'it would be more constructive to stop leaking contradictory and confusing reports to the press, and instead get on with the job . . . Criminal investigation is unacceptable to the international aviation community. The key is prevention, not punishment.'

WHOSE FAULT IS IT ANYWAY?

Accusations, condemnation, leaks, whitewash, cover-up, self-protection are but a few of the acts that ought to have nothing to do with the work of safety-oriented aircrash detectives. Yet they crop up incessantly. Which takes us back to the question of blame. Investigators and their organizations will maintain to their dying days that they are totally unconcerned with blameworthiness. But the question is: do unquestionably good intentions amount to much more than burying honest heads in the sand?

There *is* an inevitable and irritating undertone of hypocrisy. The reason is that, while investigators do not *mean* or *wish* to contribute to any crminal or blame-finding exercise, they are urged by *Annex 13*, and required by their national laws, always to reach some specific conclusions without which their work would be useless for the real purpose of the job: the prevention of more disasters.

It is relatively easy to spot the obviously wrong approach. In the autumn of 1987, Italian airline workers were about to strike for higher pay. Chaos ruled Milan and Rome airports as passengers vied to get on the last planes out. On October 8, a twin-engine ATR-42 of the short-haul ATI line crashed into the Italian Alps near Lake Como. Thirty-seven people, most of them West Germans who had been successful in the fight for seats, were killed. The strike was called off. On the nineteenth, Alitalia suspended all ATR-42 flights as a precaution that implicated the model. Three days later, the makers revealed the flight recorder read-out vindicating their aircraft: after turning on the de-icer, the pilot had failed to maintain minimal speed. Officials announced that the crash was thus caused by the crew's non-adherence to the norms for operations in bad weather, i.e. pilot error.

In 1982, a Japanese DC8 overran the runway at Shanghai. Thirty-seven people were injured. Subsequently, a report was submitted to ICAO with the following conclusions: the Chinese Civil Aviation investigation team 'determines that the main cause for this accident was the explosion of the air brake bottle damaging thirteen hydraulic system tubes and two emergency air brake system tubes, some of which resulted in the failure of extension of flaps and loss of normal as well as emergency wheel braking, thus increasing the roll after touchdown distance to a value greater than the available runway and stopway length. These factors prevented the captain from stopping the aircraft within the runway and stopway confines.'

It says clearly *what* happened – not a word about *why* and *how* it could have happened. Was the bottle faulty? Was it a bad design? Was there a maintenance fault? Was the pilot trained to deal with such problems? Did the co-pilot offer sufficient help? Was there a management problem? No other DC8 user would ever learn anything from that mishap or the report.

'Bad, but not exceptionally bad,' commented Olof Fritsch. As he points out in his international accident investigation course in Stockholm: 'Much too often we get detailed description of *what* happened, but no explaination *why* it happened. Like when it is said that the pilot came in too low. But why did he? Was he mentally disturbed? Was he an alcoholic? If so, how come the management didn't notice it? Or were his instruments faulty? Or was the altimeter of the type that can be misread all too easily? Or was he badly trained? Or misled by air traffic control? These are the answers we need.' And one could add that half an answer or a superficial one could be more harmful than admitting that for the time being the cause would have to remain 'undetermined'.

'It's finding the cause behind the cause that is the essence of investigation,' said Ira Rimson. 'We must go beyond the blinding glimpse of the obvious, and eliminate what did not and could not have happened. But having sat through several courtroom shamblings of truth, it is apparent that many potentially accountable parties are much more dedicated to hiding evidence of their own culpability than they are to admitting their responsibilities for the safety of the public.'

The investigative community has been bogged down for decades with the debate to find expressions for their findings that would circumvent all implications of blame, overlooking constantly that not the words but the uses they were put to by governments, industrial and judicial authorities were the real culprit.

In Canada, investigators used to be required to find the *probable cause*. Then it was changed to *cause related findings and other findings*. 'Then the new Board decided, by no means unanimously, to go back to the *probable cause* supported by *causal factors*,' said Ken Johnson, Executive Director of CASB. 'We're searching for the most helpful formula to prevent accidents. The wording will be reassessed eventually, and it wouldn't surprise me if we reversed our decision.'

Canadian pilots prefer the enumeration of factors to the more blameworthy pinpointing of a single cause. So does Don Cooper of the AAIB: 'We now talk of causes or causal factors rather than *the* cause of an accident. In fact, whilst Annex 13 to the Chicago Convention recommends "state the findings and cause(s) established in the investigation" in the written report, the results of all significant investigations have to be sent to ICAO separately in the form of a summary of facts, and a list of factors for entry in their database. These factors are selected to show why an accident or incident occurred. ICAO says that investigators should not concern themselves with distinguishing so-called *primary* and *contributory* factors as such distinctions are time consuming, very difficult to make, and often arbitrary. Factors are listed in the order in which they occurred.'

Similarly, the Canadian armed forces 'do not establish a prioritized list of cause factors or break them down into primary, secondary or contributing' because they define a cause factor as 'any event, condition or circumstance the presence or absence of which, *within reason*, increases the likelihood of an aircraft occurrence. So whether the cause factor is listed first or last does not matter – its mere presence indicates it was part of the causative equation.'*

* Major T. A. Bailey writing in ISASI *forum*, April, 1988.

Similarly, the German system leaves it to the reader (and the police) to pass judgment because, as Manfred Küppers explained, 'we give *findings* with a list of the chain of events. Causes appear without any weight attached, they're just causal factors, sometimes with *additional factors*.'

In Japan, the language offers an easy way out. The English translation of the law requires the establishment of 'The cause(s) or probable cause(s)'. That is because in Japanese grammar, there is no singular or plural – the original text thus allows for both interpretations.

The internationally most influential American system sticks to demanding a *probable cause* which the NTSB is to find and name if at all possible – and they almost always get the report out in twelve months! Board member John Lauber defends it: 'Before I came here, I made a few speeches in favour of *causal factors*, something like the formula used abroad and in the armed forces. But then, working from the inside, I discovered that the approach is not as narrow as it seems to be. Although we're required to end up with a single causal statement, we broaden it by highlighting the other probable causes, too. Take the Delta 191 accident at Dallas Fort Worth. We incorporated, in fact, three probable causes: the crew's decision to fly into a storm area with lightning, the training programme of the airline, and the given operations procedure for recovery. Blame? No, we don't think in those terms. But there's hardly ever a situation where we don't look for the responsibility of organizations and individuals because we'd never find the causes if we viewed all accidents as Acts of God. We may even reflect upon our own past investigations: was it merely the state of the art that prevented us from spotting a hazard behind a similar earlier accident?'

As honest and comprehensive a view as one could hope for, but does it really deny that the report will help others to apportion blame? Want it or not, the final report *will* exert serious influence even though there is a special, built-in risk of potential injustice: in their reports, investigators need not provide the kind of proof of conclusions that would be legally acceptable to the courts.

In the article already quoted, Ian Awford advocates the abolition of *the probable cause* to be 'replaced by another concept which does not have *proximate cause* or *balance of probabilities* ring to it . . . it should be sufficient merely to analyse the circumstances, list them, and make safety recommendations.' And leave it to the courts to weigh the evidence and draw their own conclusions? In an ideal world, yes. But to whom would they turn if not the best qualified experts, the investigators themselves, to give testimony (as now) and help them weigh the evidence with or without actual reference to their report?

So why not face it? Causes must be named, and, except when events were totally *unforeseeable* due to the state of the art, people and organizations must bear the responsibility – even in many so-called Act of God cases which, with foresight, could have been prevented from having such disastrous effect. Yes, the blaming process, the weighing of various aspects of reports, and the juggling with enabling, contributory and directly causative factors should be left to regulatory authorities and the courts of law, but the investigators ought to stop kidding themselves that their work will ever cease to be used as the linchpin of the wheels of justice.

The judgments of courts are subject to appeal on various grounds particularly in the light of new evidence. The findings of accident reports are no different, and this principle is embodied in the words *probable cause*. It implies, as investigators never fail to emphasize, that there always remains a tingle of doubt, that investigation techniques may improve dramatically, that no case is ever closed. But to live up to this fine principle, they ought to have better machinery for actually reopening cases whether the dispute arises on an individual or a national level.

'WHAT A WAY OF LIFE!'

'What a way of life!' a German investigator once exclaimed; I was not sure if he was complaining or showing off.

I have yet to meet an investigator, active or retired, who ever tires of discussing his work. I cannot just call it dedication or even a way of life; it is a form of addiction. They talk, dine or Sunday brunch air safety, mysteries and victories; their wives feeding the conversation with occasional lines like 'and remember when you were stuck in that swamp' or 'had to learn climbing a glacier' or 'got that call at three in the morning', as if forgetting that an aircrash detective's wife is often left alone to deal with family problems, change the fuse or unblock frozen pipes.

Hardship and surprises go with the job, and when an investigator is 'on call', his go-kit of clothes, personal equipment and favourite hand-tools is never further than an arm's length away in readiness for tropical or Alpine weather. I tried but failed to pinpoint whether their dedication is mostly to the ideal of air safety or to the work they do for it – hunters often find the chase more exciting than the kill. Probably it is a combination, and it may well be the constant challenge they enjoy most. The endless puzzles in a messy aircrash stamp out the painful sights of blood and suffering. When vital clues are missing or distorted or destroyed, the task calls for ingenuity, clear thinking, and the highest level of detective work which must utilize knowledge and understanding of science, human nature, flying, operations, air traffic control, engineering, design, maintenance, meteorology, psychology, human behaviour, the environment, medicine, drugs, alcoholism, fatigue, and lately, more and more, the role of management and economic pressures.

They will not be satisfied until they discover not only that an apple *can* fall on your head, but also the explanation *how* and *why* it fell so that next time other apples can be prevented from falling as well.

'Unfortunately, safety comes in penny packages rather than in great leaps to new standards,' said John Owen. 'Much of the work is hard slogging instead of the glamour of invention'.

Helping to face all this is the backing of firm international camaraderie within the profession. It operates behind the scenes with an assurance of total confidentiality which even some of their own employers may frown upon because such complete candour could be beneficial to safety in the long run but potentially damaging to a company short term. A shortsighted view, yet not unknown in the industry where vast fortunes and worldwide goodwill are at stake.

'If some mishap occurs, what you see in the papers is usually wrong,' said John Purvis of Boeing. 'What matters is that if, for instance, an aircraft develops a crack, I can call up any one of my counterparts in the industry, and ask for the *whole* story, and they'll tell me, no holds barred, things like how many cycles the aircraft had done when the crack occurred, how it ran, how long it was when it finally caused trouble . . . anything. And they know they'd get the same frankness from me.'

'It's the same with airlines,' said Roy Lomas, head of the British

Airways Air Safety Branch. 'And this kind of cooperation comes in handy not only in the course of investigations, but also in prevention. The regulatory authorities ought to be quicker and more positive. They have their own problems, I know, but sometimes we feel we have no time to wait for them: for the sake of safety, colleagues on the grapevine will reveal things they're not yet ready to admit in public, well in advance of legislation that may or may not follow.'

The investigation fraternity is made up of former military, commercial or test pilots, a great variety of engineers, air traffic controllers, plus a sprinkling of doctors and life-long bureaucrats with aeronautical leanings. Sometimes their views clash. A few believe that only pilots can truly understand what went wrong, while, occasionally, I came across the view: 'Crash means wreckage, so the task is technical investigation – what else?'

Like Captain Eric Pritchard, chairman of the Air Safety Group who used to fly all sorts of aircraft, many got interested in crash investigation simply because they cared. Some were induced by actual sights of tragedy. Ray Davis, now a flight recorder consultant, was once 'seconded by BEA to assist the AIB for what I thought would be a couple of hours with the Trident crash at Staines. I was intrigued by the problems, and a year later I was still there.' (Two years later, he was offered a job with the AIB – and he accepted it despite a twenty-five per cent drop in salary.)

Wing Commander Ian Hill 'had dental and medical qualifications, was to be a surgeon but turned out to be allergic to scrubbing solutions, switched to pathology and forensic science', and finally, swapped murder investigations for medical crash detection with the RAF Institute of Pathology and Tropical Medicine, a leader in this field. (It is probably to his advantage that he still does some forensic work and lectures on the subject.)

Geoffrey Wilkinson had some engineering training, was a fighter pilot, was loaned to the U.S. Air Force in Korea, crashed an F84 in a rice paddy, spent a year in hospital with numerous fractures, tried to run his own airline, then applied to be an investigator: 'I wrote to the Ministry saying that here was a man they couldn't afford not to employ. To my amazement, they gave me a job.' His rise to the top and great respect are history.

Most investigators seem to agree that whatever a man's basic training is, the real way to learn their trade is on the job. 'When we recruit, and we tend to get more and more graduates, it's essential for them to realize what tremendous impact the job will make on their social and private lives,' said Ken Smart, deputy head of the AAIB. 'Selection of the right candidates is vital. They need tact, diplomacy and a great deal of understanding sensitively to interview people in trauma. Many of the witnesses will be in a state of shock, but it's important that they're interviewed at the earliest opportunity.'

These days, most would-be professionals as well as those who assist investigations on behalf of unions and smaller airlines, tend to go on a specialist course, such as those of various duration run by the University of Southern California Institute of Safety, the Cranfield College of Aeronautics, the Swedish Institute of Aviation Safety, and the NTSB itself. Apart from techniques and procedures, these courses spread principles that are as crucial as they may sound obvious. The cornerstone of successful investigations is a mental attitude. Frank Taylor, Director of the Cranfield Aviation Safety Centre, once summed it up: 'Collect all the facts. Take nothing for granted. Never jump to conclusions. Don't be tempted to tailor evidence to your pet theory as this may lead you to disregard or even suppress evidence that doesn't fit. If you have a theory, use the scientific approach to test it, that is, try to knock it down. If it continues to stand up, it may be correct.'

'At the end of the day,' said Roy Lomas, 'all investigations require professionalism and sufficient funds. A little of each is worse than nothing. If that is the case, wise men will simply call in the experts.'

Apart from tact, diplomacy, sensitivity and experience, what investigators cannot learn on courses is integrity. Yet it is a crucial prerequisite of their trade because every accident is a battle-ground of conflicting interests between individuals with whom it is all too easy to sympathize – dissenting experts with impressive qualifications, a mighty and rich industry, the regulatory agencies. They may all strive to make greater safety the outcome of every calamity – that is why they are most unlikely even to try to bribe investigators – but it is no secret that they all hope to find that whatever happened was not their fault. If they did not firmly believe that they had done everything right, they would not be honest in their efforts in the first place.

In a major accident, particularly in America, the damages

claims can easily be in the billion dollars bracket, so as Steve Corrie said: 'An investigator is right in the middle of clashing high-powered interests with enormous stakes. Yet he must use them all, tame the aggressive, harness them for team-work, and get information out of them. Very few of them would actually distort information, but some may be reluctant to volunteer anything. The investigator in charge is the only one without an interest in the outcome, and usually the lowest paid of them all.'

In most countries, a government investigator's pay is above the national average, but in industry, with their qualifications and long experience, they would earn much more, particularly as airline pilots. 'In Canada, we recognize their almost missionary zeal,' said Ken Johnson of CASB, 'but because they must face considerable pressures, we also try to protect them with what I call an anti-temptation system. In the past, they could moonlight as consultants to the air industry. That's not on any more. They must choose. With management approval, they can do some extra work but only for companies outside our industry – if they can find the time, which is not very often. Every year they must sign a "conflict of interest" statement. If they want to transfer to jobs with a regulatory agency, the job must be cleared for conflict of interest. For example, they shouldn't go to a place where they'd be required to evaluate recommendations they themselves had drafted for their previous employer.

'Offers of favours and gifts are the main forms of temptation. We've got to reject them, no matter how insignificant they are. It's okay to accept lunch, as long as we can reciprocate. In many countries, airlines give free jump-seats to investigators on or off duty. Our people cannot accept them. They must never fly without paying for their seats.

'Against the pressures, we back them with trust. Even if they make honest mistakes, they know we'll back them. When recently we've had to change our report of a MU2 accident, there was no thought of reprimanding the investigator. Lying, laziness, sloppiness are the only things we regard as crimes.'

Wherever investigating and regulatory functions overlap, it is all too easy for the state to insulate itself from blame. But the ogre of undue influence can raise its head anywhere. In Canada, Gary Wagner of the pilots' union is convinced that, following the Arrow accident at Gander, the report suffered very long delays 'because political pressures were in play, and the Board could not get its act together'.

Chuck Miller told me that 'in 1974, there were some quite disgusting efforts by the Nixon White House to gain direct control of the NTSB all the way down to investigations level. Nixon held the view that he must control the entire bureaucracy by installing his own trusted executives, without any regard to the Board's presumed independent status, and by the vilification of some incumbents like myself in various ways that are well documented in the Watergate hearings – let alone in my files. After Nixon tightened our budget to keep us in check and limit the scope of investigations, we testified at congressional hearings that resolved some of our problems, and at least made us truly independent of the Department of Transportation. The stress, however, took its toll on my health, and I left my job in disgust.'

Sometimes the pressures are subtle and gentle. 'Some people will misinterpret a medical report, and try to put words in the pathologist's mouth, hoping that he's tired, and may just let it pass,' said Ian Hill. 'Well, we don't – I hope.'

Peter Martin has observed pressures at work many times: 'the technique differs from state to state, but governments often try to safeguard some imaginary national interest by protecting their air services, their flag-carrying vanity airline or a large airline that may provide a huge chunk of their currency earnings. Most of the examples are too defamatory to mention because in some countries the investigators are too weak to resist.

'In 1973, there was an air traffic control strike in France, and the military were roped in to provide cover. When two Spanish airliners collided in French airspace, the French made a claim promptly, without any hesitation, that the pilot was negligent because he failed to heed a direction from ATC. The Spaniards were equally quick blaming the negligence of the military controller. I was involved with the litigation for years. There is no doubt at all that all sides tried to exert severe political influence, and there were serious attempts of interference with the proper examination of facts. The worst conflict emerged between the French investigators and the prosecutors to find a suitable scapegoat.

'After the Air New Zealand crash on Mount Erebus in Antarctica, again there was great government pressure, and that's bad. There's never an absolute truth, but pressures aim to distort facts, shift emphasis and reduce responsibilities.

'Manufacturers are in a unique position to exert undue influence. Government investigators, even the British AAIB, lack resources and need to rely greatly on the planemakers' help, goodwill and honesty. Sometimes the makers are less than completely forthcoming. Once we were acting on behalf of a helicopter operator suing the manufacturer on the basis of a theory that had been propounded by the accident report. When the case came to court, we were suddenly confronted by a new version of design and construction data, a version that virtually invalidated not only our case, but also the report itself.'

General accident cases may suffer more than others from pressures: although public interest and the death toll may be low, the stakes remain high, and usually, it is a solitary (in America, a regional) investigator with limited time and funds who has to bear the brunt of it all. All traces of pressures may disappear with special ease because the eventual report will often be short, limited in scope, and, by the time it could be questioned, much of the evidence will be lost with the disposal of the wreckage.

Surrounded with all the conflicting financial and personal interests, it must be ego-boosting to the investigator to feel free from Mammon. 'Cost effectiveness is an important consideration – but not in our work,' said Terry Armentrout, former investigation director of the NTSB. 'When we make our recommendations, we can suggest the best, irrespective of the cost, even if we suspect that some of them may be unacceptable or too expensive to the FAA or the industry. Some people say it's easy to side with the angels, but they've never experienced the weight of responsibility that goes with it.'

– 5 –

THE CAN CARRIERS

'Accident investigation is the least desirable form of accident prevention', said Olof Fritsch.

Prevention deals with the future, and a report is almost useless if it dwells only on the past, failing to recommend some line of defence. It needs not say how exactly something should be fixed – that is up to the regulatory authority that will carry either the burden of follow-up action or the blame for inaction for whatever reason.

In a speech to Rotarians in 1988, Jim Burnett summed it up: 'Our recommendations . . . are advice. Some would say we regulate by the power of the raised eyebrow. There's no force of law behind us . . . But because we're non-regulatory, we have no institutional interest to protect . . . This allows us to be completely objective . . . ' And when I mentioned that the investigators use the media to highlight that 'raised eyebrow', and because of that, they are sometimes accused of blatant scare-mongering that feeds the usual post-crash press hysteria, he said 'It's people opposed to our recommendations who take that line. Because they may not like it that once we know the facts, we simply don't pull punches. We cannot afford to if we want to maintain our credibility and independence.'

But pulling no punches also implies that you stick your neck out. 'Politicians will often jump and demand to know why some recommendations were not accepted,' commented Assistant Secretary Scocozza. 'But they overlook the fact that 300 people at the NTSB cannot duplicate the work of 50,000 at the FAA. It's the high acceptance rate of the recommendations that misleads the critics: yes, the FAA makes mistakes, but that doesn't mean that the NTSB is always right.'

If the conclusions of an investigation are wrong or the reasoning that follows them is faulty by, for instance, taking a limited view and disregarding a potential domino effect, the safety recommendations can create new hazards. The risk is always there, and those who make suggestions as well as those who accept them are aware of it.

Take two recent, fortunately non-catastrophic, examples. One concerns the Boeing 737. Following some incidents of engine mounting bolts failure, the FAA *recommended* in 1982 the introduction of a secondary cable support which would hold the engine even if the bolts fractured. After more incidents, U.S. Air actually lost an engine in 1987, and the FAA *directed* all users of the 737 to carry out an ultrasonic inspection for cracks in the aft bolt after every 600 landings. In 1988, the FAA issued a further airworthiness directive: before any 737 could complete 4,000 landings, a special support structure must be installed to ensure that the engines cannot fall off.

In January 1989, it was Piedmont's turn to lose an engine during take-off from Chicago. The 737 managed to get airborne and land safely. When it was found that first the aft bolt and then the other two had broken, the shadow of doubt fell upon the efficacy of the ultrasonic tests because the hazardous failure happened just thirty-five landings after the latest inspection. A fortnight later, a new FAA emergency directive doubled the required frequency of the checks. But further investigation revealed another hazard: the Piedmont 737 was not yet fitted with a secondary support structure, but it did have the cable support; the aft mounting bolt might have fractured during taxiing, but the cable, instead of yielding to a less dangerous incident on the ground somewhere near the gate, might have continued supporting the weight of the engine – at least until take-off when, under the greater strain, the treble bolt failure occurred.

So now the NTSB recommends the precautionary inspection of all aircraft that have the cable support.

Ken Johnson of CASB gave me the other example: 'When the civil version of a helicopter, which had been used extensively by the military, came into commercial service, it was decided that an auxiliary oil tank should be fitted so that, if through some fault the helicopter started to lose oil, there would be enough reserve to get it down safely. Then we lost one in Canadian waters. We found that the crash was caused by the failure of the line that connected the standard hydraulic reservoir to the new auxiliary one! Without the *precaution*, that line would not have been there. Recommended changes need a lot of caution, not only precaution.'

Thomas Hinton, director of CASB investigations added: 'Our job is to spot the problem not to teach regulators how to suck eggs. But the Board's recommendations must be respectable.' And Bill Tucker, Director of Safety Programs, completed the CASB philosophy: 'We've found it better to distinguish between the investigative function and the drafting of recommendations: in essence, investigators are trained to find out what went wrong rather than the best way to correct safety deficiencies. That's why we have a group of analysts to work them out. These people are specialists who can devote all their time to research and feasibility studies.' It is likely that many, even advanced countries, could profit from adopting a similar attitude if they were willing to bear the extra cost.

The ICAO Accident Prevention Manual urges everybody concerned that 'safety recommendations regarding serious hazards should be made as soon as the hazards have been positively identified, rather than waiting' for the completion of the investigation. Most investigators concur: after the Manchester accident, for instance, the British AAIB issued some recommendations actually *years* before the report.

But the most worrying question is: what is the fate of recommendations? Unfortunately, their early acceptance can be interpreted as an admission of failure or negligence. Yes, that is possible, but, more often than not, it merely reflects upon the state of the art. Fearful of litigation, manufacturers who quickly comply with early recommendations sometimes claim to do so reluctantly and only as an added precaution. (I was warned that these days, under pressure from their legal advisers, the planemakers might not even admit what their product was.

Fortunately, this was not borne out by my experience. Well, not quite. While British Aerospace chose to pretend that no questions had been asked, Boeing, for instance, was most forthcoming.)

'Engineers are an honest lot,' said aviation lawyer Tim Scorer. 'Whatever legal advice they're given, safety remains their chief concern, and they may put things in writing even to their legal disadvantage. It's true, however, that generally manufacturers are not very keen on mandatory modifications that are costly to customers. But this attitude can be shortsighted and unwise commercially. A second accident of the same kind impinges more upon the reputation of the product. Admittedly, it is the regulatory authority that chooses between making a certain change mandatory or optional, but they often rely heavily on the manufacturer's views. Airlines are in an even more unenviable position. If a change is not mandatory in every country, the conscientious ones may feel that they'll lose the competitive edge to operators who are governed by weaker authorities that may work hand in glove with the national carrier. So is there a case of negligence against those authorites? Now that can be hard to prove.'

It has also been suggested that some regulators will protect their own track record by resisting improvements that may be or may be seen as corrections of their mistakes. If so, they would be taking tremendous risks: unprevented though publicly foreseen accidents may amount to criminal negligence.

In an ideal world, reaction to a newly seen risk would always be instantaneous, but the average route to safety is not a non-stop flight from Utopia to Shangri-la. After decades of great animosity, the NTSB and the FAA, foremost proponents of clashing interests, seem to have evolved a state of peaceful coexistence. This is what most investigators told me (with a fair amount of grumbling off the record), and this is what Anthony Broderick, Associate Administrator of the FAA said (with a fine understatement about a historic cover-up):

'Our relations with the NTSB have never been better. They're charged to advocate safety without any consideration for administrative, financial or practical problems. Their complete independence from us is crucial. It was brought about by the crash which killed Senator Cutting in the thirties when clearly, the FAA did a less than thorough and professional investigation. Air traffic control played a part in that accident, but that was not brought to the surface willingly by FAA folks. In fact, any country which doesn't set up an independent system like ours is really missing the boat. It's the only way. It brings about a creative tension* between the two bodies, but that's appropriate: the NTSB testifies to Congress and always expects us to do more whenever possible, and we should expect them to be somewhat critical of our lack of response or progress or speed, but that's just fine, that's what provides the checks and balances that made this nation the world aviation leader.

'Our inclination is always to accept their views – unless someone can prove why we shouldn't. Sometimes they make mistakes, sometimes their ideas are too costly or we just don't know how to put them into practice. That's why our acceptance rate is not a hundred per cent. Sometimes we get frustrated that a big report may take them a year to produce, but on the other hand, in many countries it takes much longer'.

Like all forms of friendly coexistence, this, too, needs careful watching. This is the job of the NTSB Bureau of Safety Programs. According to the law, the Secretary of Transport has ninety days to give 'a significant response' to every recommendation. '*We've got it* or *we're reviewing* it is a mere ackowledgment, not a significant response', said Safety Recommendations Division chief Rick van Woerkom. 'Until 1976, we had no specialized group to deal with safety recommendation follow-ups. This activity was left to the people who wrote the reports. If they were busy with new accidents, many things got snowed under on their desks. We've made 2,370 recommendations to the FAA since 1967. Now all those, along with newly issued safety recommendations. are computerized. Some are classified as "urgent" and call for follow-up in ten working days. Every two weeks the computer kicks out a list of items that haven't received a significant response in time, so that the Safety Board can send a reminder. Each recommen-

* In July, 1988, at the height of bitchy warring between Dukakis and Jackson for the Democratic Presidential nomination, Jackson called his relationship with his rival 'just creative tension' which appeared to be some euphemistic positivism as opposed to something British, i.e. euphemistic but compromising, such as 'there's room for improvement' in a political sense as well as in the CAA-AAIB shotgun marriage.

dation is reviewed for activity at least once a year, and the form of follow-ups is standardized.*

'Rejected or not, we keep an eye on the fate of our recommendations. The FAA knows that Congress reacts negatively to delays, and will show displeasure when the budget hearings come. On the other hand, we know that if we overlook something that has an "open – acceptable" status, i.e. the FAA agrees with us but hasn't done a thing about it, our Chairman won't enjoy facing the press with a major issue we allowed to stay dormant for years. On average, the FAA responds in seventy-seven days, and the rate of acceptance is seventy-eight per cent, which is good.

'If they disagree with a recommendation, we encourage action on review of issues, but generally keep it open for up to five years. When there's another accident or some new fact emerges, it's good ammunition for us to say "we told you so". Unfortunately the computer still can't alert us that a new case may not be the first of a kind, but people will remember, and then we can trace back all its forerunners'.

The system certainly enabled the NTSB to exert pressure on the regulators whenever they were dragging their feet. For example, as too little has been done about frequent runway incursions and about the risk with contaminated runways, the NTSB has conducted a special study of the problem using dozens of their past proposals as evidence. Similarly, a vast accumulation of recommendations is helping the fight for improved air traffic control and the introduction of an airborne collision avoidance system. And the long delayed installation of clearer CVR and multi-channelled recorders is probably the best illlustration of this point.

In the 1960s, the Americans led the world into the era of flight data recorders to preserve evidence of crew action and aircraft behaviour long past their tragic end, and to open the way to eventual invaluable performance monitoring. By the 1970s, the rest of the world (first of all Britain, pushed by a most effective

* E.g. 'Do not use "We are pleased . . . " in the introduction. This is a subjective statement and belongs in the section commenting on the response.' Or if the recommendation was rejected but the NTSB still feels that something ought to be done, 'The letter should be closed with an appropriate (depending on the tone of the discussion) tactful sentence' to say that continued efforts to solve the problem after all would be much appreciated.

AIB) overtook the United States, where inertia set in coupled with a dangerous overdose of self-satisfaction. Despite constant prodding the FAA was reluctant to force a struggling industry to spend vast amounts on more modern equipment that would not increase seat/mile utilization, cost-efficiency, revenue, profits or any other concepts the accountants and powerful lobbyists of the airlines would readily understand. The unspoken argument was that recorders would improve accident investigation reliability, and point towards long-term safety measures, but, unlike extra leg-room and leggy air hostesses, not even the neatest recorders would sell a single seat. In the face of evidence freely available in accident reports, the FAA and the airlines had their excuses on tap.

'Take our cockpit voice recorders,' said Don McClure. 'It's absurd. The British and others have hot mikes [individual crew microphones] as well as area mikes [those that pick up every sound in the cockpit]. Our FAA and airlines claim that pilots dislike hot mikes. That's rubbish. It's been ALPA policy since 1987 that hot mikes should be worn when an aircraft is flying below 10,000 feet. After all, it's our members who suffer from poor CVR read-outs which misconstrue or miss out something.'

Jim Burnett speaks with pride about the victory for CVR and DFDR standards: 'The world expected leadership from us, and yet at one stage we faced a situation that the U.S. was about to take an anti-progressive attitude in ICAO, a world forum. So the NTSB took a very aggressive position to promote improvements, and fortunately the DoT backed us. Under the weight of the massive evidence we had on record, the FAA gave in and accepted at least the bulk of our recommendation even though retrofitting of advanced recorders on all airliners was still blocked because of the cost.'

'The FAA had no case to reject recorders with a minimum of thirty-six instead of five channels', said Matthew Scocozza. 'They were wrong, and we convinced the Secretary that Jim Burnett was right. It's an important role of the DoT to act as a sort of supreme court of appeal. If the FAA became completely independent from the Department, we'd need special legislation to force their hand in a case like this, and that would take much, much longer.'

Associate Administrator Broderick refutes some of the criticism: 'It's not that we disagreed about the need of better recorders, but when your decision is going to affect thousands of airplanes, you've got to tread with caution. Now, at last, the proposals are out for comment, but it'll take another year or so to implement them. We're talking about some 300,000,000 dollars, an investment of 50,000 dollars per airplane. And don't forget, the real advantage of the extra channels showed up only in the eighties with the introduction of the digital recorders. Add to this that the process of making new rules can be quite ponderous. The FAA must listen, study the subject, give others a chance to comment.

'One of the most difficult aspects of my job is to establish priorities for rulemaking actions. Last year we proposed or adopted actual rule changes in over fifty areas which I'd classify as significant. We have a backlog at any one time of at least five years' worth of effort at that rate, and we must put priority on those projects which most directly and immediately will result in safety benefits. If a new ruling lacks such direct benefits and involves high costs it may easily lengthen the gestation period, and may even take as long as eight years to go from concept to action. When there is great urgency, things can be speeded up. Take for example ageing aircraft rule changes. The first ones for Boeing took less than two years, and I expect the changes will be completed for the entire fleet in less than thirty months. All of this is to say that the flight recorder rulemaking took far longer than one might hope, given an ideal world and no worries about competition for scarce resources.* In the four years to 1992, for instance, we're going to use all the excess capacity in U. S. avionics for the installation of collision avoidance and windshear warning systems. If we came along with some other demands for additional retrofits that needed significant wiring changes in the voice and flight recorders already in use, the airlines wouldn't have the capacity to comply, and our regulations would become self-defeating.'

But the regulations still aim at no more than thirty-six channels when all the leading airlines see the huge advantage of using many more. In effect it means that the FAA will save money for only *some* airlines, perhaps the ones that need most prodding. Or will it

* Many people in the industry refuse to accept the view that flight recorders do not provide a *direct* safety benefit, and therefore criticize the FAA for using such views as an excuse for the long delay.

all lead to yet another battle for reviewing the reviewed standards almost before they come into force? And how about the relatively smaller so-called commuter aircraft? That fleet carries a vast number of passengers, it needs urgent safety improvements that could be pinpointed by recorders, yet it took fifteen years to issue a mandatory update forcing its operators to install any recorders at all. (Recent commuter accidents like the one at Bar Harbor highlighted the problems caused by the lack of recorded data.)

In Canada, safety recommendations are made public, and go direct to the appropriate Minister, usually the one in charge of Transport Canada, who must respond within ninety days. Rejections must also be in writing. On average at least eighty per cent of the recommendations are accepted but 'these still need a lot of watching,' said Bill Tucker. 'Bureaucrats like to generate answers to get things out of the way, but people change jobs, and great ideas sink to the bottom of the pile. The aftermath of the fire-in-flight case at Cincinnati taught us the lesson. The CASB's recommendations were accepted by the Ministry as were those of the NTSB. A committee was set up to organize follow-up action, but when some key members moved elsewhere, the committee apparently just ceased to function. So now we assign each case to an individual officer to watch.'

German investigators believe that 'if too many recommendations were made, people would be too busy to look at them.' To guard against that, the Flugunfalluntersuchungsstelle keeps up a barrage of letters until they get a decision. (They investigate some 600 commercial and general aviation events a year, and make fewer than twenty recommendations.)

Many countries feel that they have too few cases to justify the running cost of a watchdog body, but Olof Fritsch's argument is unanswerable: even if a computer may be superfluous, card systems are cheap and easy, and can ensure that no valuable suggestion will ever be forgotten.

In Britain, the situation is rather ambiguous. The AAIB is firmly convinced that some eighty per cent of its recommendations are accepted by the CAA without much ado. This high success rate is based on the CAA *saying* that a suggestion is acceptable or worth consideration. Thus even a 'considered but rejected' item may become a triumph on paper because there is no follow-up system, no money for a specialist to keep track of the fruits of the

AAIB's labours. It would be comforting to believe that the lack of controls is a manifestation of perfect cooperation. If so, it would be a sudden, dramatic improvement in the secretive system. That, however, seems improbable. Barely a week before my visits to Farnborough, Geoffrey Wilkinson, the previous Chief Inspector of Accidents, told me: 'The problem is that, while the FAA may at times be in the pocket of the aviation industry, our CAA is just plain afraid of controversy. If only occasionally they were a little faster and gutsier just as they were, for instance, after the 707 accident at Lusaka!'

It is not untypical that, in Britain, many such attributable statements of criticism are voiced only after retirement from public service. William Tench, Wilkinson's long-serving predecessor, had also retired before he wrote *Safety Is No Accident*, a noteworthy book, in 1985. He called the CAA record of implementing recommendations 'lamentable'. He said 'The clearest message which comes through in my experience . . . is an irrational, though consistent, reluctance on the part of the airworthiness authorities to implement fully the safety recommendations of the accident investigating authorities. This may be a psychological hang-up on the basis of "nobody knows better than we do what should be done in the field of air safety", or an attitude that the acceptance of recommendations may interrupt or even conflict' with the normal course of their activities following their own grand design.

In the London *Standard* (25.10.85), the CAA was quick to deny the validity of Tench's allegations. They claimed that most recommendations were carried out meticulously, and stressed most emphatically that 'the CAA and not the AIB has the legal duty to decide what regulatory changes are the most effective for improving safety. This means that in a few cases the CAA may find that a particular AIB recommendation is not practical.'

Like the AAIB, the CAA continues to claim that its cooperation with the investigators is as good as could be. Much of the time this may well be so. But despite the AAIB's official assurances, several serving investigators express non-attributable dissatisfaction with the overall speed and receptivity of the CAA. Deputy Chief Inspector Ken Smart explained that, for instance, 'in the late 1970s and early 1980s it did appear that the helicopter industry in particular accepted and implemented AAIB safety recommendations on North Sea operations more readily than the CAA. However the Helicopter Airworthiness Panel's report in 1984 identified a shortage of CAA staff resources in this area. The CAA has addressed the matter, and today there's a good working relationship between the two organizations. Inevitably, there will be occasions when our views on safety differ. I believe that total agreement on every issue would suggest a far too cosy relationship between investigators and regulators, and call into question the independence of the AAIB.'

The all-round independence is, however, somewhat precarious. The CAA is independent of the industry which pays for its existence and upkeep. The AAIB is independent of everybody, and responsible to the Minister of Transport – who, in turn, also oversees the CAA. AAIB criticism of the CAA may imply that the Minister has not exercised sufficient control over the regulatory body.

It is astonishing to find that the AAIB has no computerized data base of its own to keep in mind its vast amount of invaluable collective experience. The official attitude is that they do not *need* it because the CAA has one, and it would only be a duplication. No wonder that investigators complain bitterly – off the record, of course – that whenever they need to look up precedents, references and past suggestions, they either have to go cap in hand to the CAA to ask for information (which cannot bolster their precious independence) or, particularly if the required data is likely to be an obvious embarrassment to the CAA, they must read and search laboriously through hundreds of lengthy documents as if computers were yet to be invented.

Equally bad if not even worse is that there is no organized 'chaser' system to review continuously the fate of all the recommendations that were accepted but not acted upon, no computer 'flagging' mechanism to warn the AAIB whenever it is high time to nag, beg or even blackmail the CAA. It just cannot be right to leave it to dedicated pressure groups like the Air Transport Users' Committee or SCI-SAFE *(formed by survivors and the bereaved of the Manchester tragedy)* to act as the vociferous conscience of politicians and the CAA. And it must be wrong for the investigating authority to take the moral high-ground, sit back, and wait for the opportunity to fire the occasional salvo of 'we've warned you, haven't we?' accompanied by a, usually confidential, complaint about inaction and delays.

But what is at the root of those delays?

THE STATE OF THE ART or THE ART OF THE STATE?

Decisions taken by a state's regulatory authority may reflect either or both halves of the question. When it is an American decision, it is likely to have worldwide repercussions. The regulators of aviation carry an extensive range of responsibilities from certifying the airworthiness of aircraft to licensing and watching the performance of operators as well as airports, pilots, maintenance engineers, air traffic controllers, etc. In 'a variety of safety matters' the FAA's Technical Center at Atlantic City is, for instance, a world leader, wrote Frank Taylor, Director of the Cranfield Aviation Safety Centre. Although the CAA 'has very little money available for research of any kind', a great deal of valuable work has been done in Britain, and 'pound for pound, our efforts will probably always be more effective than the U.S.A.'s, but *few would deny that in civil aviation safety, we have usually waited for the Americans to act first.*'* So have other states (most of them far more justifiably, as they have far less expertise at their disposal), and the FAA has always had to endure a great deal of flak from all directions partly because of this extra weight of international responsibility. The following is just a random sample of such criticisms:

Joseph Sutter, Boeing's Executive Vice President, was pulling no punches: 'Like that of the airlines, the technical expertise of the FAA has also suffered in recent years . . . In the past, U.S. airplane manufacturers enjoyed an advantage over foreign competitors because we could deliver our airplanes with an FAA certificate' which 'does not enhance the saleability of U.S. products today. Many foreign airlines are concerned with FAA actions which are inconsistent with the FAA's own rules or are reflective of pressures brought to bear by special interest groups.' He complained about 'governance by pressure politics rather than

* Article in *Aerogram* of Cranfield College of Aeronautics, May, 1987. Emphasis added.

technical regulation', a process that 'destroys FAA morale', and, as a result, 'many actions of the FAA are premature and inconsistent and run the risk of reducing safety margins rather than improving safety . . . '*

At a congressional subcommittee hearing in 1984, Captain Duffy of ALPA accused the FAA of covering up safety violations by airlines, and failing 'to conduct the professional, thorough investigations on which the industry has relied for so many years . . . Requests for air traffic control tapes and other data are so slow that evidence is destroyed before our safety experts are able to analyse it . . . ' Whether all this is 'happening out of design or sheer incompetence, we don't know'. And in the same year, in an Appendix to the NTSB safety study of airports, ALPA complained about 'the *total* lack of response' to navigable airspace problems by the FAA: ' . . . all the users agree changes are needed and the FAA consistently and aptly avoids dealing with it. It has always been our understanding that the FAA exists to serve the flying public. Their performance on this topic indicates otherwise.'

Gary Wagner of Canadian ALPA: 'The FAA cannot just *suggest* safety measures. If they matter, they must be mandated. A service bulletin is not mandatory, an airworthiness directive is. Some third world countries tend just to copy the U.S. – they may not have anyone even to read all bulletins.'

The NTSB keeps up a constant barrage despite the currently sweet relationship. In the above safety study several delays were recorded concerning recommendations even though they were classed as 'open – acceptable'. The tragic Midwest Express DC9 crash at Milwaukee, Wisconsin, was attributed to the flightcrew's improper use of the flight controls in response to an engine failure. But the investigation revealed an astonishing 'lack of FAA oversight of certificated overhaul facilities', said Jim Burnett.† 'At the public hearing of this accident, the testimony of a customer maintenance engineer who coordinated shop reviews for Pratt Whitney was very revealing.' The engineer stated publicly that in the previous five years he had not 'looked at a single shop that

* Wings Club: 23rd General Harris 'Sight' Lecture, New York, 1986.

† Speech at the ATA Engineering and Maintenance Forum, Cincinnati, Ohio, 1987.

would survive close scrutiny by the FAA to their requirements, particularly in the area of calibration of instruments.' Not in a single shop could he pick up 'an instrument that's not out of calibration, or I should say not out of calibration date'.

A senior British investigator told me: 'Don't quote me, but it's a fact that you only get anywhere with the FAA and the CAA by blackmail. Or by a few repeat accidents.'

Investigators sympathize with the FAA. 'We criticize them, but we know they have 50,000 people to duplicate the work of 250,000. So they must delegate. Aircraft certification can be up to one man against the whole of Lockheed or Boeing,' said Steve Corrie of the NTSB.

Two years ago, on his appointment as FAA Administrator, Allan McArtor inherited an unenviable situation. In a speech to the National Aviation Club* he pointed out the unprecedented growth of traffic ('at an all-time record level' the FAA 'handled over 96,000,000 operations' in 1987, 'and maintained a safety record unsurpassed worldwide'), but admitted that there was 'a crisis in public confidence in flying'. That was why, at his swearing-in ceremony, he called for 'immediate measures to demonstrate progress to a doubting public', and unveiled *IMPACT 88*, a programme to focus on key problem areas.† Some of these, under blanket headings, may include the art of the state in reacting to recommendations *the* right way – if such a concept can ever be agreed. When, for instance, in the wake of an accident, obvious modifications to an aircraft are called for, should they be made mandatory? The painful search for *the* right way to go about issuing ADs or SBs after the initial DC10 door problems gave an ample demonstration of this dilemma.

'I feel no animosity towards the FAA,' said NTSB Acting Chairman James Kolstad. 'When we have a competent Administrator like McArtor, a lot can be achieved by non-confrontation. Some people don't subscribe to that style. I think it's more constructive than pressure by brow-beating. It tends to yield better results because, to me, the FAA is just another member of the team. I'm a communicator: pick up the phone, and talk. Sure,

* Crystal City, 1988.

† Before he could implement his numerous ideas, the American system of political appointments proved true to itself, and he was out of the job when President Bush replaced him with Admiral Jim Busey.

jealousies and personality problems can make things difficult. We may recommend to the FAA to issue an AD, and they may say that they disagree, that we lack the background to recommend that. So conflicts may arise, but less so if we have better communications.'

The same could be true, of course, on an international level. After an engine had fallen off a DC10 at Chicago, the FAA suspended the type certificate for a couple of weeks. The CAA and the rest of the world had to follow suit. But the CAA (an independent, self-supporting body, founded by the industry *not* the government) decided that the grounding was not really necessary, and that the FAA was wrapped up in its own legal system. So, three weeks later, Britain allowed the aircraft to return to service while the FAA still kept it grounded. Who was right? Was the CAA influenced by the fact that Laker Airline had just taken delivery of some DC10s that were vital to their operations? The CAA was convinced that the engine mountings of the DC10s were safe. On that occasion, they were to be proved right. But what if they were not? What if Chicago were to be the forerunner of an Ermenonville? It would have been a heavy can to carry – a risk they, and the flying public, must live with.

In 1972, the Papa India Trident crash at Staines was a major landmark in the history of accident investigation. It vindicated CVR; led to irresistible demands for extra parameters in flight recorders (by then BEA had already installed sixty-four-channel recorders when the law required five channels, and that was a tremendous help to the investigation), tightening of crew rostering for experience, fuller incident reporting, confidential *confessions* by crews, more strenuous control of pilots' heart condition, better training for sudden pilot incapacitation, elimination of 'non-standard' procedures in the cockpit, regulation to make air safety departments compulsory for all British airlines (BEA already had one, but it needed greater authority to investigate *minor* incidents), the introduction of monitored approach procedures – to mention but a few. It also highlighted a problem with the art of the state rather than the state of the art at the time: at least two almost identical forerunners of the tragedy were seen as minor incidents; about one of these, the Staines report stated that it 'fell neatly between two stools, DTI considered it to be a mechanical defect; ARB were satisfied that it was an operational or human error.' Thus the regulatory authorities, the Air Registration Board (now CAA) and the Department of Trade and Industry, failed to get together and make a decision. As a result, 'neither took action'.

The choice between two subjects of safety research may at times be influenced heavily by the question: how many lives will it save?

'Oh yes, I know that this a big question in some places,' said Lufthansa's Captain Heino Caesar, 'but it's a sad argument, and I've never heard it mentioned in public. It would imply that a company can afford an accident in, say, once every ten years. But that's not acceptable. One death or one accident is just one too many. Sure, starting an engine is always a compromise to some extent. We must give a service and earn a living as safely as possible.' A commendable view, shared no doubt by all, but the level of the compromise is the key factor. 'The trouble is, that our evaluation will always be tested by accidents,' he added. 'It's a problem, and it'll be an even bigger problem in the future, under increased pressures of cost, when some of those nice-to-have extras may be cut out even though they're part of our safety cushion.'

The irksome *how many lives will it save* enigma is most prominent in the decision about where to invest in preventive research and measures. Earl Weener, Boeing Chief Engineer of Airworthiness and Product Assurance, told me: 'The statistics speak for themselves. If we reduced the *maintenance*-induced accidents by half, we'd achieve a two per cent improvement overall. But if we halved the *crew*-related accidents, it would make a huge impact on safety. And yet, look at the FAA research budget. About five per cent of it is related to safety, but the bulk of that is used to deal with things like crash survivability rather than preventing *operational* errors which would save more lives.'

Occasionally, the FAA rejects an NTSB recommendation saying that it had been studied but was not found to be cost effective. 'Of course, the same may become politically beneficial two accidents later,' said Rick van Woerkom. 'In about 1970, we asked the DoT what they regarded as a standard figure for the cost of a life lost in a transportation accident. We got no answer. Nobody wanted to go on record with placing a value on human life. Not on paper. Yet obviously, they must have some figures or they couldn't do cost-benefit studies. The National Safety Council has a figure, I believe, and the Insurance Institute for Highway Safety must have one, too. Or else insurers couldn't set their rates.' (In 1962, a study carried out for the FAA estimated an airline passenger's life at 400,000 dollars which included related losses and even the cost of investigation; in 1983, the value per life saved stood at 650,000 dollars; since then, that figure must have risen sharply with inflation and much increased punitive damages.)

Jerome Lederer, President Emeritus of the Flight Safety Foundation, drew a dividing line between cost-benefit and cost-effectiveness studies because the latter would not always include the dollar benefits.* But at the end of the day, it is the passenger who pays for it – with higher fares or with his life. As Martin Shugrue, President of Continental Airlines put it in *Newsmakers* (CNN TV 27.11.88): 'This a very competitive market, and there are no free lunches.'

Lederer quoted C. Dousset of Sud Aviation, who once gave a good example of numerical assessment for the most effective use of available funds – an exercise that may ignore the smaller probable risks in order to maximize the effort against the greatest hazards. Some safety equipment is unlikely ever to be used during the lifespan of an aircraft. On the basis of accident data, he calculated the number of passengers who would probably be saved by the occasional use of various, already mandatory items of safety equipment per million dollar investment:

Escape slides	7
Oxygen masks	1.5
Life jackets	7
Life rafts	0.2
Fire extinguishers	1.5
Terrain avoidance system	9

In the light of this, it is painful to note that, when the ground proximity warning device, a well proven lifesaver, was not made standard on the Boeing 727, Spain did not ask for this nice-to-have extra – with disastrous results.

* Paper presented at the Alaska Air Carrier Association, Hawaii, 1987.

On November 20, 1974, at 4.42 in the morning, the engines of a Lufthansa 747 'were started up in accordance with a revised procedure introduced by the company the previous year'. Twelve minutes later, as the aircraft was lifting off the runway at Nairobi, the crew experienced severe buffeting. Drinks in the first-class cabin were thrown at the ceiling. The pilot thought the vibration came from an engine or the undercarriage. Then suddenly, at about a hundred feet up, the rate of climb fell rapidly to zero. A former airline pilot, sitting in the window seat of Row 16, looked out and saw that the wing leading edge flaps were not extended. The aircraft sank, made contact with the ground, careered on with its tail scraping the grass, skidded, hit an elevated access road, began to break up, and spun round. There was an explosion, and a fierce fire broke out. Evacuation was limited to the left-hand side. Fifty-nine people were killed, fifty-four suffered non-fatal injuries, forty-three, including four of the crew, got out uninjured. The aircraft was destroyed.

The East African Community Chief Inspector of Accidents chose to ask for expert help, and called in the AIB. Twenty-four hours later, Colin Allen and Dave King were there. 'The scene was in a disorderly state', Allen told me. 'Crowds gathered to view it from the road above, and there were lots of souvenir hunters – pests, really – but luckily, no vital part had been looted. Much of the wreckage was still very hot. The access road was an early suspect. It had been built recently, and because of flooding in the rainy season it was some eight feet high, running at right angles to the extended centre line of the runway. If it wasn't there, the aircraft might have landed in scrubland, but it was not to be blamed: it was beyond usual safety limits, and many other airports have to live with bigger obstacles.

'Dozens of people came to help from Lufthansa, Boeing, the FAA, NTSB, CAA, and there was some minor friction because all the interested parties tried to protect themselves, but that's natural: they must believe in their own work until proven otherwise.'

Soon the leading edge flaps became the main suspect: they were found in a retracted position – just as the ex-pilot had seen them in flight; in the badly charred cockpit, three of the four relevant switches were still *off*; the motors those switches would have activated were found to be in working order – had they been *on*, they would not have failed to extend those flaps and prevent stalling. But did the pilots know that they were getting airborne in a partially stalled condition?

The investigators concluded that the crew initiated 'a take-off with the leading edge flaps retracted . . . ' and, in the short time available to them, they were unable to identify the problem. One of the major *contributory factors* was 'the lack of warning of a critical condition of leading edge flap position'. The other such factor was 'the failure of the crew to satisfactorily complete their checklist items'.

Eventually, Lufthansa dismissed the pilot and the engineer because, presumably, stricter execution of the checklist procedure might have prevented the accident despite the lack of warning. The engineer was prosecuted for homicide by gross negligence, but the court acquitted him. 'I'm glad it went that way,' said Captain Caesar. 'I'd never blame the crew alone for an accident because they're only the last link in the chain. Besides, it's often a matter of luck. At Nairobi, for instance, they had a five-knot tailwind over an inversion. Had it been a five-knot headwind, they might have damaged the aircraft but without any loss of life, and our only question would have been: would any other experienced crew make the same mistake?'

Captain Caesar was right to be disinclined to blame the crew alone. Many people claim to this day that the accident was wholly preventable! The investigation dug up six known incidents of the same kind in the two previous years.

The special hot-start and checklist techniques used only by Lufthansa were irrelevant. (*After* Nairobi, a further three cases came to light: none of those had been reported before, even though 'mandatory incident reporting was in force in the state concerned'. There might have been even more similar incidents, but far from all operators were in the habit of revealing them.)

Although Allen's report did not name any of the airlines, it was no secret in the industry that the first three fully reported dangerous incidents with the leading edge flaps had been suffered by BOAC. John Boulding, who was in charge of their investigations at the time, told me: 'That Nairobi accident shouldn't have happened. The authorities had two years to prevent it. After our first case in August, 1972, we wanted a modification on all 747s to warn the pilots that they might have misleading information about those flaps, but both the manufacturers and the CAA saw the "event" as a probable crew error, and refused to take any action. We informed an IATA safety meeting about the problem.

'Two months later we had a second incident, and we published it in our Safety Bulletin that's available to the entire industry. A month after that, we had a third one. With the help of Boeing we tried to reproduce the fault on the ground. Although we failed with that, the manufacturer issued a Field Service Memorandum about the incident, and promised that an Operations Bulletin would soon tell everybody about proper check procedures for the flap warning lights.* In January, 1973, Boeing produced an analysis showing that the standard warning light could be incorrectly illuminated because of any one of six electrical faults.

'That might have put off some people but we demanded, nevertheless, a special modification to give the pilot clear warnings whenever he had reason to believe that those flaps were extended though they were not. So eventually, Boeing okayed that the function of the flap warning horn should also cover the leading edge. That, however, was regarded as exclusive to us and Aer Lingus whose 747s we maintained, and accordingly, as Allen's report would state, they "did not notify other 747 operators" about the existence of our modification. So we decided that we ourselves would inform all airlines. I remember the day when I mentioned it at an international gathering of safety officers at Heathrow. The German colleague arrived a little late, and so didn't hear me. That wouldn't have mattered because we also publicized it through the IATA Information Exchange. Oddly enough, only KLM came back to us to ask for details.'

In May, 1974, a fifth incident was reported. By then Boulding felt desperate: 'I thought that if the FAA failed to take notice, the CAA ought to bring pressure on them to make our modification mandatory to all who flew 747s.' He felt he had to do something – something that might cost him his reputation or get him an MBE. So in August, 1974, he wrote to the CAA saying 'If we are to prevent an accident arising from a take-off with these leading edges retracted I consider we ought to try and get the authority

* Following that, Lufthansa introduced an additional relevant item in the flight engineer's checklist as a precaution.

to have another look at this matter. As the industry has already had two close shaves the next Operator might not be so fortunate. Perhaps in the light of the two events they may be inclined to issue an AD.'

The Nairobi report says: 'This letter and the two copies of it which followed at intervals of a few weeks were never received by the state airworthiness authority, who, although they knew of the general purport of the letter, were reluctant to act on the reports of an occurrence in another State without adequate evidence. Consequently no further action had been taken at the time of the accident.'

On November 13, 1974, the sixth case occurred. The crew managed to sort out the problem, and reported the incident only on their return to base on November 20. By then, at dawn on the same day, fifty-nine people had died at Nairobi.

Regulatory wheels began to turn within a few weeks, and, after consultations, the FAA issued a suitable AD in March, 1975, 'to be complied with within five months'.

In due course, Boulding received an MBE.

In *Safety Is No Accident*, Bill Tench agreed with Boulding that the Nairobi disaster could have been prevented had safety authorities been more alert. 'It's not a matter of alertness but the state of the art,' John Chaplin of the CAA argues adding that 'the Lufthansa case shows how hard it is to be clever even with hindsight. The FAA felt it was a cockpit *discipline* problem, while we thought it was a cockpit *layout* problem. As always when there's a discrepancy of opinions, the view of the state of manufacture prevails so far as the aviation world at large is concerned. We couldn't warn "everybody concerned" because we might not even know who "everybody" is. We're obliged to report to the state of manufacture, and in this case, to Boeing. A special problem was that the 747 used in Britain was a slightly different version from the rest. So it had to be up to the FAA to consult Boeing, and disseminate information, if necessary, worldwide.'

The state of the art was also emphasized by Captain Caesar: 'We did a lot of soul-searching afterwards. Did we overlook something? It's impossible to say. We relied on the Americans and the system as it was originally devised. Incidents often look ominous to one airline, but not to another. As we had different checklists and training philosophy, it seemed that the British problem wouldn't apply to us. Furthermore, our pilots stated that the leading edge green light had been illuminated. If that really was the case, even the British system would not have warned the crew for the lights could have been activated by a small electrical fault. What's the use if the pilot could be misled that way? The British alternative was open to subjective judgment – and rejection. Unfortunately.' (Lufthansa introduced many devices to combat 'unthinkable' mishaps, but Caesar would have been happier if more accidents could have been avoided by 'prophylactic anticipation rather than empiric lessons paid for in blood and tears'.)

In 1988, a 727 of Delta Air Lines crashed at Dallas. At the public hearing, NTSB evidence suggested that both the trailing and leading edge flaps must have been fully retracted during take-off. That would make the crash inevitable. Needless to say, 'pilot error' allegations began to fly about like pollen on a hot summer day. The final report may prove eventually that the pilot failed to perform an essential task. But is that straightforward? Did the pilots actually know that the flaps were not set? Henry Duffy, the President of ALPA, pointed out in a letter to *FLIGHT* magazine on February 25, 1989: 'the first officer can be heard completing the flap checklist item' on the CVR, and 'the clicks on the recording' are most probably 'the sounds of the flap handle being set; the fact that no take-off warning horn sounded' might indicate that the pilots did not know that the flaps had not been set. Duffy insists that, on the 727, there is 'a serious flaw in the take-off warning system, which fails to alert a pilot of this potentially fatal condition', and that it 'must be corrected to prevent a similar accident from occurring in the future'. Does this have a ring of familiarity? Viewing it from the old Nairobi crash site, we have been living dangerously in that 'future' for fourteen years.

Yes, it is true and understandable that the pilots' associations can be biased in defence of their members, but disguising the truth about one member could kill many others – and so could a heedless jump to conclusions.

Could ICAO act as a supranational arbiter?

The simple answer is: no. The only authority ICAO has is by the consensus of its members. It is governed by the Chicago

Convention (primarily for the free operation of aircraft), and the various *Annexes* that deal with airworthiness, airports, crew licensing, investigations, etc. All members accept (or declare to differ from) these standards and recommended practices, but, beyond that, it is a matter of trust. 'If, for instance, a member state issues a certificate of airworthiness that says that an aircraft complies with international standards, then in safety terms all ICAO contracting states will give that aircraft overflying and landing rights automatically,' said John Chaplin. 'Nobody can question that certificate without a *very* good reason, and in a way, we have to perform an act of faith. An authority like the CAA can prevent a foreign aircraft landing at or departing from a British airport if it is thought to be unsafe. This right we have exercised, but only very rarely, only when we were a hundred and one per cent sure about our suspicions, because such action can stir up an immediate international incident, with diplomatic notes flying back and forth.'

The differences in standards are acknowledged without value judgment that might be considered offensive. Ron Ashford, CAA director of airworthiness wrote in 1986: ICAO 'requires all aircraft involved in international navigation to conform to "a detailed and comprehensive national code", but the detailed provisions and their interpretation may vary from country to country'.* John Owen, who teaches investigation, put it more bluntly: 'International law is a misnomer in aviation. It's a code which is unenforceable unless each country does so through its own laws. ICAO lays down standards but has no teeth. Most things in air safety depend upon international goodwill.' At the FAA, ICAO was often called an introspective paper tiger. That was echoed by numerous regulatory authorities. Which revealed a plainly hypocritical stance because ICAO is their own creation, and none of them would want an international body to tell them what is and what is not right. That is why they do not breathe life into their Pinocchio.

'It's the fear of losing a single fee-paying member that keeps ICAO in check,' said Sidney Lane of BALPA. 'They're too reluctant to upset anyone because their finances are slender, and they have to cut their budget all the time.' The membership list,

* *Flight International*, 19.4.86

now 157 strong, shows few notable exceptions (though both Koreas are, East Germany and Taiwan are not accepted as members to avoid double representation), and non-members send observers to Montreal. Some specialists believe that it is the proliferation of countries that debilitates ICAO, and restricts the effective radius of national regulatory agencies because the first thing new countries want is an independent national airline. An official (whose name together with the name of his employer had to be omitted because his comments were felt to be out of date by now) told me: 'We can't look over the garden fence to check them – we have our hands full checking the British industry. Drake's spirit, the British mercantile tradition, is still alive, and it's like riding a tiger: exhilarating, but you mustn't fall off. For us to try watching other countries would be a daunting task. Only a much stronger ICAO could police them.' Some people see the answer in the tacit existence of elitist groups within ICAO to lead international developments, but even the advanced European countries find it very difficult to create and enforce some uniformity of standards.

An aviation bureaucrat of international repute said: 'No, you'd better not quote me by name, but yes, I can see ICAO policing come about but not before aircraft flown by third world countries start falling on big cities, be it New York or Moscow. Far-fetched? Not at all. Only nobody likes to talk about it'.

ICAO is, in fact, a club that gives some duties but many rights to its members. 'It's a heavy wheel that keeps turning slowly, and no new ideas can speed it up,' said Olof Fritsch. 'When I came here many years ago, I first looked for new ways to make quick changes. Then I discovered how it all works – the only way it can work. We try to see the world the way it is, not the way we'd like it to be. We've got to be pragmatic; take half a loaf if we can't get a whole loaf; go by precedence and routine, because then the member states can come along with us more readily. Until we get a world government, if ever, we can't expect much more unity from our members; every one of whom is proud to voice an opinion. We don't pass judgment – that's up to the Council, i.e. the members, not the Secretariat. And it can't be helped that in Council and Committees much work and debate go into discussions to decide what and how should be discussed at the next session because even minute changes could be a matter of principle to one member or another. That's how the system works. But it's no good for member states to complain about ICAO itself: if they really want something done, all that's needed is that they write a single letter about their specific concern. Then the whole Secretariat jumps to attention, and starts working towards finding a globally acceptable solution.'

Attention? All hell would break loose in the citadel of the Montreal headquarters, some people claim. There is said to be a great deal of 'whinging and whining' at ICAO about things like wet runway certification, or like too many pilots calling 'low fuel' emergency to jump the queue and get an immediate landing slot. When nothing happens, officials and delegates may grumble that *somebody* ought to do something about it, whereas a letter from a single state would indeed start the ball rolling. It would be, however, a slow roll, because committees fight over single words for ages.

I needed an electron microscope to observe a fractional turn of that 'heavy wheel' when I sat through a routine session of the Air Navigation Commission in 1988. After lengthy international consultations, the aim was 'to amend slightly' an Amendment introduced in 1981 (!) Highly qualified (and presumably, highly paid) experts delivered speeches ranging from detailed comments to numerous procedural observations. After three hours, no agreement was yet in sight.

A fine illustration was the process to produce a guest appearance of the expression Minimum Equipment List (MEL) in ICAO documents. In 1965, the Fourth Navigation Conference referred briefly to *go, no go* items on the pilot's list of essentials for a flight. They meant MEL but failed to use the expression. It took until 1972 to get a similar reference into *Annex* 6 – still without spelling out MEL. Various states of manufacture and registry issue a master MEL, but the practice is not universal or compulsory. Even in the United States a lot of mischief used to be going on, airlines flying with less than the required minimum, until the FAA stepped in. Smaller countries could benefit from some ICAO guidance, but in 1986, fourteen years after the first reference, all they got was a mere incidental mention of MEL in connection with extended range operations. MEL deficiencies have continued figuring in numerous accidents as a major factor, but it took until 1988 that at last ICAO began to produce some real guidance material. That is hoped to be incorporated into *Annex* 6 in 1990.

Generating recommendations is, however, unequal to getting them read and used. Changes in an *Annex* will probably filter through to everyone. But one shudders to think of the fate of other valuable documents that accumulate on shelves around the globe. 'Nobody knows precisely how many rules, standards, specifications and guidelines have been laid down' by ICAO. Annual meetings have been held since 1944 'to amend and update the rule book. The index to this alone runs to eighteen volumes,' wrote Robin Morgan (*Sunday Times*, 25.8.85), quoting Eugene Socher of ICAO: 'There are so many regulations and documents' that 'I don't think anyone in the world has seen all of them.'

If self-policing by an international body is impossible, what chance has the public got to keep the authorities in check?

THE SHORT ARM OF THE LAW

In most countries, a proposition of legal and financial responsibility by the regulatory authorities would be a non-starter because no state is keen to cut off its nose to spite its own face. In the communist bloc (as much as remains of it) and in the third world, often the question cannot even arise.

In the West, one could probably adapt Murphy's Law (whatever can go wrong, will go wrong) to the legal profession: if the proverbial Murphy was a lawyer, whoever could be sued would be sued – even if the chances of success were minimal.

It is widely believed that the chances of suing government bodies successfully are less than minimal. Only the mightiest – rich law firms or the industry itself – could stand up to the scientific, financial and legal resources of an FAA. So do the regulators enjoy impunity? *No*, the lawyers tell us.

'The British CAA's position on certification, unlike the FAA's, has never been established in a court judgment,' said Tim Scorer. 'If there was a case of glaring negligence in, say, classifying a safety modification- i.e. making it optional instead of mandatory – I believe there could be grounds for liability. Here is an example of some of the problems that can arise on certification. A French light aircraft was tested by the French certification authority, and was found to be unspinnable. When it was imported into Britain, it had full French certification, and in accordance with ICAO practice and reciprocal arrangements with Europe, the CAA accepted the results of French certification: it was not obliged to undertake the normal stringent tests to which the aircraft would have been subjected if it was a new British design or came from a non-European country.

'A flying club purchased the aircraft for training purposes. At that time students had to demonstrate the ability to recover from spinning. It was questionable therefore whether this was the appropriate aircraft for the purpose. However, the flying club devised a way of making the supposedly unspinnable aircraft spin contrary to all the tests. They failed to notify the CAA of their discovery. One day the aircraft got into a spin and could not be recovered. Both the instructor and his pupil suffered very serious leg injuries, and they sued both the club and the CAA. Against the CAA it was said that they had failed to properly certificate the aircraft, and had failed to discover that it was capable of being spun. The case was compromised without reaching court – leaving for another time and place the question of whether the CAA might be sued in a major airline disaster'.

Joseph Nall, a lawyer member of the NTSB Board told me: 'In English Common Law, the Sovereign can't be sued. Over here, this *sovereign liability* doctrine has been eroded. In many instances, the public can sue the government because of the Federal Tort Claims Act, but there is the exemption called Discretionary Function to contend with. A government agency cannot be sued for making the wrong judgment. But if a government employee is negligent, the government is liable.'

That is how the FAA, or rather the government which it is part of, can become a party to litigation. 'Right now, in 1988, there are about 300 aviation accident related cases pending against the U.S. government,' said James Dillman, FAA Assistant Chief Counsel for Litigation. 'They arise out of various FAA activities, mostly air traffic control functions but also from midair collisions, the collection and dissemination of weather information, and from allegedly misaligned navigation aids that sent the pilot to Timbuktu when he only wanted to go to Cincinnati. We've been sued successfully many times. If the FAA became completely independent from government, as some people keep suggesting, it would have to charge vast sums for its services only to buy itself sufficient insurance coverage.'

It used to be much easier to sue the FAA on the basis that, if it okayed the design of an aircraft, it had to be responsible for its safety. But then came a classic legal wrangle that grew into yet another wall with which governments protect themselves. It all began with perhaps the ultimate horror story of the air.

On July 11, 1973, a Brazilian Varig Airlines 707 was approaching Orly airport, the Parisian destination of a long flight from Rio. It was reported to the captain that a small fire had broken out and could not be brought under control in a rear lavatory. Foul-smelling, thick, black smoke spread fast into the cabin, and the venting system seemed unable to cope with it. It would have been dangerous to deploy the oxygen masks because they would only feed the flames. So the crew put on their own masks, which work 'on demand' only when they inhale, to begin a rapid descent. They made a successful emergency landing – quite a feat with a large jet in an open field. They quickly unlocked the doors to start evacuation, but only clouds of deadly fumes, a mixture of carbon monoxide and cyanide, began to escape. The passengers remained strapped in their seats – dead. By the time of the landing, the cabin had become a gas chamber that asphyxiated all but eleven of the 135 people aboard. The ensuing fire consumed most of the fuselage.

'The investigation confirmed that probably it was a cigarette end that started the fire in a trash-can for towel disposal in one of the restrooms,' said James Dillman. Although 'fag ends' are often blamed unfairly because it is not easy to start a fire with them, this time the conditions were particularly favourable: 'By regulation, the container must be of a design and construction that would suppress any fire that might occur there, but in this case, it apparently was not. What's worse, it had holes in its sides, almost as if designed to be a burn-can in which oxygen can feed the flames through the holes. When the 707 was certified fifteen years earlier, it probably had the right type of trash container, but in service, due to quick, frequent changes, the can aboard might have been the hundredth replacement. How it got there was anybody's guess, but Varig, as well as the victims' relatives, sued the U.S. government alleging that the FAA had been negligent in the aircraft certification.'

In 1981, eight years after the accident, the U.S. District Court at Los Angeles exonerated the FAA, and dismissed the case. Varig and the relatives appealed, and the Court of Appeal reversed the decision. Now it was the government's turn to appeal to the Supreme Court. At that time, 183 'negligent certification' cases were pending against the FAA, with claims amounting to almost 800,000,000 dollars.

Meanwhile, another case had begun in the U.S. District Court at San Diego. (Same state, California, but different District, where a precedent from the Los Angeles court was not binding.) This case concerned an accident at Las Vegas in 1968. A de Havilland Dove crashed three minutes after take-off, killing the pilots and the two passengers. The history of the aircraft was a key factor of the investigation. 'The Dove had been bought from another air taxi company that operated in a colder part of the country,' Dillman told me. 'Presumably, passengers complained about the cold, and the company had a supplementary gasoline-fuelled heater installed. The modification was approved by the FAA regional office in Texas. A gas line, running from a wing tank to the heater, was fastened to the wall of the fuselage by a plastic device, the sort electricians would use to bundle wires, something like tying up a trash bag with a twist around its neck. The aircraft was sold a year later, but it would have been pointless to remove the extra heater. According to the crash investigation, vibration rubbed a hole in the line, gasoline dripped out, and a fatal fire on board ensued.

'This case was lost by the FAA at the District Court level, the government appealed, and the Court of Appeal affirmed the lower court decision, and so this case, too, went to the Supreme Court. At that level, the two crashes were brought together and became *the Varig case* even though it covered two different certifications – one for type, the 707, the other for a modification. More than ten years after the Varig accident, the Supreme Court ruled in favour of the FAA, saying that "the FAA has a statutory duty to promote safety in air transportation, not to insure it". The crucial element in the 707 crash was that the FAA had to set down the essential safety design principles for the trash can, but had the discretionary right to rely upon the manufacturer to adhere to these principles, and not to check and certify every detail, and so couldn't be held responsible for what it didn't inspect. It's the manufacturers who are responsible for complying with all design regulations. In other words, the government rules that murder is illegal, but it's the citizen who remains liable for murder – not the police for failing to prevent it. The Court held that if the FAA could be challenged for every actual inspection mistake – like overlooking the potential hazard in the Dove heater fitting – it would only encourage the FAA not to inspect anything beyond the bare essentials, for they can be liable only if they look at something.'

The Varig crash became a turning point: from then on, it would be well nigh impossible to sue the government for any sort of 'certification negligence'. The FAA litigation case load dropped immediately by almost a third. Several other kinds of certification lawsuits – ranging from the Tristar fire at Riyadh to the licence for a pilot with myocardiac infarction – were also dismissed on the basis of the Varig principle that, according to the Supreme Court, the Good Samaritan Doctrine could not be applied to the regulatory agency. (A doctor may be expected to stop and help an accident victim, but he can be sued for negligence if anything goes wrong; if, however, he fails to stop because he considers it unnecessary, he cannot be held responsible whatever happens.) On June 7, 1988, the FAA told me that as a result, 'negligent certification' cases had virtually disappeared. But just six days later, the Supreme Court ruled in a polio vaccination case – *Berkovitz v. United States* – that 'the federal government may be sued for damages for negligence in approving vaccines for public use that are defective or unsafe.'* Will that puncture the protection shield of the aviation authorities? 'Perhaps, but the effect will be limited', said government lawyer Dillman at a Symposium of the American Bar Association, in 1988.

'These days, the biggest problem concerns the insufficient inspection of deregulated airlines,' said Philip Silverman.† 'But how can we hold the FAA liable without some restructuring of the law? The budget is allocated by Congress, the FAA submits how it will be spent – so much for ATC, so much for inspectors. Can you fault them and hold them liable? When I was a kid, I got an allowance of fifty cents a week. My friends got the same. A Maltit cost a quarter, and my friends bought *two* right away, every week,

* Wall Street Journal, 14.6.88.

† Sadly, this fine lawyer died shortly after our meeting.

while I got *five* chocolate-sodas costing a dime each. My purchase lasted longer, but can you fault my friends?'

Not unless it was proven to them that, say, their favourite could seriously damage their teeth, and, even more so, if they learned but soon forgot the lesson a painful visit to the dentist had taught them.

Long before the Supreme Court would make its historic decision, the Varig crash triggered off an intensive study of inflammable materials used in cabin construction. The FAA initiated the adoption of much stricter standards, and introduced *no smoking* rules in all aircraft lavatories. Those who then believed (and some still do) that the steps already taken were sufficient, and that the risks would forever be eliminated, must have been living in a fool's paradise. Or else they might subscribe to a conspiracy by ignorance and silence: for if it is painful that the world is often too slow to learn lessons, it implies criminal negligence that lessons taught by accidents are allowed to be unlearned repeatedly, as if they carried their own *read and self-destruct* command.

– 6 –

LESSONS THAT GO UP IN FLAMES

Whenever a flight of Saudia, the Saudi Arabian national carrier, is about to leave the ground, 'a prayer for the travellers from the Koran is intoned loudly through the aircraft.'* In many, supposedly routine situations aboard some airlines, and in some such situations on many airlines, even the agnostic may feel that a prayer for safety would be in order – if not the most positive source of hope.

In December, 1988, I had a most educative experience when flying from Bangkok to Singapore with Biman, the sizeable and growing national airline of Bangladesh. Boarding was not easy, not unless one sat back in rows of empty seats that offered a grandstand view of the scrum at the far end of the departure lounge. When after a fifty-minute delay the gate was opened, there was a surge that pinned two uniformed girls to the frame of the exit. Passengers battled with each other as well as their own numerous pieces of *hand-luggage* which included TV sets, bedrolls over shoulders, massive wicker baskets balanced atop heads, string-held boxes, some of them the size of a davenport.

In the crush, the checking of boarding passes was a Houdini act.

Aboard the aircraft, the fight for *Lebensraum* was equally relentless. The overhead bins would not take half the luggage. Some people proposed to sit on boxes or place them on their laps, others squeezed bulky items between the rows of seats or tried to

* Christopher Walker in *The Times*, 17.3.1989.

leave them in the aisles. Objections were brushed aside – or simply disobeyed as soon as the weary cabin crew turned their backs. In any case, by now it was much too late to limit the size of hand-luggage. The safety briefing was ignored – the noise level rendered it inaudible, anyway.

As soon as the DC10 was in the air, half the people left their cramped seats to stretch and smoke in large groups in the aisles. During drinks service, at least once the serving trolley had to be lifted over a box that blocked its path. Some forty minutes into the flight, shouting broke out at the back. It was the accompaniment to a fist fight. Some six more men joined in, and bodies were thrown with thumps against the side of the fuselage. The purser gave as good as he got but failed to separate the fierce pugilists without reinforcement from the cockpit. While order was being restored, a terrified Singaporean girl crossed herself, and turned to a sari-clad stewardess who was bored rather than troubled by it all.

'How can you allow this?'

'Happens on every flight.' And contemptuously: 'They come from dry country. They cannot hold the drink.'

'Then why do you serve any?'

A shrug of the shoulders: 'It's expected. Free!'

The landing was smooth. The warnings to stay seated were ignored. As soon as the wheels touched the runway, most passengers were on their feet. More luggage was piled high in the aisles, and there was a forward crush. Had there been a crash and fire with the chance to run for it . . . No, nothing happened. Not that time. The Singapore girl must have prayed well.

For situations like that, one should blame not only the authorities and airlines: passengers must bear some responsibility for their own safety when their preparedness and behaviour may become the last line of defence in so-called *survivable accidents* – probably the two most cruel words in the aviation vocabulary, because they imply the *evitability* of tragic events. The Varig aircraft,* an 'airborne gas chamber', made a safe emergency landing in 1973. Those killed in the 727 which burned out at Salt Lake City (p. 38) could also have been saved. So what happened to the invaluable lessons those horrors had taught us? Had they just gone up in flames to be forgotten and relearned again and again? One must

* Discussed from a different angle in the previous chapter.

conclude that the answer is *yes*, for the last decade alone has produced some thirty such cases, and, while the actual toll air accidents take may show a downward trend, the proportionate number of deaths in aircraft fires has increased to seventy per cent of the total.

In survivable accidents, the aircraft structure tends to remain undamaged or sufficiently intact to allow the quick evacuation of the mostly uninjured passengers on board before fire prevents their escape. If there are fatalities, *fire* is likely to be named as the culprit. But is this because of lazy use of the language or because preoccupation with the old adage of 'no smoke without fire' has obliterated the proposition that conversely 'no fire without smoke' could be equally true? For painful experience has taught us repeatedly that it is *smoke* rather than the fire itself that incapacitates and kills most victims. It travels and acts faster than flames, and fuels flashovers that may incinerate the evidence. Any delay – and a foreseeable but unforeseen disaster may follow. An almost random selection from the abundance of horrifying examples illustrates this point.

NO FIRE WITHOUT SMOKE

A Saudi Arabian Lockheed 1011 Tristar departed Riyadh for Jeddah on August 19,1980, at 18.08 GMT. Almost seven minutes later, there was a fire warning. Still climbing, the flight crew spent three minutes just 'looking for the aft cargo smoke warning procedure'. (The manuals were confusing, and the flight engineer was dyslexic.) More time was wasted on trying to confirm that it was not a false alarm, while the flight engineer was sent twice to investigate in the cabin. Once he reported fire in there, once he saw only some smoke and signs of panic. By then the captain had decided to return to Riyadh. Preparing for an emergency, the tower wanted to know how many people were on board. 'We don't know exactly, think we have a full load.'

Three minutes later fire in the cabin was reported. Following that, an attendant who had tried to take a closer look burst into the cockpit with these words: 'There is no way I can go to the back * * after L2 R2 because the people are fighting in the aisles.'

The CVR picked up 'singing' (possibly religious chants) in Arabic, and what was said thereafter.

At 18.27 the captain declared that 'as soon as possible we're gonna be down'. The cabin crew were begging the passengers in Arabic, English and Urdu to 'sit on your seats, sit on your seats, ladies and gentlemen take your seats – nothing will happen to aircraft . . . don't stand like this . . . no danger from the airplane . . .'

Well before landing, cabin attendants asked the captain several times if they should evacuate. 'What?' 'Do you want us to evacuate the passengers as soon as we stop?' 'Take your positions.' As they got no answers, they kept trying to restore calm in the cabin. 'Give me your attention please, be seated ladies and gentlemen, we are about to land, there's no reason to panic . . . Place your hands behind your heads for impact, girls demonstrate impact position . . . ' There were at least nineteen pleas from cabin and flight crew to passengers to do as they were told. Some twenty minutes into the emergency, passengers were still being urged to sit down and fasten their seatbelts. The sound of the public announcements lacked authority. Perhaps the continuous singing reduced their importance.

Seventeen seconds past 18.35, the flight engineer began to insist that the girls' questions about evacuation must be answered. The captain responded only forty seconds later, twenty-seven seconds before landing: 'Tell them, tell them to not evacuate.'

The tower noted no fire as the aircraft touched down at 18.36:24, and continued a routine *roll-out, turning off the runway and taxiing slowly for two minutes and thirty-nine seconds*. At 18.40:33, 'We are trying to evacuate now,' was the last transmission from thc aircraft, but the engines were not shut down for more than three minutes after stopping on the taxiway. Onlookers stood by bewildered: none of the exits was opened.

They saw smoke and flames through windows but only at the rear of the cabin.

Firemen who had never fought a training fire let alone a real one, and had never been given a chance actually to operate an aircraft door from the outside, managed to force the first exit open twenty-nine minutes after touchdown. 'The cabin was observed to be full of smoke and no life was observed nor were any human sounds heard . . . Flames were seen progressing forward from the rear of the cabin' *three full minutes after opening of the first door*. An ominous statement to which the final report seemed to attach no great importance.

Of the 301 people aboard none survived.

All the cabin crew and virtually all passengers were found packed tightly in the front third of the aircraft. As the flames would only have reached them well after the clouds of smoke, noxious gases rather than fire must have killed them. But much of this is no more than logical hypothesis: the pilots as well as the passengers were buried without an autopsy. The bodies that were 'viewed' at all by non-medical investigators revealed no impact or crushing type injuries. So the accident was classified as 'survivable'. To crowd the front of the aircraft, there had to be a surge forward; had that occurred before landing, the DFDR would have noted it, as well as the pitch trim reaction to it, just as movements in the cabin had been recorded during the flight. So if people were alive when the plane touched down, why was there no evidence that anybody had even tried to pull the emergency handles and operate the doors? It was thought possible that the panicky crush was such that the doors were 'prevented from moving inboard the necessary few inches prior to opening'.

Uncontrolled post-crash activities and some criminally idiotic 'tidying up' of the wreckage destroyed whatever evidence might have been left by the fire, adding to the immense difficulties experienced by the NTSB and AIB investigators who were brought in to help – too late. They concluded that the stove with a green bottle of butane found among the seats in the cabin was just a red herring, and that the fire had probably originated in the cargo area. That, however, is still disputed by specialists who believe that the findings suited Lockheed only too well. BA investigator John Boulding was convinced that the fire had probably started in an inaccessible spot between the hold and the fuselage skin where thoughtless design allowed hydraulic, pneumatic and electric controls to converge in a confined space that the extinguishers could not penetrate. (His airline removed some lining and sound-proofing for added safety, and a subsequent Malaysian accident seemed to have supported his view.)

The captain failed to prepare for evacuation, failed fully to utilize his non-assertive crew, failed to recognize the urgency of the situation, and failed to stop on the ground wherever he could, shut down his engines without delay, and order immediate evacuation instead of taxiing about.

It is widely believed that the investigators' final report was the left-over of two previous versions that had allegedly been torpedoed by Lockheed and the Saudis. Better crew training and emergency procedures were only two among the meagre safety recommendations that included stricter control of passengers' baggage, and some oddities like 'Amend Saudia's personnel policy and practices to stop the rehiring of flight crew members for a flight crew position after they have been removed from another flight crew because of substandard performance.' The disastrous effect of toxic fumes was reviewed briefly in the report, but there was no recommendation to deal with the problem. Although fumes gathered first at the top, and the report discussed the effects of venting through the floor, no recommendation was made to study and perhaps modify the system. To taxi around and clear the runway was not the pilot's eccentric invention. It was standard procedure, worldwide. Though it was seen to have wasted several minutes during which most if not all lives could have been saved, the report made no recommendation to stamp out this dangerous practice. With the omission of post mortems, due to a combination of carelessness and religious objections, yet another golden opportunity for greater safety was wasted, and the lessons from the Riyadh disaster would dissipate in the acid bath of obscurity.

In Spain, a DC10 of Spantax aborted take-off at Malaga on September 13, 1982. It could stop only in a field beyond the runway. A wing was torn off, and a severe fire erupted. The cabin was intact when the aircraft halted, and even when the fire trucks arrived. Yet the flames burned through the fuselage before evacuation could be completed. Fifty-one of 393 people on board died. Their escape was seriously impeded by a massive amount of debris and other obstacles in the aisles, and also by a form of negative panic that seemed to have paralysed some passengers in their seats.

If the sensible reader expected the investigators to have initiated a full review of passenger behaviour and escape facilities, the reader would be wrong.

June 2, 1983 is an important date in the history of aircraft fires. Flight 797, an Air Canada DC9, was en route from Dallas, Texas, to Montreal, Quebec, when three circuit breakers associated with the aft lavatory flush motor tripped in quick succession. It might have been due to some motor malfunction.

The time was 18.51. Nine minutes later, when the pilots were assuming that no dramatic action was necessary, some 'light gray smoke' was noticed in the cabin. Flight attendants found curling black smoke in the lavatory which they saturated with CO_2 from an extinguisher while one of them walked to the cockpit to inform the captain. As the aircraft was less than half full, they moved all passengers to the front section.

The flight attendants then reported that the smoke appeared to be clearing, but when the first officer went aft to investigate, smoke prevented him from opening the lavatory door which felt hot to the touch. (On his second visit he had smoke goggles with him, but no breathing apparatus which was not required to be carried on board.) Meanwhile the captain decided not to descend because he expected the fire to be put out. Sixteen minutes after the initial problems had occurred, there were further electrical malfunctions, and the first officer reported: 'I don't like what's happening, I think we better go down, okay?'

At 19.08:12 the pilot declared: 'Mayday, Mayday, Mayday.'

The sector controller would clear the path of the aircraft: 'Can you possibly make Cincinnati?' The answer was affirmative – the airport was twenty-five miles away but offered good emergency facilities.

For the fire was not going out. Its seat was probably behind panels where the CO_2 could not get to it. The flight attendants were trained to use an axe, stored behind the captain's seat, to break down certain panels and so obtain access to the fire, but they had not been taught 'which lavatory panels could be removed or destroyed without endangering critical airplane components'. (Afterwards, the NTSB would remain unconvinced that firefighting could have been carried out as prescribed 'without a full-face smoke mask with self-contained breathing apparatus'.)

Although smoke was now spreading into the cabin, several factors supported the chances of survival. There were three flight attendants to control and prepare only forty-one passengers for speedy evacuation. Everybody moved forward of Row 13. Full emergency briefing was given, but it was not heard in full by everyone: the PA system had broken down, and the busy flight attendants did not want to waste time on retrieving the megaphone. Male passengers were designated to open the overwing exit windows. As black, acrid smoke began to fill the cabin from ceiling down to knee level, people breathed sparingly through articles of clothing or wet towels distributed by the crew.

Smoke also entered the flight deck through the door (!) which remained open all the time. The captain used smoke goggles and the oxygen mask as he struggled with multiplying instrument problems. Crash-fire-rescue vehicles were waiting. The aircraft landed without any structural damage, and caused no incapacitation to anyone on board. Thus the accident was *survivable.*

The plane stopped on the runway thirty minutes after the first signs of trouble, twelve minutes after the delayed Mayday call.

The pilots tried to enter the cabin to help evacuation, but were driven back by thick black smoke that reduced visibility to nil. By then evacuation was in progress, but only the crew members and eighteen of the passengers managed to get out, in sixty to ninety seconds, before flashfire engulfed the cabin and rendered the environment non-survivable within a further thirty seconds.

Twenty-three passengers never made it though two doors and three overwing exits were fully open.

Most of the fatalities occurred up front (and were found in the aisle) where only the galley bulkhead separated them from the opened doors. (See drawing on page 138.) Though that potentially dangerous design feature might have impeded quicker and full evacuation, the report gave it no consideration.

Aircraft are certified for ninety-second evacuation using only *half* the exits. How come that so few people got out through *most* of the exits before a killing flashfire? Were they already dead from the fumes? Wet towels would filter out some deadly fumes but not carbon monoxide. As so frequently, survivors said: ‘There was no panic.’ Did they then suffer from ‘behavioural inaction’, i.e. negative panic, and wait too long for commands to move? Most of those who got out had made their own escape plans and counted the seats to the doors, but survivors testified that, due to the smoke inhaled, they ‘barely had the strength and presence of mind to negotiate the overwing exits’ that were already open. Strangely enough, *the report failed to spare a thought for smoke-hoods for passengers.*

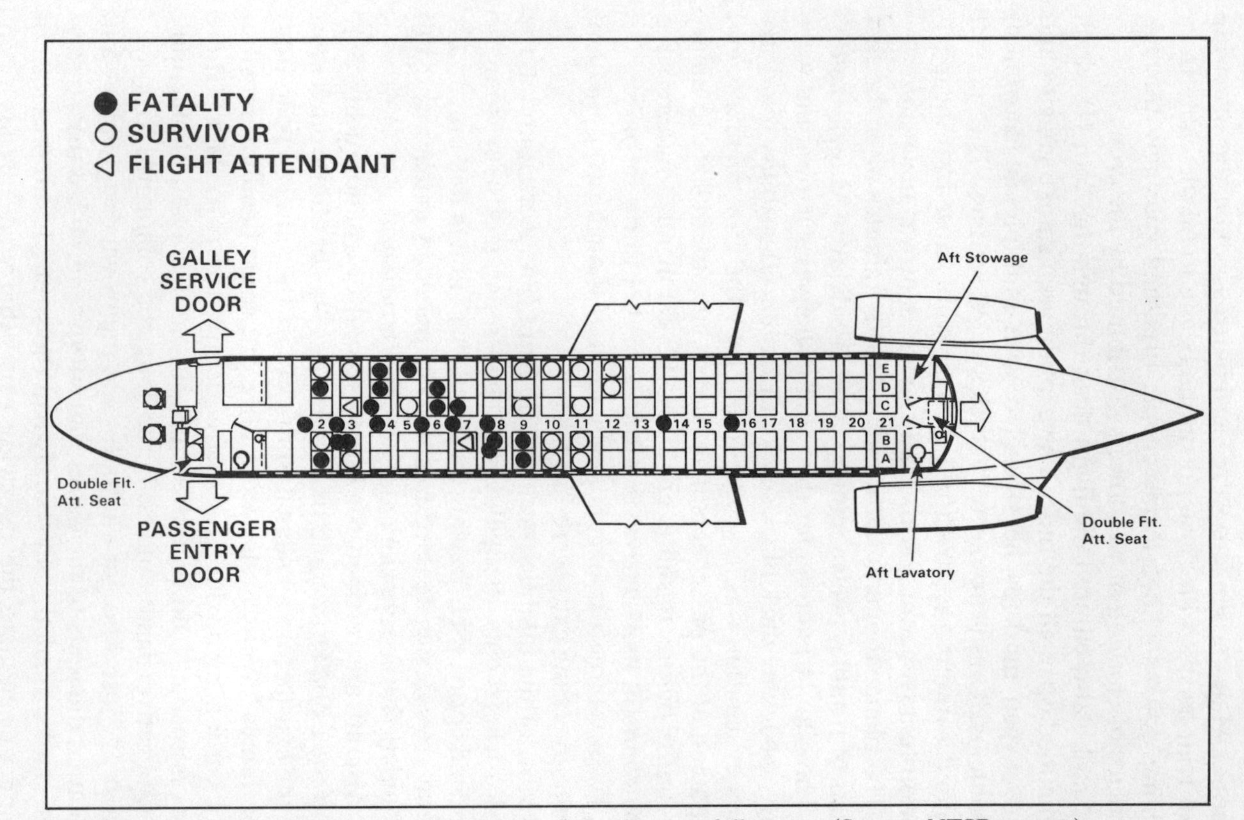

At Cincinnati 23 people failed to make it although 5 exits were fully open. (Source: NTSB report.)

The recommendations were, however, devastating: oft-repeated advice concerning lavatory fires, reiterated references to numerous accidents covering a full decade. The need for *smoke detectors* headed the list. Their installation had been resisted by arguments about unresolved technological problems. Quite miraculously, certain solutions became acceptable to some airlines which introduced them *voluntarily* within a few months of Cincinnati. Detectors became mandatory only four years later.

'We had the right idea long before Cincinnati,' said Jeremy Haines, Maintenance Quality Director of Air Canada. 'We certainly led the field in putting automatic extinguishers in the waste area of washrooms. We already had them on many of our aircraft for years – though unfortunately, not on that particular DC9. After Cincinnati, we accelerated the program, and put in smoke detectors. At first, for speed, we used ordinary household types we could pick up from local hardware stores, so to speak. We did that while the rest of the industry were still waiting for new regulations.'

Smoke masks for the crew weighed heavily among the recommendations, as if it was a new idea which it certainly was not. The NTSB had urged the introduction of CPBE (Crew Protective Breathing Equipment) in the wake of numerous crashes such as the PanAm cargo flight accident in 1973. After the Varig crash, the FAA planned compulsory oxygen masks for cabin crews, but the proposal was soon withdrawn for 'further testing . . . to establish standards'. Eight years later, when masks would have enabled the Air Canada crew to fight the fire more effectively, the standards were still being debated. Yet another two years later, in 1985, the FAA issued a new 'proposal for rulemaking'. The rule came into force at last in 1987, but implementation would be mandatory only after 1990.

THE SURVIVAL OF THE LUCKIEST

Over the last three decades, the fire hazard has been attacked on many fronts to make the not necessarily fatal accidents more survivable. Complete prevention of fire would of course be the ideal solution, but economic viability does not cater for an ideal world. As Frank Taylor, Director of the Cranfield Aviation Safety Centre, emphasized time and again, 'there're dozens or maybe hundreds of combinations of events that could cause an aircraft to end up as a smoking heap on or close to an airfield. To prevent all those accidents may be desirable but would in practice be virtually impossible, and surely less cost effective than concentrating our attention on the smoking heap.' Decisions about the most effective application of the available 'safety money' are strongly influenced by the old enigma: what is more useful, a fence at the top of the cliff or an ambulance in the valley below? As there are many 'cliffs' and even more routes to the edge of the precipice, 'we must continue trying to prevent accidents but we must also try to minimize the effect of those that will inevitably occur'.

Fires in the air provided one of the first battlegrounds, and the industry chalked up several victories. Preventive design and construction, better detection and extinguishers for engines as well as cargo holds did not eliminate the hazard but reduced its frequency. An inflight fire was no longer necessarily a foregone disaster.

The fight against *post-crash fires* has been far less successful, though not without some great triumphs. Take the struggle for safer fuel. JP was more dangerous than kerosene but it enjoyed weight and price advantage, so it took many years to produce a statistically significant number of 'clear-cut' disasters before the champions of the *save money before lives* principle were swayed. (The ASG succeeded in making the problem a public issue in the 1960s, and Lord Brabazon proposed a 'fuel-duel': he would stand in a pool of kerosene, his opponents in a puddle of JP4, and both would light a match to demonstrate their faith in their respective fuels. The challenge was never taken up. America followed the British lead only in the 1970s.) Today, Canadians are probably the only ones to use JP4, and, according to the CASB, seem to have no problems with it. Air Canada has never been convinced about the extra risk, but as Laird Stovel, Manager, Operations Engineering, told me, 'few stations rely on JP4 these days – Edmonton, Victoria B.C., Rouyn, Quebec, and Regina, Saskatchewan are probably the last where we load it'.

Though even safcr anti-misting fucl and firc-rctarding fucl additives have been available for many years, their advantage is still claimed to be 'not proven' and strongly resisted on technical grounds. But the real reason for the resistance to their introduction is probably financial. As they would increase the fuel bill and require some costly modifications, they will be researched more vigorously and accepted only if regulatory authorities grow convinced that 'enough' lives could be saved to justify legislation.

Crashworthiness is another area where advances thrive on painful lessons from many accidents. Much has been done to create a planned break-up sequence, protect structural integrity, safeguard passengers from injuries when the cabin is not destroyed, and reduce the risk of fire.

'The problem was best and most tragically illustrated for me when we were called in to help with the investigation of a Yugoslav accident at Rieka,' said Geoffrey Wilkinson, a former chief of the AIB. 'The aircraft was Russian, a TU-134, the type we called the *BAC-one-elevenski*, though all the similarity ended with the looks. I flew it once, and it was not a happy experience. It had no stickshaker, and it was very primitive by Western standards. A British test pilot would have failed it on cockpit visibility alone. It was as if you were peering through a letterbox, and you had to lean to one side.

'Anyway, the pilot got not untypically poor thrust response as he was trying to come in through isolated thundershowers, hit the end of the runway very hard – and the wing came off! After a thousand-metre skid, the plane overturned and caught fire. The pilots got out through the hatch, a courier just walked out through a hole in the fuselage, but everybody else on board, sixty-seven people, died. And yet it should have been a survivable accident. The fire took almost two minutes to break into the cabin, but most passengers were still hanging upside down like fruit-bats. The emergency exits were not marked properly, and most of them were hidden behind curtains. But the main cause of the tragic outcome was a design problem: the 134 was sort of cobbled together, without much thought for crashworthiness. Its wing was a scaled down version of the 104, a bomber, optimized for long cruise rather than airline operation. Its landing gear was much more robust than the wing. The undercarriage should have been the first to collapse, but it didn't. Then progressive failure of the structure ought to have followed, but only the wing came off. Six months later, a Cyprus Airways Trident had an almost identical heavy landing and skid on a training flight. Its undercarriage collapsed taking much of the impact. The structure failed in a planned sequence – and no one was hurt. That's what an aircraft should be designed for.'

Much has been done for the fire hardening of the fuselage to give passengers better protection but a report* by the FAA Technical Center revealed that, in a fuel-spill fire, the aluminium skin would burn through in about fifty seconds. When the cabin is intact, the sidewall insulation will maintain a survivable temperature until the windows melt in about three minutes.

The containment of fuel in a survivable mishap is therefore essential, but progress in that direction is far from satisfactory. U.S. ALPA safety specialists, for instance, have been campaigning incessantly for crashworthy fuel systems, self-sealing fuel lines (like those used on many military aircraft) and double-skinned tanks, because 'almost all of the [postcrash] fires are caused by inadequate fuel containment.'† In 1988, Noreene Koan, chairing the Air Safety Committee of the U.S. Flight Attendants' Association, welcomed efforts to force 'the FAA to at least study this area' and asked: 'Why should we have to fight to bring about a transfer of military technology to commercial aviation?'

'Sure, there're certain different standards enforced by the military,' said Boeing's Earl Weener. 'They also apply to helicopters and certain light aircraft. Now there is a move afoot to apply the same standards to all airplanes. It looks good to the public, so they say: "Yeah, let's do it . . . " but does anybody stop to think if it's really applicable to aircraft where the fuel tanks are different?'

The sheer size of the Jumbo brought only false and shortlived hopes for better protection. 'In the 1960s, the argument was that aircraft could not cope with the extra weight of break-away self-sealing tanks,' said Captain Vic Hewes of the ALPA Accident Survival Committee. 'Now we're going to have the extra quantities of fuel everywhere, even in the tail, and so may have to face blazing fires all round the entire aircraft structure. Concorde has always had fuel running back to front the length of the aircraft for trim purposes. The new 747-400 will have up to 3,000 gallons in the horizontal tail section. The MD11 and the new Airbus will also have fuel tanks in the tail. At present, extinguishing agent

* DOT/FAA/CT-83/10.

† The ALPA Guide To Accident Survival Factors, 1989, repeating the old demands.

requirements are based on fire envisaged in two-thirds of the aircraft. Now their availability will have to be upped to cope with fire in the *whole* of the aircraft.'

Is it then surprising that perhaps the most widespread fear of air transport is that we have to fly on a keg of gunpowder?

Accidents taught investigators innumerable lessons about crashworthiness inside the fuselage – but not all were taken to heart. 'Since the introduction of large jet aircraft, there has been a relative increase in the injury and fatality rate during survivable accidents. This is caused, to a great degree, by cabin interior furnishings which break loose at relatively low impact loads and injure the occupants or trap them and thereby prevent their escape in the event of fire.'*

The *strength of the seats* was always the major consideration. When seats break loose in otherwise survivable accidents, head injuries and broken limbs render their occupants easy prey for smoke and fire. In the late 1980s, FAA seat strength regulations were still based on criteria established in the 1920s upgraded by estimates of the 1950s. Manufacturers argued that the floor would also have to be strengthened at extra weight penalty, and a fine balancing act would have to be performed between seats and the human body: if seats were stronger, bodies would have to absorb more of the impact energy. Airlines were concerned about the cost benefit: the extra weight of stronger seats would impose more than five dollars of extra fuel cost per seat per annum. All interested parties – except the passengers – kept asking how many lives could any new measures save? It appears that more deaths, more atrocious injuries, were needed before a serious attempt would be made to reconcile the clashing views.

The FAA proposal for stronger seats was delayed by the need of more research. New U.S. laws – with the rest of the world following suit – would apply only to new aircraft. The fleet in use would have a seven-year grace for retrofit. Until then, passengers would have to grin and bear it. Which could be even more difficult if travelling in small commuter aircraft to which the stricter laws would not apply. In November, 1987, when the Ryan Air Beechcraft 1900 crashed in a landing attempt, it ploughed through an airport fence, and came to rest in a snowy field. The wings were

* The ALPA Guide To Accident Survival Factors, 1986.

sheared off, but the fuselage appeared to be intact. Yet most of the passengers were dead. Sergeant Stogsdill of the Alaska State Troopers said: 'There was no fire, but there was quite a mess.' He was referring to the fact that most of the seats in the aircraft had been uprooted and hurled forward. Almost three hours later, bodies were still being removed from the mangled intestines of an apparently survivable accident.

The British Midland crash in 1989 might have provided the long-awaited impetus for study and legislation. The Boeing 737 flew into the embankment of the M1 motorway. The local Trent Regional Health Authority compiled a report on the 'horrific tally' of injuries that included thirty-five serious head wounds, twenty-five pelvic fractures, twenty spinal fractures, and more broken limbs than passengers. 'An opportunity such as this, to document thoroughly the injuries of so many survivors, rarely occurs,' the report said. 'We hope to exploit it fully to obtain as much information as possible, in order to further our understanding of how the trauma of survivable accidents may be mitigated.' It might help to build crashworthy seats that would absorb impact energy through a sequence of planned, gradual collapse, but it was most disturbing to see that, in this crash, the aircraft was new and had therefore the doubly strong seats required by laws on which the ink had hardly dried.

Stronger seats are, however, only half the battle.* Infants who have no seats of their own need better protection than a parent's hug in a crash. So the CAA has commissioned the Cranfield Institute of Technology to determine how babies and small children could have seatbelts attached to adults' lapbelts. The British Midland crash confirmed, yet again, that passengers are at grave risk when freely flailing parts of the body, restrained only by lapbelts, are flung forward: between seventy and eighty per cent of aircrash deaths and injuries are the result of 'contact with structures that are stronger than the head'. (It is known, for instance, that the hard surface of trays would be much more safely stored in the arm rests than in the back of the seat in front of most passengers, but there is no room elsewhere for trays except in the *fauteuils* with extra legroom reserved at inflated prices for the first-class and business travellers.) A pathologist, exasperated

* See Chapter 16.

by decades of experience in reviewing avoidable carnage, told me that 'in tourist class and on charters there is so little space to bend forward that perhaps they ought to fly nobody taller than five-foot-two'.

Although properly anchored shoulderbelts could prevent most of the serious head injuries, we are told that we would object to using them on account of discomfort. We used, allegedly, to be reluctant to wear lapbelts, and are still thought to abhor backward facing seats. But has anybody actually asked the passenger? As a matter of fact, yes. In 1947(!) the U.S. Air Force questioned 1,020 passengers after trans-continental flights in rearward facing seats, and 'found that people were overwhelmingly in favour of the change'.*

Individually, many lawmakers would prefer rearward facing seats that could at least reduce the gravity and extent of injuries. They admit that, with modern materials, even the extra seat weight problem could be solved. The risk of loose objects being catapulted forwards and hitting passengers in the face on impact could also be eliminated by better locks on overhead bins. Nevertheless Ronald Ashford, CAA Group Director of Safety Regulations, told Kieran Daly of *Flight International* (11.2.1989) that although the CAA had 'favoured' rearward seats for many years, no 'clear case' for their mandatory installation had yet emerged, and the idea was not included in the new joint European safety requirements because 'the majority of the countries' could not justify it. In the light of accumulated recent crash experience, perhaps the CAA could make a determined move one day to bring its allegedly favoured view of many years to belated fruition.

While (apart from the military) no more than lip service was paid to reversing the seats, vast sums were spent on the development of *fire-resistant textiles and fireblocking seats* over three decades. The logic behind the idea was sound: if the conflagration of furnishings, including wall panels and curtains, was prevented, passengers would have more safety in the air, and more time to walk away from survivable accidents. The industry argued against it on the usual grounds of viability, cost and weight penalty as well as the difficulty of making the new materials hardwearing and washable, but, after each major

* James Castle, a former BA engineer, writing in *The Times* on 10.1.1989.

accident involving fire, the public outcry was such that the regulatory authorities were compelled to act.

Unfortunately, instead of pooling international resources, most of the research was left to would-be vendors. The FAA tested some 600 different materials, and found that more than a hundred would satisfy its minimum criteria. Though in 1979, an expert report by AGARD, the NATO Advisory Group for Aerospace Research & Development, did not mince words suggesting that better means of passenger protection ought to be sought because substantial fireblocking was a long way off, there was no major change in regulatory policy. Led by the FAA, plans were drawn up in the early 1980s to introduce fire-resistant covers at some 200 dollars per aircraft seat. That sounded more substantial than it was: the cost was reduced by mass production, and at today's rates, all 650,000 seats of the U.S. fleets would be refurbished for the price of two shorthaul jets. The weight problem turned out to be negligible: some new covers weighed as much as a bottle of Scotch per seat, others were no heavier than an in-flight magazine.

In the wake of Cincinnati, under huge pressure to do something about the fire hazard, the FAA launched the legislation for fireblocking in 1985, and the CAA was quick to follow suit. The directive applied to all new aircraft, and old ones were to be retrofitted. British Airways began to comply right away, well ahead of the time limit. (In August, 1985, one of their British Airtours 737s, called *River Orrin*, was awaiting its turn in the queue of the fleet for refitting.)

There were, however, very serious questions about how much safety improvement the new seat covers would bring. They would resist burning matches, and prevent fires lit by the much maligned 'fag end'. Presumably, many legislators failed to attend fire brigade tests which demonstrated how hard it was, anyway, to start a blaze by a carelessly discarded or even most carefully planted cigarette. On the other hand, no cover could withstand fuel-fed combustion, and protect forever the polyurethane foam cushions which, though difficult to ignite, would eventually burn like napalm in a 1000°C inferno. Not that such excessive heat would kill anyone – long before it could develop, that job would be accomplished by a whole range of black, toxic gases.

The new materials would certainly buy some extra time, slowing down charring and smoke generation by an estimated seventy to a hundred seconds. In 1989, a German research project* used cabin materials that complied with all the latest safety standards. The burner-assisted fire became self-sustaining after 105 seconds. In the first hundred seconds 'habitable atmosphere prevailed' in the test section of a wide-bodied jet. Then the temperature and toxic gas concentration increased rapidly, reducing visibility to zero; 150 seconds into the fire, before a flashover would occur, and at a time when most seats were yet to suffer any more than minor burns, the conditions in the cabin became completely unsurvivable.

The delay brought about by the new materials is invaluable, no doubt, but will it be enough? Certainly yes – *if we accept the regulatory authorities' assumption that every aircraft that passes airworthiness certification can be evacuated in ninety seconds.* Had those authorities queried, however, the validity of their own assumptions based on rather artificial evacuation tests conducted in clean, smokefree air, the question would have to be repeated: does the 1985 law of fireblocking add enough to the passengers' survival time? The answers would now have to be *certainly not*, if there's an inflight fire overland, when even prompt action by the crew could be thirty minutes away from safety; and most probably not in post-crash fires where people would become incapacitated by smoke and lost in the murk long before conditions became deadly. (Though cabin compartmentalization could be helpful, a five-pound fire-retardant curtain alone would emit enough cyanide to kill all aboard.)

More than two decades ago, *Aircrash Detective* presented a great deal of evidence to show that, while much had been done to prevent fires, the question of survivability had not been addressed properly. The changes since then have amounted to patching up rather than a rethink. Evacuation, passenger behaviour, the effects of smoke and panic continued to be viewed mostly through the tinted glasses of old assumptions, so it should be no surprise what Frank Taylor's statistical investigation revealed: in the three decades from the mid-1950s the accident rate had been reduced

* Dussa, Fiala, Wagner, Zenzes of the Deutsche Forschungs- und Versuchsanstalt für Luft- und Raumfahrt e. V. Institut für Antriebstechnik, Köln, Germany.

– but the actual fatality rate in survivable accidents had not. In 1985, with the new fireblocking legislation, far-reaching and costly care was *seen* to be demonstrated rather than actually exercised. Was it impossible to do more for survivability? Were there unexplored avenues of possible improvements that had been overlooked, underestimated or ignored? If yet another object lesson was necessary as a reminder of the 'unlucky few' who had not got away, and of experience that had burned into oblivion, we did not need to wait for long.

– 7 –

WASTED LIVES

Twelve minutes past six in the morning, on August 22, 1985, *River Orrin*, one of the 737s still awaiting its turn in the fireblocking refurbishment queue, was cleared by Manchester Tower to take-off for Corfu. There were on board two-pilots, four cabin crew and 131 passengers, including two infants. As the aircraft hurtled down Runway twenty-four, still well below take-off speed, a 'thump' or 'thud' was heard. It sounded like a tyre burst or bird strike. The captain chose to stop. It was a routine procedure with closing the throttles, selecting reverse thrust, and extending spoilers. The co-pilot applied maximum wheel braking but thc captain warned him: 'Don't hammer the brakes.' There was still plenty of runway ahead, and a tyre failure would not amount to a serious emergency.

Nine seconds after the *thud* a bell rang – No. 1 engine was on fire. Then the Tower warned about 'a lot of fire' on the port side. From the cockpit, the flames were invisible. The captain asked if he should evacuate. The controller answered: 'I'd do via the starboard side.'

Having decelerated gently, the captain did what he was conditioned to do: taxi off the runway, turn into and stop on a link road. Wasting some twenty seconds would have seemed a small price to pay for not messing up Manchester's single runway at the height of holiday traffic. The pilots had no inkling of the nature of the fire, and could not have guessed – because nobody knew –

that even a light, seven-knot breeze (to which they now exposed the engine) could have the devastating effect of turning the fuel-fed blaze into a flamethrower aimed at the fuselage.

Meanwhile, the pilots had to get through a busy schedule, including a written emergency drill in which the call to evacuate the passengers came as item No. 14. The captain broadcast: 'Evacuate on the starboard side please.' The purser opened the flight deck door demanding: 'Say again.' He got the confirmation as required by the manual. Eight seconds later, forty-five vital seconds after the thud, *River Orrin* came to a halt.

By then, there was pandemonium in the cabin. The *thud* had been heard. Somebody cried 'Fire!' A man shouted: 'Sit down, stay calm!' Passengers jumped to their feet. The purser, using the public address system, urged them 'to sit down and to remain strapped in'.

Smoke and flames broke through windows at the rear. As the first screams could be heard, people crowded into the aisle. Some climbed over the seats. Others seemed to freeze, and seatbacks collapsed on them under the weight of the climbers. Within five seconds, thick black smoke rose towards the ceiling, filling every orifice, every lung, gelling into chunks of filth, the size of Oxo cubes, in every mouth.

It blanked out the light and deadened all sounds.

At the rear, two stewardesses opened a door – and slammed it shut. Flames were roaring outside, enveloping the fuselage in black clouds. The two young women held their ground to direct passengers forwards. Their brave effort would be their last.

A dental technician was urging his fiancée to run. 'What about my handbag?' He shouted at her to forget the blessed thing. As people on fire were pushing to escape, the two struggled arm in arm along the aisle, but in the mêlée he lost her. Forever.

Up front, the purser tried to open the starboard door. It jammed. With great presence of mind, he checked the portside, saw that the forward spread of the fire was slow, and that firemen were just beginning to flood the area with foam, so he opened the door and deployed the slide. Another twenty-five seconds had been lost. Evacuation began at last but the flow was stemmed by the galley bulkheads: under tremendous pressure from behind, people were squashed into a human cork blocking the narrow passage. While the No. 4 stewardess pulled people free to clear the obstruction, the purser returned to the jammed door, and forced it open. Seventy seconds had been wasted before evacuation could begin on the starboard side. The two crew members helped, directed and pushed people out, until the smoke became completely unbearable and compelled them to dive out.

Halfway down the fuselage, another tragic delay was developing. The starboard overwing hatch would offer safe exit. I once saw the opening of such a hatch in an exercise: an ablebodied man, who had performed the task umpteen times, worked the handle, pulled the hatch in, and shoved it out through the hole in a single sweep in maybe four seconds. Aboard the *River Orrin*, 10F, the seat next to the hatch, was occupied by a girl. Urged to open it, she read the 'Emergency Pull' marking – and instead of the handle at the top, she pulled hard on her arm-rest that was mounted on the hatch. Another girl reached across, operated the handle – and the forty-eight-pound door fell on her friend in 10F, trapping her in her seat. A man from behind managed to drag the door further in and place it on a seat. Forty-five seconds had been wasted. A choking, panicky human wave climbed and scrambled towards the gap from all directions. Some people got out. Five minutes after the aircraft had stopped, a dead man was seen trapped in that exit. A little hand squeezed through between the corpse and the metal frame. A fireman yanked at it, and a young boy, the last survivor, came through.

Though the 'flamethrower' devastated the rear of the aircraft, no killing flashover occurred. So the accident was *survivable*. Except that fifty-five victims would never know that. The miracle was that eighty-two people survived, aided by the cabin crew's selfless dedication. But brave and noble acts (like a stewardess searching on her hands and knees for live bodies in toxic darkness) should never be called for. Air transport is a service, not a proving ground for heroes.

Instead of a complete mosaic of human suffering, *one* man's ordeal, a most educational one, should suffice.

John Beardmore is a tall, slim, rational and self-assured man. His vigorous gait reveals he could run or fight for dear life.

Half-remarks are a giveaway of the love for his family. Not the sort who could be kept away from his children in trouble. In 1985, at the age of forty-two, he was young enough to enjoy the advantage of agility, old enough to think for himself in any situation – given half a chance. He could serve as a 'most likely to survive' passenger profile. Yet it was down to luck rather than the laws of probabilities that he lived to tell his tale. There is no trace of selfpity. He speaks with the joy of having come through *unscathed*. But he cannot enter lifts or underground trains. In whatever way his recollections conflict with others' testimony, whatever he failed to hear or see around him, is not a comment on his power of observation but on yet another facet of the debilitating horror of it all.

'It was an odd morning. I felt apprehensive. We were probably the last to board. The plane was packed. I'm a six-footer, and I felt squeezed in my seat, which added to my unease. Pamela sat by the window in 13A, next to her the boys, David, aged thirteen, and Simon, fifteen, and I on the other side of the aisle. When the aircraft began to taxi, we relaxed – the holiday had begun. Everybody on board seemed to wear brand-new trainers. I was the odd man out. It made us laugh. By lunchtime we could be on the beach in Corfu.

'During the take-off run, we heard a thump. Like a muffled explosion. Was it a tyre? The fear was instant. The aircraft seemed to swerve. I could see an orange glow, almost the colour of a nice English sunset. Someone on the portside shouted "Fire!" People started jumping up. "Don't panic!" I yelled, "Sit down!" The aircraft was only just starting to slow down, there was nowhere to go. I thought if we stayed calm we'd all get out. Cabin crew were nowhere in sight. The tannoy telling us to stay in our seats was very faint. I heard screams, and turned – thick, black smoke was seeping through the window frames. A stewardess came from the front, and stopped. She saw what I had seen. I'll never forget her eyes. The horror. They just told me: We're on our own. My wife got up. She told the boys to undo their seatbelts. They started down the gangway, moving fast. A split second later I tried to follow, but I was too late. The aisle was full, the growing crush had separated us, grey smoke everywhere was beginning to turn black. And the aircraft was still taxiing. Why won't it stop? Let's open the doors and get out . . .

'The push forward was tremendous. Everyone for himself, that was obvious. No chance of women and children first. No time, no space in the cramped conditions for grand gestures. Yet there was no fighting. People behaved admirably. Nobody tried to shove someone else out of the way. Nobody tried to ruin the others' chances. You just knew that your chances were very limited. I found myself at Row 7 when the aircraft stopped. *(Six rows or six steps forward in about forty seconds.)* Lot of people in front of me, few behind. I was trying to think of my options. Somehow I never looked to the right. I saw only the portside hatch. I forced my way out of the crush, and turned back towards Row 10 with the overwing hatch exit – losing my place in the queue. I had no idea what effect the smoke might already have on me, but why else would I get into Row 9 instead of 10? It couldn't be a coincidence that another man made the same mistake by going into Row 11. Now I saw the flames licking the portside of the aircraft. In an odd way, I felt sort of relieved: instinct told me I wouldn't have the strength or the power of simplest reasoning to open that hatch.

'Back in the queue, pushing up to about Row 7 once again, I remember taking my first lungful of black smoke. It came from burning plastics and kerosene, not textiles. The immediate effect was incredible. It tasted filthy. I'll never forget the shock of how thick the smoke was. It bunged up your insides. One breath – and my legs buckled though I didn't collapse. The panic of helplessness. Fear like I've never known before. My brains . . . totally scrambled. You think in slow motion. You try to hold your breath or else you fall down, and you know that's fatal. Cleaner air down there? What's the use? You'd be trampled on. It was getting very dark but I could still see. I didn't know that my glasses were giving me *some* protection. Other people's eyes had filled up with thick layers of black goo by then. And I was still fighting to hold my breath. Nobody told us to, it just half occurred to me to cover my mouth with a handkerchief. But no. I needed my hands to hold my balance. If I managed to reach into my pocket, I could never raise my hand again in the crush. Were there screams? I couldn't hear them. All sounds were muffled by the smoke. As if we were moving in heavy fog or under water.

'The realization that I was about to die was undramatic. My life didn't flash by. My last wish was modest – to die peacefully. With dignity. Without fighting death. A few seconds must have passed. I wasn't aware of the crush carrying me. I knew I couldn't hold my breath any longer. I'd fall. I squeezed myself between two seats so as not to fall on the people in front, and took that second breath. My legs went. A seatback collapsed under my weight. I tumbled, landing on my back somehow. I didn't pass out, I think I didn't. The next thing I remember is a shaft of daylight facing me. Wasn't sure it was real. Perhaps a hole in the side of the fuselage. Seemed so small. How I got near that I'll never know. I found myself in clean air. I thought I was still way down the aisle . . . but it was the front port exit. Couldn't believe it.

'The tarmac seemed deserted. A lone fireman shouted "Get down the bloody slide!" But I couldn't move. All I could think of was that I could breathe, breathe clean air . . . the relief . . . and then the panic: I must be the only survivor! My family . . . I knew they'd been in front of me, but they must have fallen. They must still be inside. I wanted to return and die with them . . . But someone shouted "Dad!" Simon was running from somewhere . . . then Pamela and David . . . No one else. We just stood in front of the burning aircraft, all faces black, all of us motionless, speechless. Then a shout: "Get over here!" And we were seized by an awful, unstoppable coughing fit, bringing up ghastly black mucus.'

He was the last survivor to get out unaided.

Pamela Beardmore got separated from the boys, and could only hope they were ahead of her. When they reached the door, David was 'hit smack in the chest by the powerful stream of foam that knocked him right back into the aircraft'. Others fell off the slides and the wing because the foam made everything too slippery. And at some stage, when the firemen ran out of water, all the hydrants nearby were found to be dry.

The case is an endless catalogue of missed opportunities. If only the aircraft had stopped sooner, if only it had not turned into the wind, if only the engines had been shut down right away (crucial and oft-forgotten lessons taught by numerous other accidents), if only the doors had been opened more quickly, if only there was no jamming at the bulkheads, if only there was clear access to the hatch, if only people had some protection from the smoke for at least two minutes – a survivable accident could have been survived by all.

There were lessons for everybody at Manchester. Who would learn from them?

'ROUND UP THE USUAL SUSPECTS'

A massive investigation was mounted.

Like so many other foreseeable but unforeseen accidents, Manchester could not have happened at all without a whole series of incredible coincidences. The basic cause was 'an uncontained failure of the left engine'. That in turn was initiated by the explosive rupture of the much repaired, yet badly cracked, No. 9 combustion can. Similar cases were on record, none of them causing a disaster. But at Manchester, when a section of the can was ejected forcibly from the engine, it struck and fractured an underwing fuel tank access panel – a terrible coincidence in itself, considering the size of the panel that had only a quarter of the impact resistance of the skin all around it. Fuel gushed freely from the full tank, and it was ignited instantaneously.

'That engine was sick,' said John Boulding. 'With hindsight, it should have been out at least two days before Manchester, but it was treated as a *marginal* decision. The rest of the tragedy might have been down to coincidences. But that can was fatigue-cracked all round. And that's not a matter of bad luck. Such cans used to be thrown away. Then it became more economical to repair them. We know what it led to. And nobody should have been surprised. It was not the first explosion. Not by a long chalk.'

As NTSB Power Plants specialist Paul Baker keeps saying: 'The engine behaviour gives warnings. Hardware talks to you. If only people listened.' Communications were far less than perfect between engines and engineers, as well as between Pratt Whitney and the airline. The welding technique in the engine manual was different from that used by BA, and, as there was no dialogue, the manufacturers thought their messages were followed while the airline disregarded them as non-applicable. When, immediately after Manchester, BA stopped repairing cracked fire cans, not everybody listened to that message either: six months later, a pre-departure walkround discovered a hole in an engine of an Eastern 727 at LaGuardia; further inspection found excessively welded fire cans – one with more than half of it burnt away, and all the others badly cracked.

Dave King, one of the best AAIB inspectors, who was in charge of the *River Orrin* investigation, ensured that, long before the completion of the report, safety recommendations would be made and accepted. Acceptance, however, does not always mean implementation despite persistent reminders by the media and eloquently irksome public conscience bodies like SCI-SAFE* whose disarmingly dedicated frontmen, survivor Beardmore with William and Linda Beckett (who lost a daughter at Manchester), will readily make a nuisance of themselves with their powerfully argued case whenever the authorities try to sit on unearned laurels. Under pressure, following the lead by an embarrassed and repentant British Airways, new, if not always newly thought of, safety measures came into force everywhere.

'Change the emergency drills, forget about clearing the runway, stop wherever you can and evacuate at once, pay more attention to light wind, create extra room to make that overwing hatch more accessible . . . Even if we achieved no more, Manchester would already have become a lifesaving landmark,' said an expert who, though closely involved with the investigation, was muzzled into anonymity by a zealous bureaucrat of the *silence can't hurt* school. 'Unexpectedly, we didn't stop there even though, when we had begun the job, it all seemed quite straightforward. Remember the film *Casablanca*? After Bogart shot the Nazi officer, Claude Rains as the French policeman issues the order "round up the usual suspects" to cover his friend. Now I don't think anybody wanted a cover-up in our case – though the arguments some people put forward for years might suggest otherwise – but the whole thing seemed just routine. We'd round up the usual suspects, make the necessary noises, and move on to the next case.'

Yet worrying questions began to crop up, most inconveniently, from day one. Why was it, for instance, that so many people died at Manchester when barely a year earlier, everybody got out of

* Survivors Campaign to Improve Safety in Airline Flight Equipment, a fast growing and well informed pressure group created by survivors and the victims' relatives who were determined that nobody else should ever experience what they had been through.

another 737 at Calgary in Canada? In that case, too, deviations from the engine repairs manual weighed heavily among the causes. That pilot was also induced by a *thud* to abort take-off; his taxiing off the runway and doubts about the fire lasted even longer than at Manchester; evacuation was initiated mainly by the passengers themselves after long delays; the windows melted and smoke entered the cabin, but all 119 people survived. Was it a matter of luck? Was it because more doors were used or because people were not so tightly packed aboard the Western Pacific aircraft or because the uncontained engine failure only punctured the tank as opposed to the dinner-plate-size hole at Manchester or because – as the Canadian investigation suspected – almost all the passengers were very experienced air travellers, familiar with the 737?

Most of those nagging questions would soon find their way to Eddie Trimble of the AAIB, an ex-Rolls Royce engineer, who was caravan holidaying in the Lake District when the news broke. Back home a week later, he was briefed to look into the survival and evacuation aspects of the tragedy, the hazards of the tight seat pitch and narrow bulkhead, with particular attention to the jammed door – a relatively simple problem that would soon be solved by a minor design modification.

'None of us knew much about matters of survival, but I thought I'd better wrap it up in a couple of months,' he said, 'because due to manpower shortage, we always have to run fast just to stand still.' (Trimble's projected 'couple of months' was not unreasonable: in the CASB's Calgary report, for instance, cabin safety received merely a mention and the suggestion that ongoing efforts to reduce the survival risks should be monitored.) 'I began by examining the survivors' statements time and time again in order to grasp their problems in escaping. But a shock was in store for me. What stood out in all interviews and statements taken by the police and our Ops. investigators was the effect of smoke. Okay, we knew that smoke was a problem, but it surprised us that heat and fire were much less of a problem than toxic gases. We heard repeatedly what an instant debilitating effect that smoke had. It made them groggy, weakened their legs, made them feel that a second or third breath would knock them out. They couldn't see their hands inches away from their eyes, and as people kept falling in the aisle, no wonder they pushed, jostled, and tried to climb over seats. It began to look to me as if the provision of fast egress could be dependent on or even secondary to better protection of people from smoke.'

Observations and thoughts like that brought *cabin safety* firmly into focus. By the end of the year, the worst in aviation history to date with more than 1,300 fatalities, questions began to demand some answers. Has enough been done for survival? Should even more money be sunk into fireblocking? What else could be done?

Those who favour instant solutions – and which airline does not? – dredged up a promising idea that had been around for a long time: *floor level lighting* would be relatively cheap, easy to install, and demonstrate loving care for the travelling public. Based on the success of simulator trials of the late 1970s, a variety of such devices were now introduced by everybody in a great hurry, beating mandatory deadlines (1987 in America, 1988 in Britain). That just ignored the fact that real passengers' eyeballs would be covered by thick layers of hurtful smoke deposits. To see those lights in low visibility, people would have to follow instructions such as those on the emergency cards of Monarch Airline's new 757s: a woman is shown escaping on all fours. It may work in an hotel fire – near the floor there is also less smoke – but people who tried to crawl along the floor of the aisle on an aircraft would probably be trampled on by a panicky crowd. True, such lights would have helped at Cincinnati where the aircraft was only half full, but would have been of no use at Manchester. In the first survivable accident with *ground proximity guidance* already fitted, no survivor stooped to utilize those lights in his escape.*

Meanwhile, interest was revived in another old concept of great potential: an *airborne sprinkler system* could retard the development of heat, fire and smoke. It could lengthen the period during which the atmosphere remained breathable, and multiply the chances of escape. As an additional bonus, total hull write-offs could be prevented in some cases of post-crash fire. Ever since the mid-1960s, the idea had been considered repeatedly, and the FAA evaluated it in the early 1980s. Every time the installation cost, weight penalty (water and plumbing) and the risk of malfunction were the main objection, and every time more lip service was paid to the need of further expensive research – leaving it to

* Delta Airlines 727 crashing on take-off from Fort Worth airport, Dallas.

struggling inventors and would-be vendors actually to do something.

Manchester gave a tremendous impetus to the project. A company, aptly named SAVE, offered a cabin fire suppression system it had been developing; high-profile tests were organized; and the world could not afford to ignore them. Everybody was duly impressed: atop a blazing pool of kerosene, cabin furnishings remained unscathed. If the *River Orrin* carried such a spray system, most if not all people would have walked out of the fire. Forty-five gallons of extra water on a 737 could have suppressed the risks for at least three minutes. By then the fire brigade would have been in action. At the moment, firemen cannot fight the blaze inside the cabin during evacuation when people need their help most. But if the design allowed firemen to connect their hoses to the system, they could maintain the spray indefinitely from ground-supplies – provided the hydrants did not run dry, as they did at Manchester.

The weight penalty appeared to be minimal – the pipes and the on-board water would reduce the number of passengers by no more than two. But the tests also revealed and re-emphasized some serious unsolved problems. The airlines would face a huge expenditure if the world's commercial fleet had to be retrofitted with sprinklers, because the cost could be a quarter of a million pounds for an old 737, more for a Jumbo, double if a fail-safe duplication was required, but considerably less if the system was designed and built into every new aircraft. Would it be worth it? Yes, if certain questions could be answered satisfactorily.

Would it work in the air as well as on the ground? What about malfunctions? Would an accidental discharge merely drench the passengers or ruin the electrical systems especially on aircraft using fly-by-wire technology? What if, in a survivable crash, the fuselage is fractured – would the water mist still function? And what would it do about the biggest menace of them all – toxic fumes?

The tests succeeded in washing most gases out of the atmosphere, but carbon monoxide remained present in large quantities. As some killer fumes would be absorbed in water, particularly when the heat turned it into steam, passengers could be poisoned by swallowing it. Much could be learned from the technique used in mines and the petrochemical industry, non-toxic additives in the water might deal with carbon dioxide, but a vast amount of research had still to be done. At least four new water-mist systems are being developed, and there is talk about in-flight trials. At long last, five years after Manchester, the FAA, CAA, Transport Canada, French and other authorities show willingness to fund the work even though voices of substantial opposition continued to be heard.

Ben Cosgrove, Boeing vice president in charge of the Engineering Division, told me: 'The authorities are pressurized by politicians and the public, and in turn, we're accused of dragging our feet. Until now it was all about non-flammable furnishings. I've spent millions, hundreds of millions on changing my airplanes to reduce burn-through, and now the British come up with something else. Politicians want to mandate us to put some spray system on board. Sure, they give us very impressive demonstrations. It doesn't work so well on airplanes, but it's a very impressive show. Think of the money we're wasting all the time only because somebody wants to sell something or get publicity or build a political career! No, I'd prefer to do something rational even if it takes time.

'Now if they really want to achieve an optimum low burn rate, I'll first insist that all my passengers enter the airplane naked. Because their clothes are a serious fire hazard. Maybe men should fly in one airplane, and ladies in another. But seriously, I'd want to prevent accidents rather than apply band-aid afterwards. Take Manchester. The welding of burner cans and the inspection of engines were a sad story. That has now been dealt with. The location and weakness problem of the wing tank access panel has been solved. So if we can prevent a repetition, there won't be another fire. I don't dispute it, fires are horrible, but making emotional decisions won't save lives. We must use our money to the best effect – and that's never chosen by emotional speeches and politicians.'

Despite such strong views, the bandwagon is rolling, and sprinklers may provide added protection – one day. It is seen by many (including Ron Ashford, Safety Director of the CAA) as the most promising long-term prospect for saving lives in post-crash fires. The AAIB's Manchester report also sees great potential in the system. The FAA Technical Center in Atlantic City is running full-scale tests, conducting long-term evaluation with the CAA in a big way. But how long is long-term? And how long away are we from its beginning? The variety of estimates, ranging up to fifteen years if ever for retrofitting all aircraft, is not a harbinger of profound reassurance, particularly not if even Raymond Whitfield, chairman of SAVE, feels that water spray systems could not be mandatory before 1992 for new aircraft and 1997 for old ones . . . allowing for further research, depending on sufficient support, and barring political football being played with the project.

So where does that leave those who want to fly to Disneyland or a business meeting today or next year? That was yet another question Eddie Trimble had to contemplate when, back in 1985, he embarked on the longest 'two-month job' that would take well-nigh three years out of his life.

'I listened to the survivors, saw the victims, and knew we'd have to do something *now*, not in the twenty-first century', he said during lunch break at the Cranfield course for accident investigators. 'The pathologist confirmed that at Manchester eighty-three per cent of the fatalities had not been *burnt* to death – they had succumbed to toxic gases. And that was not untypical for such survivable accidents. I then happened to see an ad. for some sort of smokehood. I went to look at one out of curiosity, and began to read up on the subject.' He would soon find yet another warren of missed opportunities and lessons that had been learned only to be forgotten.

'TOMBSTONE OR MILESTONE?'

The smoke hazard so painfully illustrated by the Manchester disaster was no stranger to the industry. In 1965, after the Salt Lake City accident, a great deal of American research was initiated to see if some form of passenger protective breathing equipment was a feasible proposition. By 1966, the U.S. ALPA had seen enough to recommend smokehoods for passengers. Various types were tested, problems of application were tackled with some success,* and, by January 1969, the FAA could publish a 'pro-

* Much of the work was done by CAMI, the Civil Aeromedical Institute, at Oklahoma City. Its Protection and Survival Laboratory manager, Arnold Higgins, presented a paper at the AGARD symposium, 1989, offering a fine, twenty-year-long slow-motion picture of grinding stop-go efforts by authorities and researchers, including a crashworthiness committee that had the support of the major U.S. aircraft manufacturers.

posed rulemaking' for their introduction. There were twenty-one major responses: ten were in favour and three were neutral. Seven months later the proposal was withdrawn because, allegedly, 'hoods might cause delays in evacuation'. It satisfied the powerful aviation lobby, but there was no escape from the problem.

In 1982, a comprehensive cabin safety report was completed for the FAA. It demonstrated that survivability would be enhanced equally by zoned water spray and smokehoods, but the latter would be almost ten times more cost-beneficial than the former. The NTSB urged that the report should be acted upon. Three full years later, in 1985, the FAA answered that 'if the industry developed suitable devices that showed promise, the FAA would evaluate them and develop criteria for their approval'.*

The move was well timed: twenty years after Salt Lake City, just seven months before Manchester, nobody could say the FAA paid no attention to the risk. After all, others did not even pay lip service to it.

Having ploughed through mountains of documents, Trimble soon came to the conclusion that the question had not been addressed properly: did the authorities fail to read the available data or were they afraid to face facts that might not be to the liking of the industry? 'In a gas-filled environment, time was obviously the key factor in determining the chances of survival,' he said. 'Because even in a bad case like Manchester, there's at least *some* time during which people can escape.

'There was, for instance, a woman with a baby in Seat 21F. She managed to get out of that black inferno at the back, and struggle all the way to Row 10. We know that for sure because hers was the only baby on board, and a woman in the area of Row 10 was begged by a stranger: "Please take my baby." With that the mother succumbed to the fumes, and perished together with her baby and husband. The point is that she did have some time to escape. I felt sure that with some protection from smoke, she could have survived. It was suggested frequently that donning a hood would waste time. Had she wasted, say, twenty seconds, in that way, it would have gained her minutes of survival time, enough to get out. Then there was the man who died *at* the overwing hatch. Why didn't he get out? It's no good telling people

* *ibid.*

"get out quickly" if they get incapacitated on the way. And they do. Fuel and the great variety of materials burning in an aircraft fire produce a deadly cocktail of some 300 toxic fumes. A lungful of those, and the person collapses, then expires. It suddenly struck me that there was a basic mechanism at work – something I then referred to as DIES, for Debilitation Induced Evacuation Suppression. It's a rapid, closed-loop process that can slow down evacuation to the point where it ceases altogether.'

Evidence in favour of smokehoods accumulated not only from obvious sources like researchers and hundreds of autopsies in reports on survivable accidents, but also from unexpected quarters such as CAA executives and other air safety specialists who would admit, strictly in confidence, that, for many years, their hand-luggage contained, as a matter of routine, their own personal smokehoods. Yet the same people kept up a barrage of objections to the general introduction of such devices. 'What would they do if they found themselves in a fire on a holiday flight with their families?' Trimble queried. 'Would they stop to help their kids if they themselves had no protection? Would they if they did have a hood? Would they let somebody else wear that single hood? Nobody could answer that. The reaction to life-threatening danger is primeval. People cannot help wanting to survive. The most harrowing aspect is when a father gets out without his family. He immediately tries to return to get them out. He's not allowed to do so.'

A Manchester survivor, a thoroughly responsible businessman, said: 'In the mêlée, the cabin crew would never get the chance to don their hoods. We would. For our children. People would snatch them away from their heads. The basic animal instinct for survival doesn't stop to consider priorities. That's why it wouldn't help to let people buy their own or hire from the travel agents. If, after Manchester, the airlines think that we can still be regarded as self-loading cargo, they're in for a big surprise.'

By the end of 1985 the AAIB knew enough to urge the CAA that smokehoods should be investigated thoroughly. From experience they also knew, however, that sympathetic noises would be the only likely positive response, and the recommendation would gather dust on shelves – unless even more substantial evidence from actual tests were available. The AAIB had no funds or traditional authority to initiate tests. Research was not their province. It was, therefore, an unspoken yet devastating comment on all regulatory authorities and other guardians of air safety that they felt compelled to go ahead – for nobody else would do the job.

Trimble was lucky. He got extra time from Ken Smart, his immediate boss, and full backing from Dave King, who was in charge of the investigation. Chief Inspector Wilkinson allocated £50,000 to the extra-curricular work from the very meagre resources of the AAIB. (Neither he nor anybody else could have guessed that the ultimate cost to them alone would quadruple.) Now Trimble was free to travel up and down the country and put the cat among the pigeons. U.S. research establishments, like CAMI, joined forces, specialists like John Stewart (Royal Albert Edward Infirmary, Wigan), John Evans (British Coal Laboratory, Edinburgh), Keith Paul (Rubber and Plastic Research Association), and others, all dedicated to air safety, volunteered their tireless assistance. SCI-SAFE drummed up public support, and Dr James Vant of the Air Transport Users' Committee produced funds and cooperation out of some magician's hat. For those involved it became an all-consuming way of life, with their spouses contemplating divorce when the telephone would not stop ringing even at night and over Christmas. Their research ranged from smoke analysis and smoke chamber tests to experiments in theatrical and real smoke with seals, filters, oxygen consumption, mechanical lungs, dummy heads, and various smokehoods. At one point, when human tests became desirable, but the AAIB was warned off because of the hazards, Dr Vant organized tests at the Offshore Fire School, and a massive exercise aboard a Trident with 800 'naïve' volunteers.

Scientific proof would soon back the survivors' experience that a few lungfuls of gases present in aircraft fires would incapacitate or kill anyone, and that with some form of protection, particularly from cyanide and carbon monoxide, far fewer lives would be wasted in survivable accidents.

Forced to face the amassed data, the bandwagon was now rolling worldwide, politicians were quick to jump on it, further research was picking up everywhere, and smoke masks were provided for many executives flying on corporate aircraft. By 1988, the CAA changed its position from 'no' to 'maybe' if new products met its specifications. (Influenced by the FAA and kowtowing to the industry, some authorities would prefer the slow evolution of *international standards*. To research and agree those in Europe alone could take several years.)

The CAA specifications called for hoods to protect people from in-flight fires for twenty minutes – when the crew hoods were for only fifteen minutes. The explanation was that during fifteen minutes' firefighting everybody aboard would be protected. Then passengers would have five more minutes to escape. The crew? Well they must be at the exits, they don't need the extra five minutes. To which survivors answered: 'Like hell they don't! At Manchester two of them died near unusable exits, and two forward cabin crew, helping the evacuation, were about to succumb when they just managed to get out.'

Objections to smokehoods come in many guises, but they tend to have one common factor: they lack evidence. One of the greatest achievements of Trimble's campaign is that opposition can now be countered by facts from actual tests.

Hoods make you deaf. Disproved in tests of the 1960s. It is the smoke that suppresses all sounds.

Hoods are associated with 'plastic bags' and people wouldn't don them because of claustrophobia. Not so: Dr Vant polled 1,300 people including business travellers at Gatwick and Luton under CAA scrutiny, and ninety-eight per cent said they would use them.

Hoods themselves may burn. Eventually, yes, but long before that hair would burst into flames because it has a low flashpoint, and hoods protect against that.

Hoods would be difficult to don. Tests proved the opposite, hoods were much easier to put on than life vests.

Hoods would delay evacuation. Untrue: the seconds used in donning would buy minutes of protection.

People wouldn't understand instructions how to put them on. Easier than those instructions regarding life vests or how to open exits.

People may don them back to front. Some designs will work even then – others make it almost impossible to err that way.

*Hoods would be stolen.** They could be protected by a simple device

* Fritz Hofmann, emergency procedures specialist of Swissair told me: 'We lose about 500 life vests a year. People would pinch hoods as souvenirs or for protection in hotel fires. A good one would cost a lot, otherwise it's no better than a wet towel.'

that sounds a buzzer when removed. It would be cheaper than a greeting card that plays music when opened.

Hoods are too expensive. Initially, that was true; new designs cost about £50 but would be cheaper if mass produced. A cost analysis showed that a good smokehood could be provided for the price of a buttered breadroll per seat per flight. Too much? They now put personal TV sets in seatbacks for a thousand pounds each.

Each authority, each airline, seems to have its pet objection. Lufthansa, for instance, would prefer a combination of smoke and oxygen masks. That would require a complete redesign to serve both purposes and be detachable. Yet the emergency training manager of the airline told the AGARD symposium in 1989 that, though he thought it was uncertain that hoods would be introduced or would save lives at all, he, as an individual, 'being familiar with several kinds of emergencies', would be 'in favour of such equipment'. The pro-hood lobby is often accused of putting forward an emotional argument. 'They may be right,' said Linda Beckett of SCI-SAFE. 'Losing your loved ones and trying to save lives *are* an emotional issue. But does that reduce the weight of logic and hard facts?'

Facts like the survival of Alf Morris, MP, who volunteered to be a guinea pig: a kerosene fire was lit under the fuselage in which he sat donning a hood; after sixty-eight seconds the atmosphere became non-survivable; he stayed aboard 139 seconds, until the heat became unbearable, and left this 'hell on earth' a bit shaken but smiling.

John Stewart FRCS summed up an arduous investigation:

'There is evidence that fireblocking and other strategies relied upon by the regulatory authorities to protect aircraft passengers from smoke may have little or no effect because kerosene smoke alone may incapacitate passengers . . . The physiological, pathological and toxicological evidence can no longer be ignored. Measures to save lives, by protecting passengers from smoke, should be introduced as an urgent priority'.*

The basic causal report on Manchester could have been completed in about a year. By the time it was published in 1989, it had grown

* The Importance of Pathophysiological Parameters In Fire Modelling Of Aircraft Accidents, paper presented at AGARD symposium, 1989.

into a massive tome to accommodate data from several major research projects that, according to its critics such as Boeing, should have been published separately. 'True, strictly speaking, it was none of our business', said Trimble who has learned to live with snide remarks about his so-called obsession with smoke-hoods. 'But we have no other outlet, and this time we simply had to ensure that this report would become not just a tombstone but a milestone of safety. We had to show where we thought the industry was going wrong. We'd have been glad to leave all this work to others if they were willing to do it. Now at least if anyone wants to argue, they must produce data that's as good as ours'.

That could be wishful thinking. The CAA, its own specifications notwithstanding, trotted out unsubstantiated statements such as *smokehoods might cause more loss of life than they would save, and passengers, with varying degree of dexterity, might place their own and others' lives at risk by lingering in the cabin trying to put on the hood correctly.** They also claim that passengers who leave their seats without their hoods will try to fight their way back to get them.

'We know that prevention is better, and hoods are not the ultimate, ideal solution,' said William Beckett, 'but they could save lives *now*. As the cost begins to dwindle, a Mark I type could be made mandatory until something better is found.' The SCI-SAFE representatives' forceful argument at a BA share-holders' meeting received due applause, and the airline declared that, if hoods to meet its own specifications were available, they would be made standard BA equipment. By 1989, twenty-seven companies had submitted their designs, but BA invited more to compete for places on its shortlist. In May, 1990, however, the airline called a halt to its own costly and time consuming search for a satisfactory product, and decided to await CAA advice – a major setback to the pro-hood lobby – even though the pledge to be first with smokehoods could have become a marketing tool to induce competitors to follow suit without waiting for the rest of the world to make it mandatory.†

Although all the investigators' recommendations were known to

* CAA presentation at Sci-Safe seminar, Manchester, June 1988.

† See: 'Compete on Safety' in Chapter 9.

and mostly accepted, at least in principle, by the CAA long before the report was published, only some six out of thirty-one safety recommendations were implemented by 1989.

The need for more research, further consideration, and international action is said to be the reason. But will such reasons stand up in court if another Manchester happens? For experts like the Manchester fire chief and Eddie Trimble have no doubt that it can and probably will happen, that when it does happen the outcome will be just as disastrous; that there will be more major in-flight fires like Varig, Riyadh and Cincinnati, and that we may yet see 300-400 corpses gassed behind the closed exits of a virtually intact aircraft after it lands. Some lawyers hold the view that, if much publicized and available preventive measures were not applied, not only the regulatory authority but also its individual officers could be held liable for breach of their statutory duty to protect air travellers. And while the government washed its corporate hands – the CAA is independent, isn't it? – the process might not stop at that.

In most cases, airlines and others are only sued for damages. In France, after the Air France A320 airbus crash at Mulhouse, *homicide involuntaire* charges were brought. And now that a U.S. court has accepted Don Madole's brilliant argument, and awarded *punitive damages* to victims of the Korean airliner that had been shot down for straying into the airspace of a Soviet missile base, lawyers representing the Lockerbie sabotage victims have tried to show that PanAm was also guilty of *wilful misconduct* – the term that unlocks the constraints of the sixty years' old Warsaw Convention, the internationally agreed and ridiculously low limits of carrier liability. *Wilful misconduct* findings have broken those limits some twenty times in France and America. So now it is not inconceivable that even more wide-ranging serious charges could be brought.

In 1989, two years after the *Herald of Free Enterprise* ferry disaster, Britain's Crown Prosecution Service charged P & O Ferries with corporate manslaughter. Could the same happen in the wake of an aircrash? The answer is probably yes. Directors may be brought to bear responsibility. The possibility of criminal charges would help some individuals concentrate more on safety. But if higher insurance premiums and compulsory passenger disclaimers will cater for this contingency, it might prove that even costly insurance could be cheaper than precautions, and that accidents could be cheaper than safety – even if some individuals must go to jail or take the rap, while lawyers argue all the way to the bank. The legal profession would have a windfall if, for instance, it had to fight out whether a gap is a trap or a door.

A GAP IS A TRAP OR A DOOR

Even the best planned evacuation will often deteriorate into a mad scramble to escape. With some notable exceptions, it is not the passenger who should be blamed for that but the fact that those plans are not all that well laid in the first place.

Fast evacuation depends basically on three factors. Provision built into the airframe (exits, slides, width of aisles, lights, signs, fire and smoke protection), crew training, and the passengers' preparedness and behaviour.

The Manchester crash brought the entire question of egress and evacuation tests for airworthiness certification into focus. It demonstrated that in a smoke-filled environment, the '90-second evacuation with half the exits operational' concept was much too fanciful.

The *River Orrin* had four doors and two overwing hatches, half of which were opened and operational after some delays. But which half? The certification process assumes that one side of the aircraft will be safe – all starboard or portside doors, that is. At Manchester, 'half' meant two doors up front and one overwing hatch, an eighteen-inch hole that became the nearest exit for a hundred people, some three quarters of all those aboard.

Access to the hatch was limited. Access to the two wide open doors was not much better: to reach them, passengers first had to wriggle and squeeze through the gap in the galley bulkhead. Due to BA specification, that gap was twenty-two and a half inches wide – two and a half inches wider than the legal minimum, but five inches narrower than the minimum for doors in British buildings – and created a bottleneck in the fully expectable crush. While the cabin configuration for such a crammed holiday charter flight must have been a special hindrance, it was clear to the investigators that a complete rethinking was overdue.

The worst of it is that this recognition was not new. At least the CAA, everybody's punchbag after Manchester, now yielded

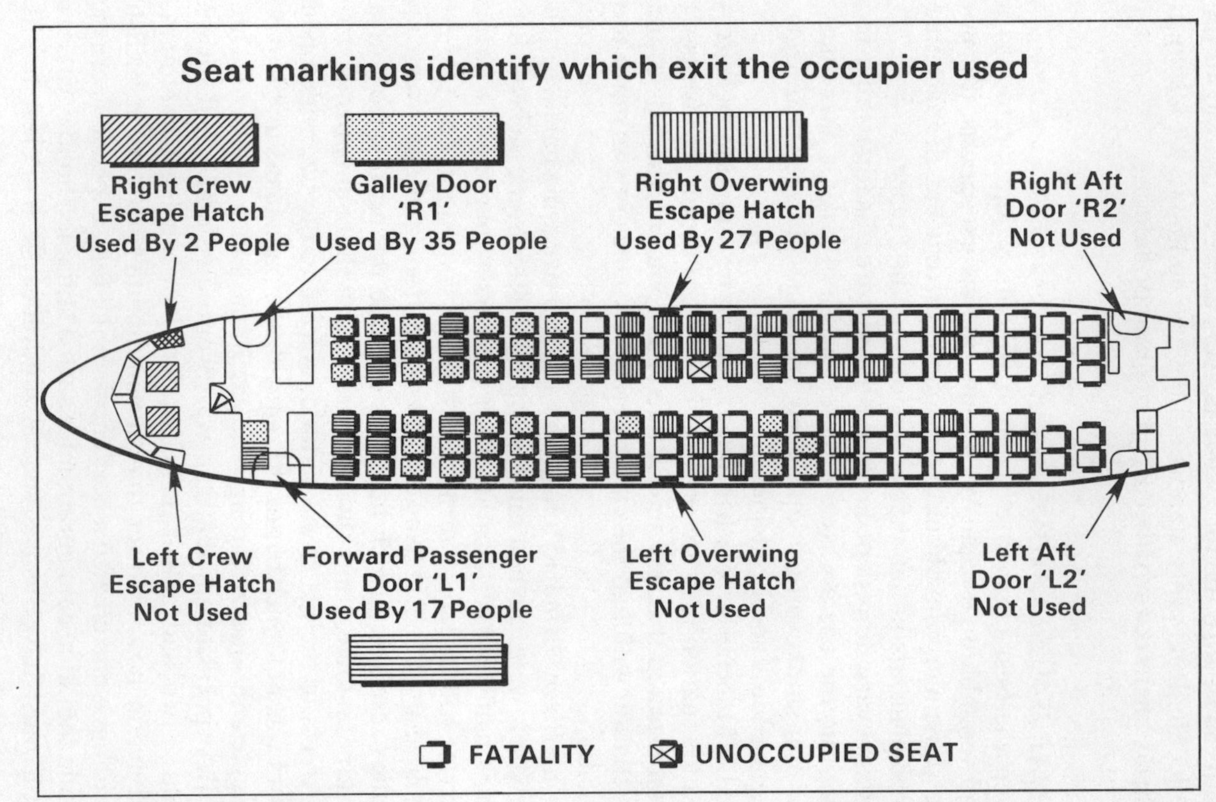

The Manchester evacuation chart offers some valuable lessons. (Source: AAIB report.)

to pressure, and after hardly excusable delays, provided some funds for experimental evacuation tests. (At the House of Commons Transport Committee hearing into cabin safety, CAA Safety Director Ron Ashford argued that there was no delay: a lengthy library search was essential, and first it had to be proven that such tests would be useful.) The work is being done at the Cranfield College of Aeronautics which can combine essential disciplines like engineering and psychology with investigators' and survivors' observations.

Even the initial results were startling, mainly because it should have been seen if not foreseen long before that assumptions are no match for facts. In order to increase the realism of the tests, an ingenious competitive element was introduced: People who might be vulnerable because of age or infirmity were excluded; more than 2,000 selected volunteers would receive £10 attendance fee, and a £5 bonus if they were among the first half through the doors. The video film of seventy-nine tests is a frightening eye-opener. There is no smoke, no danger, but people are pushing, battling, climbing over seats which collapse creating more obstacles. As Helen Muir, head of the Applied Psychology Unit, has been quoted: 'If this is what they'll do for a fiver, what would it be like when everybody's running for dear life?' Five of the tests had to be halted because, under pressure to get through an exit, 'individuals were physically stuck in the aperture', or the safety officer had to stop the exercise because 'a volunteer fell into the galley vestibule area, after being pushed through the galley restriction, and was at risk of being trampled upon . . . '*

One of the early discoveries was that a maximum opening may not always be the optimum. Theoretically, the wider the gap in the bulkheads, for instance, the faster the evacuation flow could be. That was borne out by the tests that increased the gap from twenty inches upwards, and the researchers found that a minimum of thirty-inch openings in bulkheads should be created. But when the galley was removed completely, new problems occurred: the cabin crew on duty lost their protection from a surging human wave that pinned them to the exits; if, somehow, they could still open the doors, people would be forced through, dropping to the

* Human Factors in Cabin Safety by Helen C. Muir, MA, PhD and Claire Marrison, BA, MSc, *Aerospace*, April 1989.

ground, before the chutes could inflate properly. Similarly, the lack of legroom particularly on charter flights provides not only cramped accommodation but also far too limited access to the overwing hatches. Yet when maximum space was allowed by the complete removal of a row of seats, a new set of difficulties developed in the crush, and a certain amount of flow control was proved to be vital.

A chart of exit utilization at Manchester surprised everyone. Some people got out of the worst affected areas, others died in their seats just one row away from the nearest exit; some took what would seem the obvious way out, others bypassed an available exit for no apparent reason. In the tests, the volunteers were numbered to show which seats they came from, and their use of doors as well as the sequence of their egress showed a similarly 'haphazard' pattern. Why did some individuals choose the long way out? Why did others stay in their seats 'to die' – did they freeze in panic, did they try to rise too late to beat the crush, were they trapped under climbers and collapsing seats?

Every new test seemed to expose the need to conduct five others with more refinements. What if there was smoke, even just innocuous, theatrical white smoke in the test cabin? What if groups, like families, joined hands to get out together? What about children and disabled people? (In standard certification tests volunteers must carry three large dolls, but that is barely a gesture to the reality of a mother's determination to save her baby.)

NEGATIVE PANIC

No volunteer in the Cranfield tests was *instructed* to obstruct the evacuation, but a woman in the aisle seat of a row adjoining the hatch just froze when the door was opened. She had to be persuaded to move and let others pass. The five seconds she wasted could have cost many lives. Hers was one of the cases that emphasized yet again how little is known about human response to fear and stress, 'behavioural inaction' or 'negative panic'.

A Hawker Siddeley 748 once completed three somersaults, and landed upside down. Fifty-one people were hanging strapped in their seats, but the *Daily Telegraph* reporter first at the scene was told 'there was no panic'. Aircrash survivors often make that claim. A flight engineer told Wing Commander David Fryer, a physiologist, that when his crashing aircraft came to rest, with nobody injured, 'there was a burning rivulet of fuel flowing slowly down the aisle. We all sat and watched it in silence. And it was long before anybody moved.'

Ed Slattery of the NTSB once told me about a Viscount accident: 'The stairs opened normally, but people wouldn't get going. The stewardess begged them, screamed at them, fought them' to no avail. People would not leave without their luggage, a woman refused to go out into the cold without her fur-coat, and a man standing in a growing pool of fuel under the fuselage insisted on lighting 'a cigarette to steady his nerves' before moving on. Madness? Probably yes, in a way. The mind is maimed by negative panic.

Apart from the healthy urge to do something for survival, the effects of panic create a very wide spectrum ranging from freezing (irrational in face of danger), to heroic acts of self-sacrifice (admirable but equally irrational in the cold light of the morning after). One would think that, these days, the more sophisticated air traveller just knows that emergencies are not to be trifled with. But that is no guarantee that, when the moment comes, logic and clear thinking will not desert him or her.

In August 1989, the pilot of a Britannia Airways 737 (the type that crashed seven months earlier) reported some hydraulics problems. With smoke pouring from the wheels, he made an emergency landing at East Midlands Airport. There was no damage to the aircraft. All the doors were opened, and the chutes were deployed. Nobody could tell if the aircraft might burst into flames, but many passengers would not leave without their duty-frees. They crowded the aisles, removing bottles from the overhead bins, ignoring the stewardesses' pleas and shouts, blocking all routes to the exits, and, eventually, breaking bottles as they came down the chutes. With no smoke, no fire, the evacuation took three minutes – as opposed to the ninety-second certificate for the aircraft.

'Fear will heighten autonomic activity in that heart, respiration and sweat rates will all increase, but, as well as this, there is reasonably good evidence that attention will narrow so that only matters directly associated with physical escape are dealt with, and any intellectual activity – such as following the simplest instructions – will be materially impaired.'*

So was it that numbing negative effect or the unnatural complete lack of panic that killed just a bunch of people at Athens in 1979? It was yet another of those fully survivable cases. A Swissair DC8 failed to stop on the runway, crashed, and fire broke out. The portside doors were opened, and there was ample time to escape: 140 passengers and the crew took almost five minutes to get out. Fourteen passengers were, however, burnt to death (not asphyxiated) in their seats in Rows 21-26. There was some suspicion that their path might have been blocked, but that was soon disproved: many people passed them from behind without any problems. So why did they not move? Were they too stunned? Swissair Emergency Procedures Instructor Fritz Hofmann is still puzzled:

'Among the passengers on that flight, there was a large group of European physicians going to a convention in Peking. Had they been a bunch of Sicilians, there could have been a stampede, but physicians are less panicky. They might have thought: "There's plenty of time", and waited patiently for the aisle to clear. By the time they were ready to move, there was only one door available because of the fire, and they must have been overcome or at least slowed down by smoke. But then why didn't the rest of the passengers suffer negative panic, why weren't they also too stunned to move fast or as *cool* as that group? I think that someone in that medical group must have assumed authority, and said: "Let those fools rush". And that can happen all too frequently. If the crew is not ready to scream at them, and be aggressive, somebody else might take charge. At Athens, the crew at the overwing exits were unable to give appropriate orders because of the thick smoke, and confusion was created. They weren't sure what orders to give when people converged on them from both ends. Repeated loud shouts from the usable emergency exit: "Come this way!" might have saved even that group of cool customers.

'At Los Angeles, a Continental DC10 had an overrun accident. One of the slides was burnt, another got wedged and wouldn't

* Passenger Behaviour In An Emergency by Roger Green, RAF Institute of Aviation Medicine, *Flight Safety Focus*, August 1987.

inflate properly. The hostess said: "Don't go out that way," but a passenger yelled: "I'll deal with that!" People followed him out on to the wing – and two of them fell straight into the fire.'

SERVICE WITH A SMILE

Those two accidents made a major impact on Swissair emergency training: 'From then on we assumed, and now it's proven, that, contrary to ICAO thinking, in more than half of cases less than half of the exits are usable,' said Hofmann. 'In our cabin evacuation trainers, we drill the crew to be bullies in emergencies, and be ready to force passengers out if necessary. Because cabin crew are trained to be polite, shouting at people may seem to them unnatural. The evacuation trainer has a red light that comes on only if the orders: "Follow me!" or "Come this way!" are loud enough.'

The same voice-activated light is now used in training by some countries whose Oriental and Muslim crews face an even greater barrier. Where women are still brought up to be subservient to men, and elders as well as those of higher rank must be venerated, it is a culture shock to shout or take the initiative and start evacuation without the captain's orders.

Flying with various airlines one gets the disturbing impression that the cabin crews' clothes have become a slave more to fashion and national pride than to purpose. More distinctive uniforms would lend extra authority to their wearers, and would help passengers to identify who is in charge. Pencil skirts, saris, kimonos and cheong-sams with revealing slits may perk up jaded travellers, but how would they score as fire retardants, and what hindrance would they cause in moments when a non-swimmer needs help not titillation? How many of those petite and lithesome stewardesses could shift bodily a panic-stricken passenger?

The principles of cabin crew selection have now also come into question – but only just. Few if any airlines carry more than legally required minimum cabin staff, like British Caledonian used to. Though their presence is primarily for dealing with emergencies, even the better airlines devote no more than one-fifth of their training time to safety, while the rest is to prepare them to look good, serve food and drinks efficiently and with style, be friendly and chatty at all times.

'Selection processes need more research,' a psychologist told me, 'because certain types are naturally more able and authoritative.' In the course of doing personality tests for some airlines he noted that 'scheduled services tend to go for slick, sophisticated types who are attractive to businessmen, while holiday charter operators prefer the mumsie type who can be trusted with kids, and will be a natural friend who is full of reassurance. The difficulty for both these types is to make a switch if need be, and take command.' An older person may carry extra authority, but one wonders what reassurance British passengers would get from staff in their sixties if some union demands for standard retirement age for everyone were met?

However careful the selection is, nothing can compensate for the lack of thorough training and crew coordination. Perhaps because not many accidents highlight this problem, numerous lessons taught by the few tend to be ignored. The better airlines provide 'hands on' practice in opening doors, but Jim Powell of the Flight Safety Foundation observed that many flight attendants have only a vague idea where emergency equipment is stowed, and 'may not have done more than close the main entry door a couple of times'.*

In March, 1986, a TWA 727 landed at Boston's Logan International Airport. From an overheated air-conditioning pack smoke entered the cabin. An announcement was made from the cockpit that there was no need to worry, but smoke filled the fuselage within seconds. The captain was obviously unaware of the situation for eventually he told the cabin crew to send everyone forward. The aft stairs were opened – by someone on the ground. The crew had failed to take the initiative: they 'made an announcement to evacuate – after people had already begun evacuating'.† An added hazard in that situation must have been that all three flight attendants had been hired, in reaction to a strike, just nineteen days before the incident.

The NTSB found it disappointing that the FAA issued only toothless advisory circulars to advocate improved training. In the same year, 1988, Noreene Koan of the U.S. Flight Attendants'

* *Flight Safety Focus*, No. 1, 1986.

† NTSB Chairman Jim Burnett at the International Aircraft Cabin Safety Symposium, Oakland, California, February, 1988.

Air Safety Committee revealed that 'serious problems remain in flight attendant training. Drills and actual operation of equipment are becoming less frequent, and at some carriers operating a fire extinguisher means passing it around the class.' At many U.S. airlines it had become a rarity, she said, to get 'into the water to learn about vests and rafts'. FAA inspectors often allowed the reduction of training from twelve to eight hours a year, 'and on the job training, the actual experience requirement for flight attendants, remains a pathetic two and a half to five hours.' And Captain McBride of the Canadian ALPA told me that 'there's not enough cross-training between pilots and flight attendants. They all have their little booklets, but coordination depends on the captain. Pacific Western Airline, for instance, had a policy to lock the cockpit door for take-off. If the intercom broke down, there would be no way to inform the captain about, say, a fire in the cabin.'

A DANGEROUS GAME OF CARDS

Communication with the passengers is another hazard that may be glossed over in crash reports. A special NTSB review of accidents revealed a seemingly minor but potentially disastrous malpractice: 'megaphones were rarely used in evacuations' and that created special problems when the engines were shut down because of actual or erroneously assumed engine fires. In 1980, at Phoenix, the PA and interphones were thus muted, and, instead of using megaphones, 'the second officer had to give evacuation orders to passengers and flight attendants by *word of mouth*'. While up-front evacuation began, the crew were still ordering passengers elsewhere to remain seated. Five years later, at Manchester, when the power of the PA system was reduced, some emergency equipment for the crew, including their hoods and two loudhailers, were stowed 'in overhead bins in the passenger cabin, not at the cabin crew stations'. The crew would have to go against the flow of escaping passengers to reach any of those vital aids.

But even if public announcements can be made one way or another, it is not all that easy to communicate. The language of the air is English, but it is a far-fetched assumption that everybody on every flight speaks or understands that tongue. The briefings may be multilingual, but which language will be used in an emergency? Floor lights guide people towards exits, but do not tell which exit is safe. When instructions are given, will everybody know what aft, left, right, starboard or portside means? Will every stewardess have the power to communicate with the tone of her voice, like the one aboard a BAC1-11 that crashed at Milan? When the engine was chopped and the lights went out, she faced a shrieking crowd, and a Turk who blocked the aisle refusing to move in any direction. The stewardess shouted at him firmly: 'Shove off, horrible little man!' Without understanding a single word, the man was so startled by her tone that he began to help others, and was almost the last to leave the aircraft.

Language problems notwithstanding, the basic concept is that passengers should be herded like sheep by telling them what to do when. Yet regulators and airlines know only too well that in an accident there may be no time to give detailed instructions, that cabin crew may be incapacitated, and that people who succeed in leaving an aircraft on fire often name their 'mental game plan' as their key to escape. That is why the briefing of passengers is crucial.

Pre-departure briefings are a bore to the crew. Those who hear them repeatedly may easily get the feeling that they know it all – even though they do not. Many people flaunt their lack of interest as if to demonstrate what cool and seasoned travellers they are. Where available, a video presentation is a considerable improvement on the dull spectacle of a live performance, but there is no guarantee that people will listen.

A self-confident Swedish executive once explained to me in all earnestness that 'we don't need to know any of that garbage, because we have herd instinct to follow others, and besides, it's the airline that's responsible for our safety.' Truly experienced fliers and air safety specialists could not agree less: everybody must take *some* responsibility for all passengers' fate. A pilot went as far as suggesting that 'air travel being a routine part of modern life, passenger education ought to begin at school.' Roy Lomas, Air Safety chief of British Airways, often feels like shouting at people: 'Your lack of interest could endanger *my* life. We're all in this together. In an emergency, everybody must know what to do. We carry you – and you must pay attention to briefings. It's part of the deal.' Thomas Hinton, investigation director of CASB, put it this way: 'I'm no hero. I fly a lot, but I listen to and read instructions, and count the seats between me and the nearest exit every time. I wouldn't mind if they introduced a five-minute written exam before take-off, and those who fail – out.'

Counting those seats is not a part of standard briefings. The airlines used to hide Emergency Exit signs behind frilly little curtains, and they still do not like to remind passengers that they might not reach their destination safely. Marketing executives have often fought to soften the impact in video presentations that now tend to be all smiles with a sprinkling of flowers in the background. 'Okay, say it with flowers if you must,' an American safety officer exploded, 'but the use of an oxygen mask is no joke. One day somebody will pay for treating passengers like shit-scared blockheads.' (He asked to remain anonymous for, in his current battle with his airline bosses, his job is at stake, and 'going public' is his last resort.) If his case does not apply to all operators, the experts and airlines must certainly take complete responsibility for failing to answer a question of considerable concern: how much of a half-heard briefing will be remembered by the end of a long flight and intermittent sleep, just when it may be needed most – during a landing emergency? The NTSB recommended in 1983(!) that some of the briefings should be repeated during descent in the final phase of each flight, but nothing happened. Could it be because repeats might use up some valuable minutes of selling more duty-frees?

One of the key factors in a survivable accident is the prompt availability of safe exits, but neither the airlines nor the manufacturers make it easy for us to be fully prepared. Not to worry, the opening of doors is the duty of the crew, they tell us. But then why give us hints – certainly not more than that – on those plastic cards they invite us, mostly in vain, to study? Well, because the crew may be incapacitated. Precisely. So we need to know. But my contention that most of us, non-technical, ignorant passengers would not be any wiser from those neat little drawings, was rebuffed almost every time. So I collected a random sample in 1988 and 1989.

First to be noted is that although most doors open outwards, some open inwards and others slip overhead.

At Manchester, if you are to believe the emergency card, passengers aboard the BA 737 should have experienced no problems in opening the overwing hatch. The card shows a heavy 'plug' being removed, in comfort and with plenty of room to manoeuvre, by a woman, who then neatly deposits it with ease on three seats in front which happen to be conveniently *empty*.

The card of Continental Airlines reveals that their 727–200 has no less than nine exits – to be operated in four different ways. Even if you study the drawings on take-off, will you be able to remember three hours later which way to open each type of door? And, don't forget, the overwing hatch is to be pulled *inwards*, as depicted, and must be placed on the seats in front which the artist has conveniently again kept empty.

The task to decipher the Biman Bangladesh DC10's 'safety instructions' seems to call for technical drawing specialists.

The Malév TU-134 has two doors and four overwing hatches. It has different unlocking mechanism for each door, and the card offers no suggestion what you do with the hatch – deposit it somewhere in the plane or throw it through the opening?

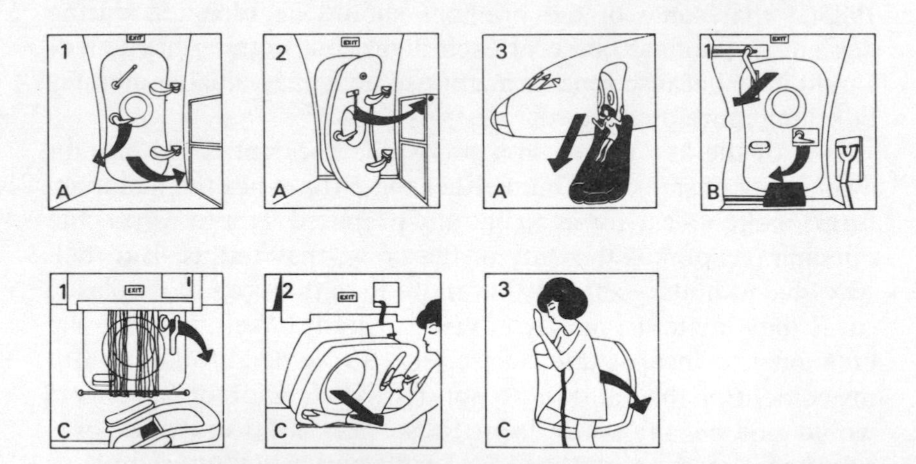

The Lufthansa 727–B card is clear about the overwing exit: pull the hatch inwards, and, if you can handle the weight in the limited space, throw it out through the opening, which is much better than trying to manoeuvre inside the 'plane and to drop it on seats usually occupied by other passengers. Unfortunately, only an arrow implies that there *is* a rear exit, and there is no instruction on how to operate the door which must be there.

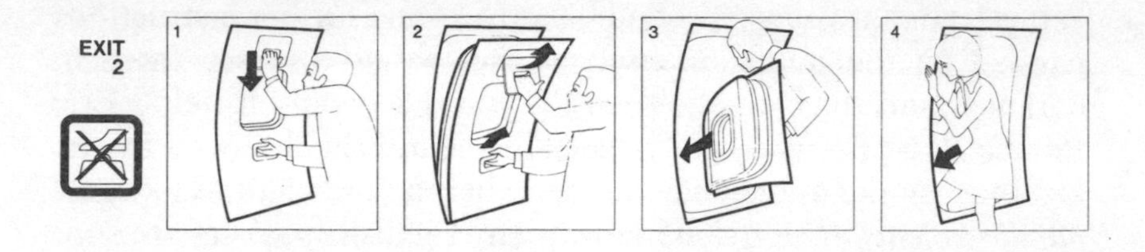

The TWA DC9-80: The three doors are identical – two red arrows in the instructions offer you guidance, but there seems to be only one handle to operate – what do you do with the second arrow? Same problem with the rear door exit, though in the drawing the tail cone seems to have fallen off miraculously to give way to a chute. You are instructed to pull the window hatch inwards. But what do you do with it? It seems to be allowed just to disappear.

So now you know the DC9? Well, try the Air Canada DC9. The door handle works, so the arrows tell you, in the direction opposite from the TWA DC9. For the removal of the overwing hatch you seem to need *two* empty rows of seats: one to kneel on for access, and one on which to deposit the cumbersome object . . . or else it would crush the people in front. The operation of the rear exit is a long story which – in an emergency – would keep passengers bemused for some time. The strip cartoon says open the door, enter a tunnel, follow some sort of rail, pull a handle to activate something, then, seeing daylight, locate, uncoil and drop a rope, lower yourself, Tarzan, to the ground, and, if you follow what picture number eleven seems to illustrate, *walk away with the tail cone on your shoulder*.

The Air Canada 767: Helpful, as the doors have identical *modus operandi*. The hatch, however, seems to come inside to land . . . on what? On top of the collapsed seatbacks which have already trapped two passengers in front?

Illustrations at the doors show how to operate the escape chutes. On most aircraft, if you use a hatch exit, you jump from the wing. Not so if you fly the Air Canada 767: the drawings show a jolly yellow slide sitting on the wing. How did it get there? Is it always there even in flight? If not, how do you put it there? Perhaps you ought to tell the fire on board to pause while you return to the artist for clarification of the emergency card that is supposed to be self-explanatory.

It is true that you would find big arrows and further instructions painted on the doors themselves. But would you see them in darkness and thick smoke? And if you did, would it help? Eric Pritchard, a pilot and safety specialist, is utterly dismayed by the fact that 'on the doors you still see unhelpful instructions like *pull handle to jettison*. Jettison? How many English speakers are sure about the true meaning of the word?'

When one floats some ideas, the airlines' and regulators' reaction is often enthusiastic: Yes, if others did it, if ICAO recommended it, we'd do it, most certainly. Why not show the opening of doors on the video briefing aboard widebodied aircraft? Could do, possible. Why not standardize at least the style of the cards that are graphic art rather than engineering drawings anyway? No reason why not, at least it would eliminate the meaningless red dots and the like on some cards. How about allowing the public to take a close look or even have a go at opening exits which could be displayed in departure lounges? Oh no, not that, that's impractical/inconvenient/dangerous/might give people the idea to try, even if it is impossible, to open the hatch in flight. How about standardizing the operation of all doors and hatches on all aircraft? Could be done, might be helpful, *somebody* ought to start getting the cooperation of all manufacturers. (At Boeing I was told that there was no real obstacle to a uniformity of latching mechanisms on future aircraft, and later engineering innovations could also be standardized. It could also help cabin crews.)

The last resort of asking untrained people to open exits for themselves will soon be an even graver risk because not only aircraft but also the passengers are becoming more geriatric. (The fastest growing segment of the airline market is in the fifty-five to eighty-five age group – people who have the time, income and motivation to travel. Dan Johnson, a cabin designer, urged manufacturers at the FSF seminar in 1986 to be mindful of older passengers who may move slowly with impaired alertness, sight and hearing.)

Why not seat able-bodied passengers next to overwing exits, and brief them individually each time? This idea is gaining popularity, but some American organizations have started a campaign for reserving those seats for the blind. (Yes, it would be good to have them close to the exit, but no, it would not help anybody if the hatch had to be opened in a hurry.)

We all see those *'Vest under your seat'* markings all the time, and almost unconsciously note those vital instructions throughout the flight. Why not a permanent display of *placards* on the seatback facing the person who sits next to the overwing hatch? A novel idea? Far from it. But, to the best of my knowledge, no airline used it before the CAA made it mandatory for all unmanned exits in the wake of Manchester. (Despite widespread criticism, some airlines like BA and KLM have seen hatches on the old 747s as surplus to requirement, and welded them shut long ago. That helped to gain extra seating. The doors in use on their Jumbos are always manned by crew, and satisfy legal minima – though not the FAA. BA has now ordered new aircraft with extra exits.)

'We can see far too many problems concerning the operation of exits,' said an evacuation specialist. 'I'd be surprised if, in an emergency, people would follow orders to throw those cumbersome hatches out. They'd just let them drop inside, and go. Yes, it would help if at least some people got a chance actually to practise opening doors. But airlines still tend to treat passengers as children, though in an emergency they expect them to behave as adults. That's why many new-style tests will have to be devised.' One of the most important of those will be to discover the rate of egress per door. So far only the flow of evacuation has been tested with the doors already opened by the 'crew'. Volunteers will soon have to try opening a hatch after a few glances at safety cards. How will they cope under pressure? How much will they remember if there is a few hours' lapse of time between scanning the card and doing the exercise? How will they cope with putting on life jackets and adopting brace positions according to the drawings? Do those strip cartoons enable people to take care of themselves, or are they a form of lip service to safety?

An excessive amount of hand-luggage carried into and cluttering the cabin can prevent smooth evacuation, yet it seems to be another area of responsibility 'delegated' to the passengers. It should not be so, but competition and the profit motive are the decisive factors.

The airlines do not worry about the potential extra weight of cumbersome hand-luggage, though that surplus could easily be put to better use in, say, stronger anchorage of seats, but the American Flight Attendants' Association has long argued that, in windshear or engines-out, it could make the difference between life and death. Carriers, however, interpret the rule as one bag per passenger (if and when they pay attention to it at all), claiming that, if they were too strict, customers would just switch to other more flexible competitors. Unfortunately, they do not check even the size of carry-on baggage – including memorable oddities like tractor tyres, a four-foot fig tree, a slot machine, a twenty-five-gallon aquarium, a BMW driveshaft, and at least once a lavatory pan.

Every item ought to be examined before boarding to ensure that it will fit into an overhead bin or under a seat. Simple measuring devices do exist, but only the Australians use them, even though they are available at some airports, such as Gatwick. Certain destinations in the Far East, Spain, and Lagos Nigeria, for example, may be notoriously lax about hand-baggage, but luggage blocking the aisles and exits or piled high on oxygen bottles and fire extinguishers can be seen on many flights everywhere.

Dangerous goods such as butane-powered hair-stylers have always been prohibited in the cabin. Security checks, if nothing else, ought to spot all such items every time, but this is not the case, and it is a rarity, indeed, that one hears about penalties for endangering the safety of an aircraft. (In 1987, for instance, on a flight from Milan to Manchester, passengers complained of painfully itchy, watering eyes which seemed to require urgent medical attention. The BA pilot diverted to Heathrow where it transpired that, in an Italian wine merchant's bag, there were leaking canisters of a cleaning fluid that contained a highly flammable and dangerous chemical. He was lucky: he lived to be fined £3,000 with £1,000 costs.)

Bottles could become dangerous missiles, slash escape slides, impede evacuation and sprinkle everything with alcohol to feed the flames. Yet, *duty-free booze*, the perk of flying, is not seen as 'dangerous goods'. The experience of some accidents and hazardous incidents do not seem to weigh heavily against the massive profits – the British Airport Authority alone takes almost

£100 millions of revenue from duty-free sales a year. If the EC countries choose to designate all traffic among members as *domestic flights*, their airports will lose an annual $1.2 billion from duty-frees. So, arguably, scheduled and charter fares would rise by some ten and twenty per cent respectively – *before* they would suddenly come under standard Value Added Tax. In the light of such figures, it seems obvious why the safer idea of selling duty-frees *at destination* is resisted. Singapore, Reykyavik and Luxembourg are proud to have introduced the system, but remain too shy to mention that they now benefit both ways, selling on arrival as well as on departure.

Under the heading of *additional hazards* the Manchester report referred to some cabin areas where the fire burned with exceptional ferocity. 'The burn characteristics were indicative of the [known] presence of flammable agents' including 'duty-free spirits, therapeutic oxygen and aerosol sprays . . . The early involvement of such materials would add significantly to the transfer of fire from the ceiling region down on to seats, carpets and other materials in the lower levels of the cabin.' It appeared specially hazardous to stow oxygen in the overhead bins, and to carry aerosols in hand baggage. It was therefore 'recommended that these materials are subject to the same controls as other flammable gas cylinders'.

Though the aircraft was American, the *River Orrin* case was an exclusively British accident. But if it fails to have a multi-faceted effect world-wide, those fifty-five lives will not have been just tragically lost but criminally wasted.

– 8 –

LESSONS THAT MELT AWAY

On January 13, 1982, Air Florida Flight 95, a 737 from Miami, could consider itself lucky to escape a diversion: barely nine minutes after it had landed at Washington National, the airport was closed down for snow clearance. The aircraft was scheduled to depart for Florida as Flight 90 at 14.15, but delays were now inevitable because several aircraft were already in the queue. While seventy-one passengers and three infants obeyed the Final Call to board, snow and ice built up all over the aircraft. That was no cause for concern – routine de-icing would soon begin – and in a light-hearted banter, the crew cracked some jokes about the conditions that were rather unfamiliar to them.

The de-icing was carried out by American Airlines providing the service under contract. AA's *Maintenance Manual* had specific instructions for de-icing 707s, 727s and DC-10s, but nothing for 737s. The port and starboard sides of the plane were de-iced by two different operators who did not follow the standard AA or Air Florida procedures. That might not have mattered much if the aircraft departed right away. But there were more delays, and it was still snowing. The Boeing *Manual* for 737s was either not consulted or disregarded. And that was more ominous because it cautioned against de-icing solutions as their dilution 'with melted snow can result in the mixture refreezing and becoming more difficult to remove'. (Only French airports have de-icing gates through which aircraft *must* pass just before take-off.)

At 15.40, thirty minutes after the completion of the job, the

CVR recorded the pilots' comments on another aircraft going ankle deep in the snow:

'Boy, this is shitty, probably the shittiest snow I've seen.'

'* * go over to the hangar and get de-iced.'

'Yeah.'

'Definitely.'

'* * de-iced * * (laughter).'

'Yeah, that's about it.'

'It's been a while since we've been de-iced.'

'Think I'll go home and * *.'

And after more laughter:

'I love it out here.'

'I love it out here.'

'It's fun.'

'The neat way the tire tracks . . . '

Another thirteen minutes later:

'Boy, this is a, this is a losing battle here on trying to de-ice these things, it [gives] you a false feeling of security, that's all it does.'

'That, ah, satisfies the Feds . . . '

'In Minneapolis, the truck they were de-icing us with, the heater didn't work on it, the glycol was freezing the moment it hit . . . '

'Well I haven't seen anybody go around yet, they're doing good . . . '

Taxiing in a sequenced departure queue, they were aware of snow on the wings. The captain hoped to remove it by using the heat and blast of exhaust gases of another aircraft which he chose to follow closely. He did not know that fresh snow, which could be blown off during take-off, would now be melted into a slushy mixture which would freeze on the leading edges and the engine inlet nosecone. Meanwhile, there was heavy incoming and outgoing traffic, and various onlookers commented on the condition of the 737. A pilot in the queue told his crew: 'Look at the junk on that airplane!' He noted that snow and ice affected the entire length of the fuselage.

The captain, as was his right, assigned the first officer to fly the sector. Throughout final preparations and the take-off, communications between the two men remained tentative, even when anomalies occurred in the engine instrument readings. (Failing to realize that icing was the cause, they never switched on the anti-ice device.) At 15.58, the first officer asked: 'Slushy runway, do you want me to do anything special for this or just go for it?' The captain answered: 'Unless you got anything special you'd like to do.'

At 15.59, the cabin speakers came to life: 'Ladies and gentlemen, we have just been cleared on the runway for take-off, flight attendants please be seated.'

A few seconds later they began their take-off run. 'God, look at that thing . . . That don't seem right does it?' said the first officer, and repeated: 'Ah, that's not right,' four seconds later.

'Yes, it is, there's eighty.'

'Naw, I don't think that's right.'

The first officer was still not happy with the acceleration. They reached V_1, the decision speed: *go – or reject take-off.*

'Easy,' said the captain, announcing V_2 to rotate. The stick-shaker sounded. 'Forward, forward,' called the captain, asking for more power, but the warning sound would not stop. 'Come on, forward . . . Forward . . . Just barely climb.'

Somebody said something about *stalling* or *falling*. The crew had fifteen seconds to gather more speed by applying full power and lowering the nose. That might have saved them.

'Larry, we're going down, Larry.'

'I know it,' said the captain.

One second past 16.01, the CVR recorded the sound of impact as 'the aircraft struck the heavily congested northbound span of the 14th Street Bridge', killing the drivers of four cars, tearing through huge chunks of wall and railings, and plunging into the frozen Potomac river.

It seemed unlikely that anyone aboard would have survived. Brave attempts to find out for sure appeared to be doomed. A boat was carried and launched by hand (the ramp was just an iceberg) but it would never reach the sunken aircraft.

Miraculously, five passengers and a flight attendant were alive in the river.

Their chances were bad. In icy waters, their hands would soon be paralysed, and they would lose consciousness. Life vests were floating everywhere, but beyond their reach. The only vest they retrieved would not come out of its wrapping. They chewed through the plastic cover, and helped a seriously injured woman.

The first rescue helicopter arrived at the scene twenty-one

minutes after the crash. It had its job cut out. At considerable risk to its crew, it rescued all but one survivor who was pulled ashore by a civilian who had dived in between drifting blocks of ice.

Eighty-two specially trained divers were flown in to begin salvage, a mammoth task. The flight recorder was soon recovered. It was only an old foil type but, together with the CVR, it helped the investigators to trace icing, the elusive culprit that just melts away.

Barely a fortnight later, the NTSB issued a massive list of recommendations for operations in snow and icing conditions. The FAA responded within twenty-four hours by calling telephone conferences and spraying the industry with telegraphic instructions. Boeing telexed warnings to the airlines, and reviewed its 737 manual.

Seven months later, the NTSB report was issued with more recommendations. The 'probable cause' was the pilots' 'failure to use engine anti-ice . . . their decision to take-off with snow/ice on the airfoil surfaces . . . and the captain's failure to reject take-off during the early stage when his attention was called to anomalous engine instrument readings.' The long delay between de-icing and take-off was one of the contributing factors; another was 'the known inherent pitchup characteristics of the B-737 aircraft when the leading edge is contaminated with even small amounts of snow or ice'.

Known characteristic? So was there a *new* lesson learned in the icy waters of the Potomac? Or was the crash a cold reminder of repeatedly forgotten old lessons?

It was no news that engines and openings that serve instruments must be covered or plugged as protection from de-icing solutions, yet the airline maintenance representative testified that he had 'never seen airplanes de-iced with covers on them'. It was no news that the heaters to de-ice the sensing probes ought to be *on* in such conditions. It was no news that ice on the wings would reduce lift. And it was no news that the 737 was particularly sensitive to icing problems: since 1970, the type had accumulated twenty-two serious icing incidents. Boeing had issued several warning bulletins. Britain had suffered a more than fair share in those incidents, and the CAA, having suggested some precautions just a few months before the Washington crash, now mandated extra take-off speed in icy weather with immediate effect.

The third 'contributing factor' was the crew's 'limited experience' in 'winter operations'. The FAA regulation was clear: 'No person may take-off an aircraft when frost snow or ice is adhering to the wings, control surfaces, or propellers.' The Air Florida Operations Manual stated equally clearly that no aircraft must ever leave with ice on the wings and tail surfaces, but the captain could disregard snow if, in his opinion, it was 'of such consistency that it will dissipate or blow off' during the take-off run.

The NTSB found that, despite their training, many pilots underestimated the risk, and that 'this crew's decision to take-off with snow adhering to the aircraft is not an isolated incident, but rather is a too frequent occurrence'.

The relatively inexperienced captain of Flight 90 had failed a line check (unsatisfactory performance in several key flying techniques, and breaking regulations) in 1980. As a result, his licence had been suspended. Three months later he passed the test. In 1981, he failed his recurrent proficiency check, but passed it a few days later after a simulator course. Most of his flying was done in non-jet aircraft in the warmer, southern parts of the country, and he had landed or taken off only eight times in icing conditions. The first officer, an ex-fighter pilot with only a thousand hours on jet transport, had experienced such conditions only twice. It emerged that, due to the company's rapid expansion, Air Florida 'pilots were upgrading faster than the industry norm to meet the increasing demands of growing schedules'. This was clearly a symptom of what, in 1987, BALPA's *Airline Pilot* called an 'erosion of the entire safety system'.* And it was also a painful illustration of the notorious 'resource management breakdown, a failure of communications on the flight deck, where the captain was clearly not listening, and remained oblivious to the problem', said NTSB Board member John Lauber.

Whatever else this accident was, a *new* lesson it was not. Here are some non-737 examples from the randomly chosen year of 1987:

* See: Deregulation and Economy v. Safety.

A Continental Airlines DC9 crashed at Denver, Colorado. Fifty-four people survived, but twenty-eight, including the pilots, were killed. The probable cause? During a snowstorm, delays between de-icing and take-off. Contributing factors? Both pilots were inexperienced on the type; the first officer handling the take-off rotated too rapidly; Continental did not know that he had been discharged by another company for failing to pass a flight check.

The ATR42, an Italian aircraft, was grounded temporarily after a crash killed thirty-seven people near Milan. The Registro Aeronautico Italiano ordered that the operation manual should increase the recommended minimum flying speeds in icing conditions. Hadn't the FAA mandated American users of the type to do so almost a year earlier? It had, but were the Italians notified? In Italy, nobody seemed to know.

A World War II Spitfire ace, with 25,000 hours in the air, crashed a Cessna 404 Titan on take-off. He and his eleven passengers were fortunate to escape serious injuries. A court found him guilty of reckless flying because he had ignored the snow and ice accumulation on the wings in the two hours between a thorough visual check and his departure.

In the same year, a CAA *Occurrence Digest* contained dozens of similar incidents involving a variety of aircraft and situations. Such diligent collection and dissemination of data must alert operators by the weight of its sheer repetitiveness, but it depends on frank and unrestricted reporting from airlines, even though in many countries an incident is regarded as a shameful episode to be covered up or a matter of secrecy or 'just one of those things' best forgotten.

Frank and fearless reporting of incidents must come from the pilots in the first place, but sometimes they may be embarrassed by their 'stupid mistakes' or afraid of consequences that may affect their reputation or even career prospects. That is why NASA in America and the RAF Institute of Aviation Medicine in Britain run systems that are guaranteed to be confidential. When a pattern of repeated mistakes emerges, revealing that perhaps it was all too easy to be trapped into making those 'stupid' errors, the 'feedback' report calls attention to it. Sometimes even a single confession could save lives – if management is quick to search for solutions rather than culprits. When, for instance, a 747 pilot confessed to CHIRP (Confidential Human Factors Incident Reporting Programme) that he had once taken off in freezing weather with his pitot probe heaters *off*, 'we immediately warned his company,' said Roger Green, Principal Psychologist, who runs the British system. 'This was paramount because an iced up pitot probe deprives the pilot of an essential instrument, the airspeed indicator. Although we'd never reveal the pilot's identity, we passed on his concern about the checklist that didn't place sufficient emphasis on that item. There were no questions asked: that company just amended the checklist right away.'

Some cases may spotlight completely unknown hazards. In 1988, a Mitsubishi YS11 of Japan Airlines System aborted take-off after reaching V_2 speed because, allegedly, the 'elevator control was too heavy to rotate'. It ditched in shallow waters near Japan's Miho airport. Eight passengers were injured. Investigators found it was the co-pilot flying the aircraft which was an irregularity: the captain was not authorized to delegate because he himself had been in command of the type for less than six months. He let the junior man do the take-off to give him some practice. That, however, did not cause the accident, and icing became the chief suspect although it happened on a not particularly cold January day, when the captain did not deem de-icing to be necessary. After tests, the investigators concluded, with considerable amazement, that icing could occur even at the relatively high temperature of 1°C., and the accumulation of slush and snow could affect the performance of the controls. Consequently, all Japanese lines were ordered to change their manuals to make anti-icing treatment compulsory. Yet some Japanese pilots seem unable to escape some nagging questions: 'Was it really icing? Do we need yet another set of regulations, when a stroll in the biting six-knot wind and visual inspection of the aircraft would reveal the hazard anyway?'

In 1989, Japan set up an air safety research centre, to be financed by the major Japanese airlines. Its first task will be to investigate the hazards of winter operations, and, specifically, snow and ice clearance from runways and aircraft. Better late than never.

Because the evidence tends to vanish, icing accidents are often disputed.

When, in 1989, the AAIB decided that a British Midland Airways Fokker-27 accident had been due to airframe icing which the pilot could not have been expected to foresee, it recommended a change in the flight manuals. Fokker, however, disagreed, claiming that the aircraft must have been flying below the already prescribed speed, and therefore no modification would be necessary. But few such disputes were as bitter as the five-year controversy that had been stirred up by a crash at Gander, Newfoundland.

On December 11, 1985, Arrow Air Flight 90, a DC8 Super 63, left behind the warm night of Cairo long after its planned departure time. It was a charter flight for 248 members of the U.S. Army 101st Airborne Division who were returning home from peacekeeping duties in the Sinai Desert. There was a stopover in Cologne where the crew of eight was changed. The DC8 landed at Gander to refuel at daybreak, and took off again an hour later. It seemed to climb hardly at all, was seen to cross over the Trans-Canada Highway at a very low altitude, and crashed into trees some 3,000 feet beyond the end of the runway. The tremendous impact was followed by fire, the flames were fed by full tanks of fuel, and the aircraft was destroyed. There were no survivors.

Some disasters offer no obvious clues to the investigators, others churn out too many.

Three witnesses noted 'a yellow/orange glow emanating from the aircraft' which might have indicated an engine or cabin fire in the air; two of them said it had been 'bright enough to illuminate the interior of the truck cabs they were driving'; the third saw the glow as merely 'the reflection of the runway approach lighting on the aircraft'.

The recorders, the modern investigator's best friends, turned out to be enemies: the old-fashioned foil recorder gave only sketchy and unreliable information, the CVR was inoperative. (At the time the American Minimum Equipment List permitted certain aircraft, including the DC8, to fly with unserviceable recorders. This attracted severe and immediate criticism from the investigators, and, two years later, the FAA modified its policy.)

In 1986, after a great deal of research, testing, reasoning and soulsearching, the conflicting views of meteorologists as well as pilots who had also used Gander on the morning of the crash were rejected, various potential scenarios were eliminated, and the somewhat inexperienced investigators drafted a report to blame icing for the disaster. It only fuelled the already raging controversy. When *Flight International* published some information received from the Canadian Air Safety Board, the 'allegations' were refuted vehemently by Arrow Air. The airline was not alone in disputing the draft. The argument, combined with leaks and accusations of a cover-up, continued inside and outside the Board for yet another two years until, at last, what was intended to be the final report was published. It acknowledged that the 'Board was unable to determine the exact sequence of events' leading to 'a stall at low altitude from which recovery was not possible', but with some emphasis on two key contributing factors, the investigators stuck to their guns: 'the most probable cause of the stall was determined to be ice contamination on the leading edge and upper surface of the wing'.

This was signed by five members of the Board, including its chairman. The remaining four members disagreed. And their thinner yet weighty report of dissent was issued simultaneously in an identical format. It dissected the majority views point by point, and offered a diametrically opposite interpretation of most facts in the light of some circumstantial evidence. It criticized the technical and medical findings, the 'incomplete' wreckage and crash site examination, as well as the entire process of analysis. Rejecting both the cause and the contributory factors, and implying sabotage or the accidental explosion of some passengers' weapons, the minority report argued with great conviction that there had been an in-flight fire and a massive loss of power in the air. 'We could not establish a direct link between the fire and the loss of power,' they admitted, but concluded, nevertheless, that 'an in-flight fire that may have resulted from detonations of undetermined origin brought about catastrophic system failures' (i.e. the accident).

Being many things but final, the two reports only opened up yet another can of worms. Various specialists, including experienced investigators of the Canadian and American pilots' unions, rejected the majority 'theory, based on, at best, incomplete research', as well as the 'unfounded speculation' by the minority.* As a result, the Board's powers were suspended temporarily, and a retired Supreme Court judge was appointed by the government

* CALPA *Newsletter*, 31.7.1989.

to review the case. The outcome was a quite devastating comment on the investigation. In 1989, four years after the accident, the inquiry simply threw out both reports. It criticized the majority for reaching conclusions without proof: 'A string of possibilities, however long, cannot form a collective probability . . . In other words, a suspicion, however strong, may not without direct support from real evidence be made into a cause.' It found 'almost no evidence which supports *any* of the conclusions of the minority'. This unique saga threw North American aviation into a turmoil, and led to the reconstruction of the CASB, but it did nothing to allay suspicions of a cover-up. The inquiry could not recommend any further Canadian investigation into the case because it saw no hope for uncovering the true cause of the accident. At the time of writing, a U.S. Congressional inquiry is contemplated, and that could be invaluable because, as CALPA says, the investigators' 'emphasis on a specious probable cause overshadowed some certain contributing factors to the detriment of corrective action and improved air safety'. That batch of such potential factors and non-contributory malpractices, unearthed by the investigation, go indeed far beyond a possible matter of icing.

One of them was the already mentioned loose and very unsatisfactory state of the regulations governing voice and flight data recorders. Another intriguing yet rarely mentioned point in the majority report concerned the weight of passengers. In his calculations of minimum take-off power, the captain used the standard average weights at the time – 170 pounds per man (150 per woman) including five pounds of carry-on baggage. Arrow Air manuals required that *actual*, not average passenger weight should be used for flight planning, but gave no specific instructions for the determination of the actual weight. On Flight 90, the passengers were not of standard civilian build. The average for soldiers of the Airborne Division should have been 220 pounds. The error was compounded by disregarding the weight of some items like cabin equipment. As a result, the actual take-off weight was underestimated by some 14,000 pounds – exceeding the authorized maximum by 8,000 pounds. In a crucial moment when icing (if there was any) might have reduced lift, and when the No. 4 engine had possibly suffered a loss of power, the extra weight might have had an adverse effect on the pilots' predicament if they were applying engine thrust calculated for a lighter aircraft.*

But would that be the pilots' fault entirely? Not infrequently, company practices – as opposed to printed manuals – and lack of stringent supervision might be conducive to negligence or reduced appreciation of risks such as icing. It is the management which creates the environment in which pilots may find it easier to err, and it is the managers who may shut their eyes to work habits based on the false economy of some *let's just get on with the job* principle.

It was up to the management of Arrow Air to issue specific instructions for weighing passengers, and ensure that all aircraft were flying with adequate records and load documentation. It was not only Flight 90 that disregarded the armed forces' published average weights. The original contract to transport observers and units of the Multinational Force contained a discrepancy that was known to the management: The contracted payload per flight exceeded the payload capability of the aircraft. The majority report opined that the 'load sheet calculations performed by the flight crew were planned to demonstrate adherence to the maximum allowable'. As this could not have been an isolated technique, the management was or ought to have been aware of it.

Before the introduction of bigger and so more forgiving aircraft, such practices used to be widespread. The flight engineer of, say, a DC3 Dakota, would not have been shocked or even surprised if his captain asked him to 'shed fifty pounds of fuel' – from the loadsheet, not the tanks. Of the 11,000 Dakotas that have been built, more than 300 are still in service. There is no reason to believe that weight irregularities ceased to occur wherever such smaller, older aircraft continue to fly. It must be done with the management's tacit connivance and, fortunately, most pilots get away with it most of the time. In commercial operations, an adequate safety margin will normally allow for a rugby's team's surprise appearance at the check-in counter. What happened in March, 1989 was unusual: an Olympic Airways flight was about to leave Samos for Athens when twelve overweight passengers

* The minority report denied that the weight consideration was a factor in the accident.

were asked to disembark because *their presence would have prevented a safe take-off*. According to Reuter's, Maria Lafi, one of the luckier passengers, said: 'I was spared despite my extra weight . . . I'm eight months pregnant.'

Although crew fatigue was not thought to be responsible in any way for the Gander case, some medical evidence was recorded by the disputed majority report saying that in the fortnight before the crash, the 'crew had been consistently exposed to work patterns . . . which were highly conducive to the development of chronic fatigue'. Arrow Air 'scheduling procedures did not address flight crew fatigue factors. No maximum duty-day limit was established.' It emerged there had already been some known cases of Arrow Air crews exceeding Federal flight time limitations, and now the CASB recommended more stringent regulations to the FAA.

Substandard maintenance by Arrow Air did not seem to be relevant to this accident, but disturbing facts came to light. A year before the crash the company was fined $34,000 for maintenance violations. (Two days after the crash *The Times* quoted an engineer saying that he had once refused to sign off the maintenance log because 'the plane was in such a bad condition that I did not want the responsibility of putting my name on it'.) The Canadian report revealed that the airline carried numerous, long overdue deferred maintenance items, and aircrews accepted 'aircraft that exhibited anomalies in the operation of the flight control systems'. Although the FAA claimed that the irregularities were record-keeping and paper violations and 'not all that terrible', the report criticized the agency for failing to spot them all.

And finally, the grave implications in *Communique #15/88* went far beyond the Gander disaster: 'The CASB is concerned that weaknesses in the regulatory control may affect the margin of safety in other air carrier operations.' Thus many apparently 'simple' cases of icing could be seen as symptomatic of a new and growing serious hazard – management.

– 9 –

FREEFORALL?

Take a mishap at Frankfurt. On March 12, 1983, a Kuwait Airways 747 was taxiing to Runway 07. Before contacting the tower as instructed, it tried to pass a PanAm 737 'holding' at an intersection. As there was not enough room to do so, the pilot left his guideline to give the 737 a wide berth, misjudged the distance, and his wingtip hit the tail of the PanAm. The damage was serious, but there were no injuries.

On the surface, an obvious pilot error of breaking airport regulations by leaving the guideline. But the crew's written route and aerodrome 'briefing by company dispatch was found to be incomplete' – it did not contain the relevant NOTAM 2112/82, says the German accident report. Thus the captain 'may not have been aware of the fact that observance of the guideline was compulsory' at Frankfurt if moving without special clearance or guidance by a ground vehicle. A more cautious manager might have prevented a potentially fatal accident by spotting the omission.

Management is probably the most underrated hazard of modern aviation. An ambiguous policy decision, a marketing push, a misplaced slip of paper may kill by creating foreseeable but unforeseen gaps in our defences.

It was a piece of well-meaning legislation that allowed American management to proceed to the edge of the precipice, but failure to learn from its consequences would be a grave error everywhere, particularly when Europe will follow suit, rather prematurely, with 'liberalizing' the aviation industry in the 1990s.

In 1978, President Carter signed the Airline Deregulation Act. While it emphasized that safety standards must not be eroded, it opened the way to immediate fare reductions of up to seventy per cent, the gradual phase-out of central authority over fares and mergers, and 'open skies' to let any line fly anywhere if it owned 'gates' to pick up and unload passengers. It was 'just the ticket' for the Reagan administration which intended to 'get goverment off the people's back'. To the passengers' initial delight, it led to an explosion of air travel through the proliferation of airlines, cut-throat competition and predatory fare wars fed by much increased load factors. (To compare two of the busiest short routes, it costs some sixty per cent more to fly from London to Paris than from New York to Washington.)

Some people consider deregulation an unmitigated success. Others call it a 'mixed blessing' with vast delays, overbooking, lost luggage, a deluge of complaints about worse, sometimes abominable, service. Other critics see it as an outright menace to U.S. aviation: some 150 smaller cities and unprofitable destinations lost their scheduled services, fares on less popular routes increased, and although passengers have saved an annual six billion dollars in fares, nobody is willing to question seriously who or what will eventually pay for that. Will safety foot the bill or will the savings be clawed back from the passengers themselves?

Many companies expanded by taking disastrous leaps in the dark to compensate for suicidal fare-slashing, and ran out of financial steam. Braniff, for instance, having lost forty million dollars in the first six months of 1989, filed for bankruptcy for the second time in four years. Pesidential went the same way. Of the 215 airlines started in 1978, only fifty-nine were still in business in 1989. Those that did not go bust were swallowed by the mighty opposition. Monopolies returned with a vengeance, and began stamping out competition in many markets. Unrestricted mergers and take-overs created the mega-carriers that ruled the airports, their hub and spokes system serving their profits rather than their customers. By 1989, just eight of those mammoths carried ninety-four per cent of passengers on domestic routes. In the same year, a study by the U.S. General Accounting Office revealed that, at the so-called hubs, dominated by one or two airlines, fares were already twenty-seven per cent higher than at other airports where there was still free competition. In September, 1989, led by American Airlines, the major U.S. operators announced massive fare increases – all within the same week. Was that a case of connivance at price fixing? At the time of writing, the U.S. Justice Department Anti-Trust division is investigating the 'coincidence'. (In 1982, the chairman of American Airlines telephoned Braniff, its direct competitor at Dallas, suggesting an agreed twenty per cent rise in their fares. The Braniff chairman taped the call. The case was investigated. AA 'settled the case with a consent decree' without admitting or denying 'any attempt at collusion.'*)

Finances dominated the operations, and problems concerning safety began to multiply. Joseph Lapensky, a former chairman of Northwest Orient and supporter of deregulation, admitted at a conference in The Hague that safety corners had been cut. In September, 1989, Samuel Skinner, U.S. Secretary of Transportation, told the International Aviation Club in Washington that 'debt-ridden airlines, forced to adopt sweeping austerity measures, may unwittingly threaten the safety of the travelling public . . . Airlines today can't afford to scrimp on the new costs of maintaining the safety of their senior aircraft.'

Praising deregulation in several respects, Jim Burnett of the NTSB has some serious reservations: 'We've not yet adjusted our thinking of safety to the new situation. We have competitive pressures on airlines, tempting them not to observe all safety regulations. In the old, regulated environment, if the FAA demanded something, the operators would simply pass the cost on to the passengers. So the FAA could relax and play it cool in a gentlemanly relationship. Now, with the fare and service competition, their role is like a referee's at basketball: if two laid-back, quiet teams play, he can let the play flow and have a slow whistle, but if he gets two really competitive, physical teams, he must be sharp. And here's the problem. Deregulation gains ought to be ploughed back into safety. Safety must not be deregulated and, if anything, we now need more such regulations. Economic deregulation demands sharper enforcement by more inspectors. But when, for instance, all airlines want to fly to a popular airport, the density will require flow control, and the carriers will cry

* *Flight International*, 3.1.1990.

"that's re-regulation!" Yes, it is, in a way, but a horse of a different colour.'

FLYING THE FLAG OF CONVENIENCE

It is difficult enough to achieve consensus for safety in the face of mounting losses, but it could be well nigh impossible on a world-wide basis if international deregulation, in whatever guise, ever comes. If some countries' regulatory authorities will be too demanding, while others relax in the name of encouraging competition, one can foresee aviation taking a leaf out of the logbook of shipping to introduce *flags of convenience* registrations, with Panama or Liberia offering anyone the right to fly – at a price. An outrageous proposition? Yes, but feasible.

'It could happen,' said Captain de Silva of Singapore Airlines, 'because under ICAO agreements, receiving countries cannot check other airlines' maintenance or training standards and controls before giving them landing rights. If Panama is a member country, we must blindly let in any aircraft registered over there. Unfortunately, there're always some unscrupulous people out to make a quick buck, and, what's worse, people will fly with them if they can save a few dollars that way. Would ICAO then get the right to enforce safety standards? If not, we'll be in trouble. Without strict inspections you cannot maintain any semblance of sanity in the air.'

Small countries which make a fortune on licensing ships and shipping companies could extend their facilities to airlines and aircraft. That in itself may not endanger safety directly if the licensees are truly responsible operators, but it can open the floodgates to others. The air operator certificate can become meaningless without advance examination and continuous inspection. Pilots and engineers can be licensed wholesale without meticulous evidence of technical, medical and mental fitness; and cut-price cabin crews can be recruited by dodgy agencies 'certifying' experience and training to deal with emergencies. Aircraft maintenance requirements can be slackened, and the befogged dates and proficiency of mandatory work items can disappear conveniently in the black holes of international paperwork. And there can be worse to follow. As the state of registration has the right and duty to make a major contribution to the investigation of accidents, those who may start a trade in flags of convenience can abuse the system to cover up their own malpractices and irregularities.

Unless the aviation community applied foresight and overhauled the international regulations to close the gaps, it would be left to individual states to withdraw landing rights from airlines and aircraft registered in countries of substandard controls. And that could lead to diplomatic and public outcries of discrimination because the question would be: where do you draw the line? There are already suspect airlines and states, yet the leading nations of aviation and their regulatory authorities remain silent. Not forgetting the *minor* problem of libel and international conflict, they feel bound and gagged by the spirit of ICAO pacts.

Some people I interviewed hinted that passengers have the final say. They do. We already see them perpetuating the risk by allowing fares to dictate their choice. The lure of *fly-by-fare* is borne out by the American experience, by the European charters' success, and also by the sharply competitive airlines of the remains of the Soviet bloc and the third world, where labour costs may be a third of those in the West, where losses are acceptable because the need to earn hard currency overrides other considerations, and where *glasnost* about safety and regulations may not always coincide with the truth. Airline safety is expensive everywhere, and even the introduction of vital extras affecting safety fails to sell tickets – as yet. Who wants to fly with XYZ Airlines *only* because it has some obscure, allegedly lifesaving piece of equipment on board? Who would know about it anyway?

A report by Harvey Elliott, *The Times* aviation correspondent, described on April 15, 1989, how Airtours, a young Lancashire company, operates a commercially ingenious scheme. Its tours are carried by aircraft that are owned by French banks, leased to a Luxembourg airline, maintained in Paris or Amsterdam, certified in France, and subleased flight by flight, just one at a time, to a Caribbean country, whatever the destination happens to be, in order to take advantage of, say, the Dominican air charter licence. It is a Monaco agency that recruits the international crew whose competence is controlled by Luxembourg – the country of registration. The result is huge success quickly. Their half-price Caribbean holidays are very popular. Everything is perfectly legal, and, by the look of it, perfectly safe. But interlocking responsibilities are so divided that many people – though not the passengers – view it with some unease. Were this technique to be utilized by some unscrupulous operator under the banner of misinterpreted deregulation, the result could be horrendous.

WHAT IS SAFETY?

Does deregulation affect safety? First, define 'safety'.

Wilbur Wright in 1901: 'If you are looking for perfect safety, you will do well to sit on a fence and watch the birds.' (Yes, if we forget that the fence may collapse or that you may overbalance and fall.)

American Congressman Norman Mineta: safety is 'the public perception of whatever the traffic will bear'.

Chuck Miller: 'a relative thing . . . a judgment call . . . Passengers may delegate the decision to someone else, but in the final analysis, we all have to decide just how much safety we want.'

John Chaplin of the CAA speculated in a lecture: 'The ideal requirement is perhaps that "the aircraft and its operation shall be safe". But as no form of transport is completely safe we must rephrase it to read: "The aircraft and its operation shall be safe *enough*". This is still begging the question how do we interpret these words. Indeed, who decides – or can decide – that such and such a level of safety is *safe enough*.' John Chaplin found that air transport must be *safe enough* because the public does not seem to be 'unduly worried' about its record.

Captain McClure of ALPA: 'We know what's safe enough. As a colleague said, we took the undue risk out of flying. So the general risk level may be acceptable. But what we don't want to see is a reversal.'

Others talk about varying *degrees* of safety as if forgetting the well-known formula – you can't be a little bit safe any more than you can be a little bit pregnant or a little bit dead. Boeing's Earl Weener put it this way: 'Pour a glass of wine into sewage, the mixture will stink as before, but a few drops of sewage in a barrel of wine . . . well, just try to taste the effect. It's the same with safety: a single moment of carelessness can ruin the bouquet.'

At a Northwestern University conference, David Hinson, chairman of Midway Airlines (the first post-deregulation carrier) called it 'first and foremost a moral issue . . . But safety is also a business issue, especially for smaller carriers' which 'may not have the resources to withstand the stigma of being labeled a *marginally safe* or *unsafe* airline'. Does that mean that a mega-carrier can withstand it? Up to a point, probably yes, for contrary to popular belief and many airline executives' misleading claims, the travelling public's combined memory is very limited, even in Japan, as long as the fares are low. The effect of a major accident is ephemeral.

'An accident shakes the world only for a few days,' said Fritz Hofmann, Emergency Procedures Instructor of Swissair. 'Some companies have accountants who argue that definitions are too vague, we just *know* that air travel *is* very safe – so why spend more on it? That's the philosophy of negligence.'

John Prescott, Labour spokesman on Transport: American deregulation has 'reduced fares but affected quality and safety'.* 'Absolute nonsense,' said Anthony Broderick, Associate Administrator of the FAA. 'Deregulation was an economic measure, nothing inherently unsafe in it. People talk about its effect on safety, but where is the statistical evidence? Sure, we seem to have a problem with the commuter airlines, and we're investigating it, but that's mainly because 1987 was a bad year for them. One or two accidents make a hell of a difference. That's statistics for you. If you have one accident one year and two the next, you have a hundred per cent increase, and everybody is outraged at the *huge* increase. 1987 was followed by fourteen months without a single fatality. That's what I mean by statistical fluctuations.'

True. But do we need to wait for such shaky statistical evidence? Would it bring back the dead if we told their surviving relatives that the victims were 'statistically safe'? Our statistically proven extremely high level of air safety over the past couple of decades was not built merely on regulations. Good airlines, good manufacturers have always done more – sometimes a great deal more – than the minima required by law.

A British Airways maintenance engineer's remarks were typical: 'Whatever the job in hand, I add my pride to the rulebook because I was trained to a standard – what we call good engineering practice. Nothing less will do for me.'

* Commons debate, 13.3.89.

CUSHIONS AND THE ALMIGHTY DOLLAR

Good engineering, flying and operational practice and management pride give us safety cushions over and above the regulations. So, does economic deregulation deflate those cushions? Does it create a financial climate which favours shaving the safety margin thinner bit by bit?

The Flight Safety Foundation studied the effects of deregulation, and found that there had been 'a reduction in the financial ability of air carriers to undertake safety initiatives in excess of FAA minimum requirements'.*

It is not deregulation *itself* that causes a drop in safety standards. But it *can* create critical financial conditions through fiercely competitive fare-cutting. Everybody loves a bargain but fares are not furs, bucket-shop ticketing is no 'liquidation sale' when 'every mink coat must be sold'. If a furrier cuts his prices too severely and goes bankrupt as a result, his customers do not crash with him.

At the time of writing, European Community Commissioner Sir Leon Brittan is fighting to foster greater competition and cheaper fares. He told the *Sunday Times* (19.3.1989) that 'there is still a long way to go towards liberalization of air transport'. The plans of small and emerging European airlines are frequently thwarted by the big names of the industry. 'If people in the street ask, "What has the Community done for me?" I cannot think a more vivid answer can be given than to say it has brought down air fares and given a greater choice of flights.' A noble programme that is also overdue – up to a point. Because if liberalization will bring cushion deflation, then the EC has not a day to lose in planning supra-national safety standards and their strict enforcement irrespective of cost.

'If the American experience is anything to go by, one must hope that 1992 will never come in Europe,' said Allan Winn, the editor of *Flight International*. 'The quality of service people get over there is reducing all the time. It's a good indicator just to see which airlines can afford and have the will to *clean* their aircraft. It doesn't directly affect safety, but it shows care. Every time they wash an aircraft, it's an extra opportunity to take a close look. And only the better-class operators do it. If you look at parked aircraft at any American airport, you can see that there're

* *Aviation Week & Space Technology* 13.7.1987.

two levels of major carriers. Those who care, will also give better maintenance. Safety cushions? Sure, the FAA may be right saying that there're no statistically significant differences. But it's like an MoT: you can slip an old banger through the annual test, but it won't be as safe as a well-maintained new car.'

The old adage is that 'if you think safety is expensive, wait until you have an accident'. But even that is becoming debatable, and Captain McClure has some serious doubts about it: 'My impression is that accidents don't cost too much to the airlines. At the end of the day, insurers must cover most of the losses, and a crash becomes real costly only if the airline, the manufacturer and others can be sued in American courts. Provided that an airline hasn't got too many mishaps, the occasional loss can be more acceptable and cheaper than safety. Beyond some short-term dollar losses, they still save. To Americans, motherhood, apple-pie and Chevrolet are paramount – except for two things: air safety and the almighty dollar. And the dollar comes before safety. It's true that beyond a certain point, you can spend all the dollars in the world, and you won't make it significantly safer to fly, but do we really know where that certain point is?'

'Safety is always a matter of degree, and so it's arguable what's worth what to whom,' said Bill Tucker, CASB director of Safety Programs, letting his hand rise like an inflating airship and plunge like a kestrel on its prey. 'That's why financial stability is crucial. When the fares are cut to the bone, erring on the side of safety is less likely to be permitted. It used to be a joke among pilots when loading fuel to ask for an extra couple of hundred pounds "for wife and kids". Not any more. It's all programmed for minimum to meet requirements. But now and then, mistakes will occur, a fuel cap will be left loose, there'll be a higher than expected rate of burn, both the destination and alternate airports may be hit by bad weather, and problems can occur. How often will they occur and how often will they lead to accidents? Well, how many angels can dance on the head of a pin? All one can say for sure is that excessive cost cutting is not conducive to safety'.

Despite resentment (a pilot exclaimed 'I don't want bean counters in *my* cockpit!'), there is an inevitable trend to replace pilots and engineers by financial managers at the helm. Although they all pay their respects to safety, the situation at some airlines may be compared to Hollywood at the time when bookkeeping tried to play the role of talent, and banking moguls, dictating artistic policy, crashed much of the film industry.

John Bibo, engineering director of the Australian Ansett, maintains 'there's no budget on safety', and his company's record attests to that principle, but aviation executives of the deregulation era are 'compelled to dedicate more time to competitive problems, less time to normal management concerns' *(Jerome Lederer)*. And Anthony Broderick told me: 'No manager will admit that he doesn't care about safety. But there're pressures, no doubt, because they're not in the business of making safety. They're in the business of making money. It just happens that you can't make money if you run an unsafe line because "safety before all else" is the ground rule of the business.'

A sound financial basis is so essential to safe aviation that punters may be wise to say: 'Show me your balance sheet before I buy your tickets.' Financially secure companies, such as Qantas, Lufthansa, Singapore Airlines, United, American or lately British Airways (turned round by brilliant management shake-up and marketing skills) could benefit from that attitude.

Without secure finance airlines cannot 'spend money on safety and maintenance', said Heinz Ruhnau, chairman of Lufthansa, in 1985. But is it a *Catch 22*? If you do not cut fares, you lose customers; if you do cut fares, you may lose money on every ticket you sell. A choice must be made, but you cannot choose to cut safety.

Management skills alone will determine success in competition as long as everybody is compelled to maintain high safety standards, which cost the same to all.

Airline finances are a matter of serious concern for the CAA: 'Financial adequacy does not guarantee a safe airline, but without the resources there could be a deadly temptation to cut back or take short-cuts in the essential safety functions. Although the link between safety and profitability cannot be demonstrated, experience suggests that it is there.' Similar concepts guide the U.S. Department of Transportation when examining an airline's fitness to fly: 'Where the carrier does not possess a strong concern for safety . . . financial stress can serve as a motivating factor for the carrier to cut the margin of safety.' In Germany, according to an article in the *Zeitschrift für Luft- und Weltraumrecht* in March, 1984, company finances must be sound or the licence is withdrawn. Between 1979 and 1988, eight small carriers had been 'requested to abandon their licences'. Nancy L. Rose, assistant professor of applied economics at the MIT, analysed the effects of deregulation, and came to the conclusion that, while the safety record of U.S. airlines was still *superb*, 'the financial condition may be correlated with accident rates at individual airlines'.

It is never easy to pinpoint that correlation between shaky finances and the deterioration of safety, but the signs can often be found, even without actual accidents. The case of PanAm, a huge, well known and respected carrier may be a good example. Or was it just a coincidence that, in the 1970s, when the airline was riddled with debts, their record was badly dented by a spate of accidents? All involved 707s, and pilot errors were suspected. The FAA reviewed PanAm operations, and found that cost-cutting measures 'severely impaired the airline's ability to train pilots properly'.*

The Shah of Iran bailed them out, they sold their New York headquarters and some of their routes, but their troubles were not over. 'Yes, under the financial pressure they endured, they had to do something,' Geoffrey Wilkinson of the AIB said. 'They were too good an airline actually to cut corners, but they would certainly reduce the costly safety margins, and I saw some of their aircraft with real tatty repairs.'

When the RAF bought two PanAm Tristars to be used as tankers, a British airline was asked to carry out the basic checks and maintenance. Allegedly, the work took twice as long as usual because mandatory modifications had not been done, and 'the planes were a bloody disgrace', said an engineer.

PanAm continued to operate in the red. In 1981, it was losing a million dollars a day. Hopes were pinned on a new manager who had 'worked wonders' with Braniff and Air Florida. (By then Braniff was bankrupt, and Air Florida's underlying problems would soon grow visible on the ice crust of the Potomac.) In 1983, PanAm was one of the four airlines (with Eastern, Western and Republic) which chalked up $400 million of net losses. For survival, airlines tried even more devastating fare cuts (up to

* John O'Brien in *Airline Pilot*, August, 1987.

So near, yet so far: just cross the M1, hop over the embankment, and there is the runway . . . but the 737 (*bottom centre*) at Kegworth was flying on a sick engine with the healthy one shut down.

A major accident may not leave much of the aircraft in one piece. Painstaking collection and sorting of the wreckage (*above*), as done here by Canadian investigators, and the reconstruction of the aircraft on a frame, may help to reveal the cause of the crash and the sequence of failures such as a break-up in the air.

Flight reconstruction by computer (*below*) – a new tool in the aircrash detective's armoury.

Obstacles and water near the end of a runway can be serious hazards. After an abandoned take-off, this 737 (*above*) skidded into New York's East River at LaGuardia Airport.

An avoidable accident? Despite extensive post-crash fire damage, the picture of the wreckage at Nairobi (*below*) still tells experts that, unknown to the pilot, the leading edge slats were not deployed. The risk was known – but not to everybody concerned.

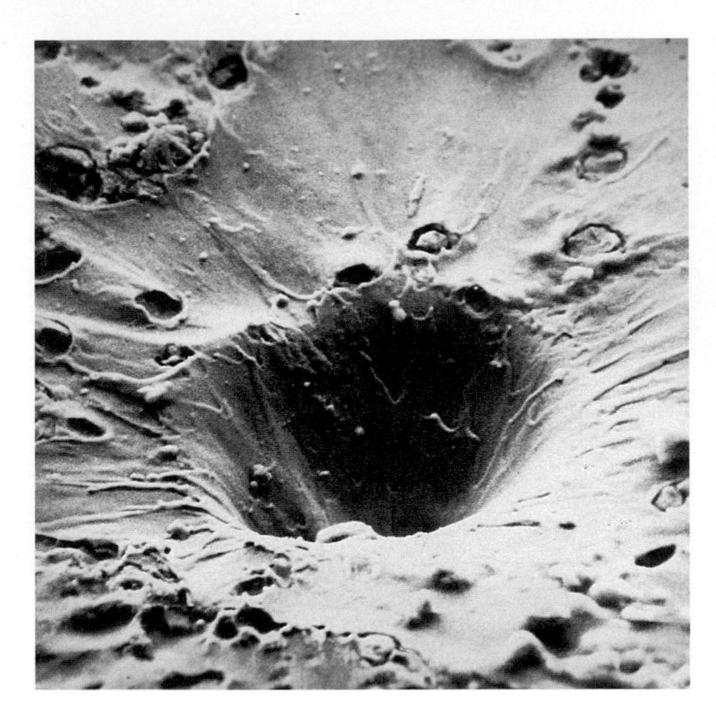

The electron microscope reveals minute 'cratering' in aluminium, the material evidence of high explosive damage.

Typical small charge explosion damage. Small bombs, like the one under a passenger seat of this 727 at Mexico City (*below left*), would have required great expertise to bring down a modern jet.

Pathologists found bullets *and* cartridge cases in a dead soldier's leg, so the ammunition must have exploded without being fired. The X-ray picture (*below right*) helped to disprove rumours that shooting on board a DC-6 killed Hammarskjöld.

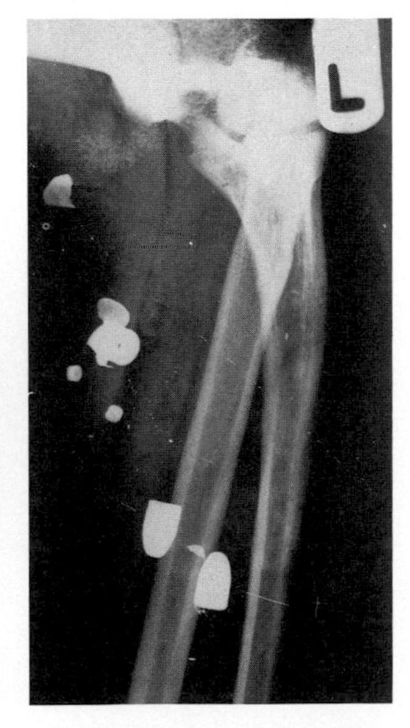

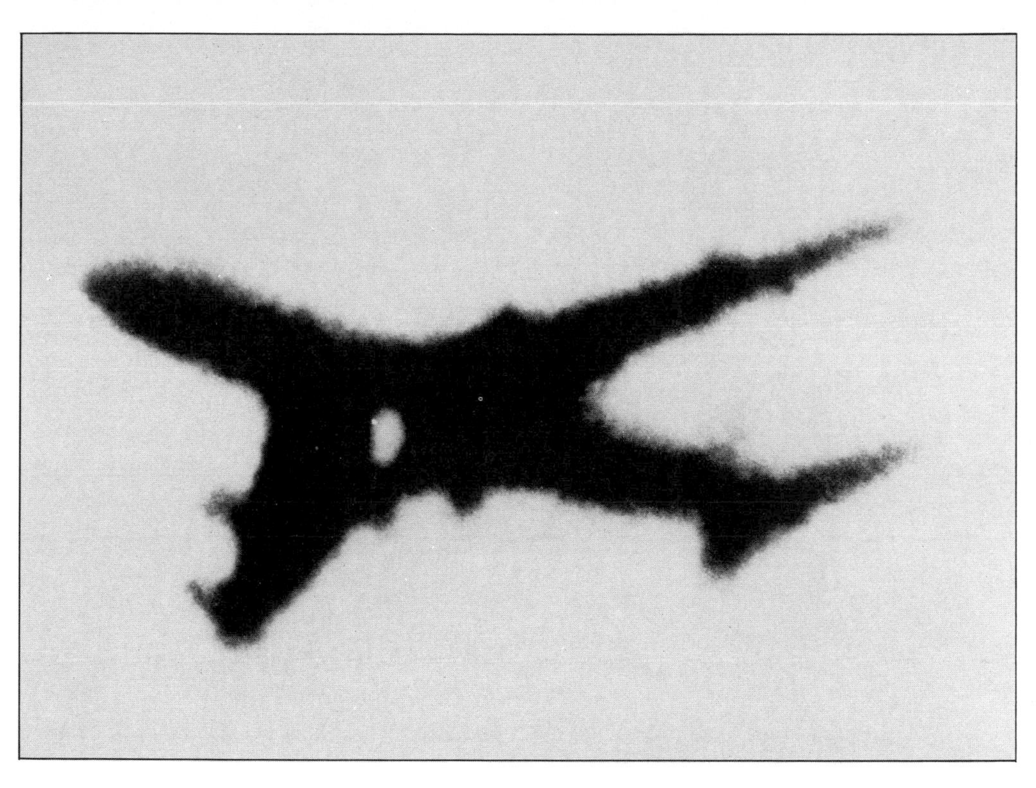

An amateur photographer's snapshot (*above*) reveals that 520 people are doomed: the Japanese Jumbo is still flying, but its vertical tail fin is already missing.

Miracle in Hawaii (*below*): although part of the fuselage was torn off, the old 737 with the 'sunshine top' landed safely. The pilot helps to evacuate his shocked but grateful passengers.

Danger on the ground: a Piper crashed into a bus and a restaurant near Munich (*above*), killing 6 and injuring 15 people.

Murphy's Law I: the famous case of the non-return fuel valve that could be, and therefore was, fitted the wrong way round (*below right*).

The sweet face of a killer (*below left*) might have helped to allay suspicions: she planted a bomb on a Korean airliner.

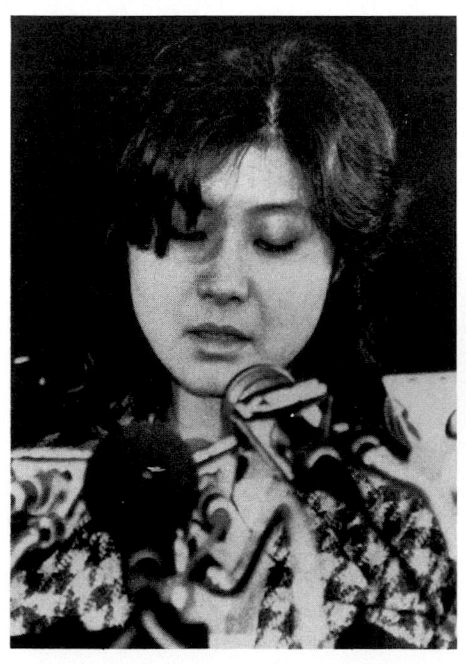

This DC9 (*above*) was on fire for 30 minutes in the air. It landed at Cincinnati with no structural damage. A survivable accident. The doors were opened, yet 23 people never got out. Why?

When icing is the cause of an accident, the evidence tends to melt away. In Washington (*below*), proof was found – mostly in the cockpit voice recorder which was recovered, like this tail section, from the icy Potomac river. (In the background, the 14th Street bridge which was hit by the jet on takeoff.)

Broken parts of a Jumbo lie at the bottom of the Irish Sea. Wreckage recovery is an enormous task. The underwater camera detects a part of the fuselage, showing the faint outline of Indian motif around the windows (*above*).

The grin of despair: a helicopter crewman recovers a dead child's cabbage patch doll after the Air India Jumbo crash (*right*). Every scrap of evidence can be vital to the investigators as the sea yields very few of its secrets.

Lockerbie is burning. The flames light up some leftovers of the deadly aluminium shower (*below*).

sixty-five per cent by Continental), and forced their pilots to accept pay cuts of up to thirty per cent. As even such extreme measures did not induce profitability, was it just another coincidence that in 1985 PanAm was fined almost four million dollars for breaking safety regulations, including the use of non-authorized substandard parts for repairs? (One of the few to exceed that record was Eastern with a nine and a half million dollar fine in 1986.) Further management changes failed to solve the problems. Though the unions agreed to cut another 110 million dollars off the wage bill, PanAm lost a billion dollars between 1986 and 1989. It then sought to take over Northwest, a robust operation, to which PanAm's losses might have appealed as tax write-offs. The bid failed and PanAm may one day be up for grabs yet again.

Like mega-carriers with apparently inexhaustible credit, national airlines with unlimited subsidies can survive precariously for decades. BOAC used to be a prime example. It lived on vast subsidies, and yet the lack of funds might have periodically affected its safety. Captain John Owen recalls 'an apocryphal account of the pioneering days of the first generation jets when BOAC was castigated by "the man from the Ministry" who was concerned about the recent spate of near disasters – one of which almost caused the premature demise of a Cabinet Minister. The letter went something like this: "Dear Sir, with regard to your recent safety record and the number of incidents, if you were an independent company, you'd be given three weeks to justify your air operator's certificate." Mind you, sometimes it's not only money. It can be staff unrest that reduces safety. Many airlines went through it. Air France, Aer Lingus, Aeronaves de Mexico . . . pilot troubles, then a big crash or increase of incidents. You may be unable to put your finger on it, but you can sense the cause.'

Iran's national carrier was propped up for prestige. Dozens of managers, pilots and engineers left the country under political pressure. A pilot, who sought asylum in Holland in 1982, revealed that the once profitable airline needed an annual 150-million-dollar subsidy because key jobs went to people with political connections, there was a shortage of spares, only a third of the fleet was operational at any time, and 'ordinary' passengers were often 'bumped' when government officials needed seats.*

* *The Times*, 25.10.1982.

Subsidies and government guarantees (whatever they are worth) keep numerous third world flag carriers flying on a shoestring. Captain Blevins, Secretary of the U.K. Flight Safety Committee, commented: 'There are many governments that feel: "We have no nuts, no bolts, no sanitation, no food for our starving millions, but, by golly, we have a big jet flying in our national colours." That's the vanity airlines' philosophy.'

The problem is not related to the size of the country. While tiny Singapore, with an excellent safety record, has shown healthy profits year after year, the Indonesian Garuda ran up almost a billion dollars in foreign debts. Air Zaire was in serious financial trouble for years. Its situation became untenable in 1986. The staff, unpaid for some time, was reduced by two-thirds to two thousand; foreign pilots were paid only after long delays; only a few of the aircraft were airworthy; but the flag was still flown. (Eventually most expatriate pilots were replaced by cheaper natives, and a French airline took over management duties.) In 1990, a Paraguayan flight to Brussels planned to refuel in Africa. Its crew mistook Conakry for Dakar where the airline's credit card was unacceptable. The aircraft could take-off only when the crew borrowed $7,000 from a passenger to buy fuel.

Nigeria Airways, unable to pay long-standing inter-airline debts, was expelled from the IATA clearing house system in 1986, and had to pay cash for all services abroad. (Nigerian Central Bank policy was blamed for the problems, exacerbated by cases of fraud – a director leased a 747 singlehandedly, without Board approval, and had it painted in the airline's livery.) Air Afrique also ran into trouble with the IATA clearing house, and retaliated with warnings of serious political repercussions. The airline, owned by ten African states, some of them among the poorest in the world, ran up large debts, had to give jobs to political appointees and free flights to special friends. In 1989, it needed a salvage plan for a huge cash injection, staff reduction and stronger monopoly position in its area. Ironically, at the same time, more subregional airlines were set up by partnerships of small states.

In Laos, in addition to the state-owned loss-making Lao Aviation, another small airline is about to be launched to serve the neighbourhood and France. Cash-starved Air Vietnam, the flag carrier, was criticized by its own state radio saying that its planes were riddled with mechanical breakdowns, service was bad, and fifty-four flights had to be cancelled in 1988 due to technical hiccoughs or lack of pilots.

Financial hardship can lead to laxity of control or outright negligence. Short of spectacular accidents (and few of even those hit the headlines if the victim is a third world carrier), those concerned with international safety hardly ever see proper evidence in proper time. In 1987, a Ugandan 707 could not leave London for Entebbe in time to beat the night curfew, and had to be parked until the morning. Another line's engineer passed by and noticed serious bird-strike damage to all its engines. He alerted the airport authority and the CAA. An urgent inspection declared the aircraft unfit not only to fly out its hundred stranded passengers, but even to attempt a ferry flight with empty seats. Uganda Airways had the aircraft on subcharter from African Express Airways, which had dry leased it from its Egyptian owners. Who ought to have done something about those engines? Uganda? The operator? The lessee? The owner? The state of registry? Time they had had aplenty: the critical damage to the engines was not new at all. In the absence of stringent supervision – it is immaterial whether it was due to lack of funds or lack of expertise – that aircraft, as well as all towns under its path, must have been protected by luck alone every time it flew in that condition.

Smaller companies and private enterprise may get caught out more quickly, but 'more quickly' may mean several years. An American DoT spokesman said: 'We keep checking their financial fitness. If we see symptoms like mounting debts and non-payment of landing fees, we initiate an informal investigation and, if necessary, ask the FAA for a special safety check.'

The murderous saga of Ryan Air Services gave an ample illustration of the combination of poor finances, poor management and poor supervision.

On November 23, 1987, scheduled Flight 103, a Beech 1900, carried nineteen passengers and cargo from Kodiak Island to Homer, Alaska. As it was coming in to land, it suddenly veered to the left and struck the ground, killing both pilots and all but one of the passengers. There was no DFDR or

CVR on board, nor were these required by regulations.

'New NTSB members are trained by going on check rides as observers,' said acting chairman James Kolstad. 'I was new here in 1987, but I had already done my check ride, and the Ryan crash was my first accident. I went to Homer with our GO team. The intense cold at that distant, desolate location kept most of the press away, so I was under reduced pressure when disturbing facts emerged right away.'

James Kolstad's word *disturbing* was a measured understatement. It transpired that the aircraft had carried more than twice the amount of permissible cargo, and its centre of gravity had been too far aft for safe operation. Company employees, worried about their mortgage payments, talked about the financial difficulties of the airline: payrolls had not always been met, caterers had not been paid, there was a badly depleted stock of spares. 'Then we began to find some gaping holes in the FAA supervisory performance,' said Kolstad. 'How could Ryan still be operating when there had been a series of serious and even fatal accidents, and a long history of management and training non-compliance with regulations? The dead captain was one of three Ryan pilots who had approached the FAA several months before the crash complaining that the management was forcing them to fly overweight aircraft. Offering suitable documentation, they were willing to testify to that effect if the FAA gave them immunity. That was understandable: they didn't want to lose their certificates by owning up to breaking the law under pressure on some occasions. The FAA refused their request, filed no action against Ryan or the pilots, and didn't even check out their stories.

'After a few days in Alaska one could sense that things don't work there the way Washington may imagine it. Ryan was the largest among lots of small local commuter lines in the area. The FAA found it difficult to hire enough people adequately to oversee the volume of operations up there. A special problem was that inspectors were left there in the job for too long in one place, and, living in virtual isolation, a chummy relationship would tend to develop between inspector and inspected.*

* Ryan's connection with the FAA had to be good: the man who became their Director of Operations in 1987 had just retired from the FAA after eleven years as an Air Carrier Operations Inspector and Airman Certification Inspector right there, in Alaska.

'I was concerned the situation might be allowed to continue by the FAA, but I'm glad to say that, following the revelations, the Administrator replaced many of the top people up there without any delay.'

Compliance disposition of managers and airlines is a key part of licensing by the DoT. The history of Ryan's non-compliance makes terrifying reading, so it is amazing that the FAA allowed the airline to operate all that time. Ten serious incidents and twelve accidents with fatalities had produced a catalogue of three dozen previous violations, and cases of inadequate maintenance and/or operational error. Some pilots' licences had been suspended and, on one occasion, Ryan had been fined a mere nine thousand dollars for pilot competency violations. In February 1987, an FAA main base inspection found numerous violations of pilot training and monitoring for which a civil penalty of 16,500 dollars was imposed. Only five days before the Homer accident, another Ryan aircraft crashed: it was found not to be in airworthy condition.

In the wake of the Homer crash, operations were suspended, and the Air Carrier Fitness Division of the DoT stepped in. 'The FAA indicated to us that at least two top Ryan executives, the President, who was the major owner, and the Director of Maintenance would have to be replaced, and, apparently, that was done,' Carol Szekely, the Assistant Chief of the Division told me. 'But when they asked us for permission to resume operations, we had to turn them down. We didn't believe that they had made sufficient changes – the family still owned the company, and all the executives who had failed to object to past practices were still there. As they didn't have enough new financing, the old pressures would re-emerge, and people who didn't have the strength to say "no" before, would still be in charge.' The wrangling continued because the family still owned the company, and the 'introduction of new blood' was not much more than a reshuffle of personnel. Eventually, the idea was to put all the family stock into trust, but the Fitness Division 'would want to see who the trustee was because it would have to be someone of sufficient standing whom the family couldn't influence in the way the line was run. The case is unusual because we issue few outright refusals. Usually the process is more subtle. Seeing our questions and objections, newly formed or temporarily grounded problem-airlines tend to withdraw their applications rather than risk a refusal.'*

THE FIRST CASUALTY OF COST-WARFARE

Few people are aware of an airline's safety record when booking a seat, and even fewer question whether the flight is backed by a virtually invisible feature – a safety department.

Some twenty years ago, Conservative MP Enoch Powell argued: 'The supreme function of statesmanship is to provide against preventable evils . . . By the very order of things such evils are not demonstrable until they have occurred (and so remain open to doubt). They attract little attention in comparison with current troubles, which are both indisputable and pressing. Hence the besetting temptation of all politics to concern itself with the immediate present at the expense of the future.'† The same could be true about the aviation industry.

Safety specialists need universal support to help them to drag themselves away from yesterday's mishaps and to hunt out the killers of tomorrow. This is the view the ICAO Accident Prevention Manual (an excellent document that, unfortunately, enjoys no more than recommendatory status) urges everybody to adopt, and this is what Roy Lomas, head of the British Airways Safety Branch wants to pursue: 'From our investigations of misfortunes and mistakes, we must try to predict the next threat and adverse trends, not in panic, but while our safety level is still high.'

The most dangerous adversary of this objective is the oft-flaunted maxim that 'safety is everybody's business' – a fine principle . . . if only it worked. In practice, matters that are seen as everybody's business, in general, tend to become nobody's business in particular. Chuck Miller did not even pause to find a classic example: 'Allegedly, a top executive of American Airlines once said: "Why do we need a safety organization? That's what we have insurance for." But in 1985, after the airline had been fined one and a half million dollars for lack of safety compliance

* At the time of writing, I understand that Ryan has replaced its entire top management to start very much scaled-down operations, within limited markets, and under strict FAA supervision.

† The quotation comes from the so-called 'Rivers of Blood' speech.

across the board, particularly in the maintenance area, he seemed to have an immediate change of heart.'

'It's true that the fine was effective,' said Anthony Broderick of the FAA. 'They decided that nothing like that must ever happen again to AA, and now they have a great reputation for an excellent attitude to safety. And it is the corporate attitude that breeds safety. It's much too easy to throw a few people into a room and say: "Look, we have a safety department." I'm not even sure that you need one at all. Having a safety department is not necessarily something that proves that you make a real contribution to safety.'

Corporate attitude is the ultimate ingredient of safety. But who will promote, bear responsibility for, and feed that attitude with relevant information? A part-timer? An overworked manager in whose in-tray that extra burden happens to land? Where will he find the time to read the flood of information arriving daily from every corner of the globe, to evaluate, select and communicate all the data, to attend international conferences and to learn from others' misfortunes?

'No chance,' said Captain de Silva of Singapore Airlines. 'In addition to all my duties as safety manager, I and my staff need two or three hours each day just to read the lifesaving news and warnings the entire industry offers to share with us. When it's all assessed for our needs, I put it to our Air Safety Committee and every interested party. That's the only way to make safety everybody's business.'

Ferdinand Füller of Lufthansa agreed: 'Our various specialists who are drafted into the safety department's work share out the relevant data. Some days so much comes in from IATA alone that, for example, a very senior maintenance man on the committee must devote on average a third of his time just to reading and channelling it all in the right direction. How anybody without a safety department could cope with it all I cannot even imagine. Some small airlines would simply have to ignore it – and hope for the best.'

'Remember the tyre explosion accident I told you about?' asked Brian Richardson of the NTSB. 'Recently a new safety guy at Avianca was shocked to find that after the Mexicana crash they still used air down there. Why? Because too many companies operate with a skeleton staff and have no-one to read, no system to disseminate even manufacturers' advisory notices, so crews and mechanics may never hear of developments.'

Advocating the introduction of a focal point of safety and prevention, and the establishment of a top level prevention adviser, there is a subdued cry for sanity in the ICAO Manual when it says: 'Most major airlines employ some of the accident prevention activities outlined in this Manual, while many of the smaller airlines and operators may not employ any.' Engineering aspects of safety are often dealt with by quality control managers, accident prevention programmes by the flight operations side, but safety 'must embrace the total organization and it is essential that a close working relationship be maintained between all parts of the organization'.

When I visited a major Canadian airline, various engineers were assigned to talk to me. 'Oh, yes, we do have a safety supremo, but he's not very good at communicating with people.' If so, he cannot be much good at his job either, for safety *is* communication. 'But we all think of safety all the time,' they emphasized, 'we need no company encouragement. Our maintenance bulletins always have a *safety corner*. Our suggestions boxes are just all over the place, collecting some 500 ideas a year, and I guess ten per cent of those are safety related. So we do not need a specific programme.' And indeed, one of them must have had safety (of his job?) on his mind when he added: 'Quote me on this, and I'll deny it'.

In some countries, such as Germany and Japan, safety departments are compulsory. JAL, ANA, Air France, all Australian airlines have massive safety outfits. Hugo Muser, former Swissair safety chief, and Manfred Reist, his successor, warned in unison that 'aircraft have become dangerously safe. It may cause complacency. People don't expect trouble, so they must be kept on their toes. So our section has expanded to five full-time specialists and a fuller safety programme, while, in America, money is being diverted from this function. We used to get safety films and guidance from over there. Now hardly anything comes. Some airlines like United still have fine safety set-ups, but others seem to think that if there are fewer fires, they can disband the fire brigade.' An article in the *Flight Safety Digest** used some DC10

* Safety Versus Economics by A.W. 'Tony' Brunetti, October, 1986.

incidents as key examples of the lack of proper safety department function: the FAA was not the only one to blame, crucial incidents went unreported or were considered to be mere 'service difficulties' and their significance was not recognized or emphasized. The recipients of some information also overlooked the hazard in mildly cautionary reports: could it be that 'the management hierarchy chose to take the risk to save manhours, time and money?'

'Insurers can play a vital role in air safety concerning particularly the smaller countries and airlines,' said Captain Blevins. 'They can say "you're virtually uninsurable unless you take certain precautions" and the creation of a safety branch or the appointment of a Flight Safety Officer could be Number One on their list.'

Some insurance brokers do that by examining the applicants' performance. When Peter Spooner, head of the Aviation Risk Management at Willis, Faber & Dumas, visits a company with his extremely detailed questionnaire to conduct a 'safety audit', senior executives tend to welcome him because they realize that it could well be an exercise that may help their profitability as well as safety. 'Usually there's an excess on claims like there's one on cars', said Spooner. 'A Jumbo may carry a million-dollar excess, a 737 or a DC9 half of that. That's why an underwriter may know nothing about minor incidents simply because there's been no claim for them. Yet they may be forerunners of accidents. When we examine a line's incident record, we can often convince them that a full-time safety officer could help them cut their delays, aircraft unserviceability and cost of mishaps. A South American company with three dozen aircraft, for instance, needed a lot of persuasion, with facts emerging from the audit, that even a small safety department would earn its keep ten or twenty times a year. So when I hear that such a department is disbanded, I find it a very odd way to save money.'

And yet this is exactly what happens when short-sighted management tries to make excessive in-house savings. In the early 1980s, when BA was still in the red, and cutting costs, the staff of the safety branch was reduced in the name of streamlining, and there was a big library 'clear-out' to save office space. More than thirty huge filing cabinets full of technical papers going back to 1945 were destroyed. Fortunately, the changes had no direct effect on safety, and the efficiency of the branch was maintained, but the shredded archives, particularly those of historical interest, were, of course, irreplaceable. Now the safety branch is back to full strength with a staff of eight people, and under Roy Lomas it continues to play an important part in international efforts to improve safety throughout the industry. In addition to the branch, BA now has a flight safety director, who sits on the Board and has direct access to the entire hierarchy. Without these specialists, BA could not publish its own *Safety Review*, a full and frank admission, evaluation and record of all the incidents the airline suffers. It is circulated freely in the industry for the advancement of aviation safety worldwide.

But will those who need it most have the time to read and make use of all that? As always, America is the chief trendsetter. U.S. airlines are keen to 'save' on safety branches these days. Many executives demand statistical proof of why accidents did *not* happen, and would deny that their excellent airlines may *still* be riding on the back of the fine practice and record – of their past. The effect of losing safety and prevention specialists can take years to manifest itself. Take NASA. The post of their director of safety had remained vacant for thirty months, and the number of employees involved with safety, engineering and quality assurance had declined from thirty to twenty prior to the *Challenger* space shuttle tragedy. Meanwhile the PR director's job had been filled as soon as it had become vacant, and the staff of the publicity department had been increased to 175.*

'They [the management] tell us that ours is an *emotional argument*. But what's emotional about pursuing safety?' asked Captain Patterson, ISASI Executive Assistant in Washington. 'They rely on past safety, and seem to prefer blood response when it becomes inevitable. That's exactly what a safety department could prevent.' Captain Caesar of Lufthansa agreed: 'While our Board gives me full job security, American safety staff seem to have a vast turnover. Many are dismissed or put back on line for disagreeing with someone like the Chief Pilot. In my opinion, you make a big mistake by regarding safety departments as luxury items you can't afford. The fact is you can't afford not to have them.'

* ibid.

Jim Burnett of the NTSB expressed great concern in 1987 about 'a growing tendency of the airlines to downgrade their aviation safety departments. The system needs this safety check more than ever.'* The airlines' loss may well be the NTSB's gain because several specialists from Eastern, American Airlines and others have joined them, but they would much prefer the avoidance of the accidents to the availability of more good people to investigate them.

Barry Trotter of the NTSB used to be Eastern's Safety Director. 'Eastern never liked to invest much in safety beyond lip service, but we had five full-timers – now they have two people who look into major incidents, and do some statistics,' he said. 'We used to investigate all accidents and incidents, and run full preventive programmes. Now I hear from pilots that investigations serve only to check on and discipline employees. An executive sent from Continental once made this statement: "If we fire a few pilots, the rest will get the word and straighten up – that's our prevention programme." That sort of view would set safety back fifty years.

'When I was on the IATA Safety Committee, PanAm and TWA used to be very active. Now they seem to find it more economical to stay away. PanAm kept its department going, but cut it by fifty per cent at the time of deregulation, and now it has only one man in safety. He does his best, but it's a huge organization, too big for one man. At NorthWest a training-pilot was asked to do the job. He's trying to learn from others because he's not trained to investigate or run prevention. Many airlines just send letters about accidents to their pilots: you call that a safety programme? American Airlines and United have learned from their mistakes, and increased their safety staff after a period of incidents. Piedmont and U.S. Air have no specific safety department. Others have part-timers, pinching a little of the time of, say, a quality assurance engineer or training-pilot, and hope it's enough if occasionally he's sent to some international conventions. The trouble is that a safety department costs money, and it's hard to measure its value in dollars. Remember Eastern's case of losing power because of the missing O-rings? Couldn't that have been prevented? Wasn't that a good illustration of value for money or

* Speech to a meeting of the Allied Pilots' Association, Arlington, Texas.

rather of the misers taking huge risks? No, it's no good to leave the problem to management alone because they say that, ultimately, safety is up to the government, and if the authorities tell us what to do, we'll do it.'

Chuck Miller concurred: 'Sometimes litigation can do the job. In a current case, which is *sub judice*, the airline involved reviewed its safety programme only after depositions were taken – and having some very critical near-accidents made them sit up and take notice. But, overall, history teaches us that most airlines don't do anything unless they're forced to do it. Since deregulation, if they spend on something voluntarily, they worry about losing a bit of the competitive edge. Only the introduction of a federal requirement could enforce an effective safety programme across the board.'

Aviation lawyer Peter Martin raised the fact that U.S. laws themselves may be responsible for some of the problems: 'It is believed that certain American airlines have disbanded their safety departments because of fear that, if they had an accident, they'd have to disclose their records and so some of their truly preventive measures might be interpreted as awareness of the risk and an admission of responsibility.'

MINIMUM, MAXIMUM AND THE GIMLI GLIDER

The deflation of the safety cushions is probably the most contentious issue of cost-cutting warfare. This deflation is a long-drawn-out process because it is caused by 'slow punctures', but those who deny that it happens at all and demand statistical evidence may soon be getting just that in the most painful way. For even the statistics began to have their brutal say in the second half of the 1980s. There came 'an exceptionally bad year' in 1985. Then another in 1987. And another in 1989. The cry went up: that had nothing to do with the cushion! And although some symptoms are visible, the need for such cushions is doubted because the regulators consider their minima 'safe enough'.

In America, numerous tales circulate about pressure on mechanics to delay non-essential maintenance, about operations being pushed right up to the legal buffers, about fuel tanks of an *old crate* being sealed with superglue in a hurry. In 1987, a near-comic incident occurred: an Eastern Airlines DC9 made a heavy landing at Pensacola, Florida – and snapped neatly into two. The roof cracked open, the tail section flopped to the ground. (Luckily, only three passengers were hurt sliding down the emergency chutes.) In Miami, passengers staged a riot when a flight was cancelled and the pilots complained that they were forced to fly shoddy aircraft.* When, in the same year, a DC9 of Continental crashed on take-off in a snowstorm at Denver, Colorado, there was no comic element for there were twenty-eight corpses among the wreckage, and pilots were quick to claim that the airline was pressurizing them to fly in dodgy weather.

In July 1987, *Newsweek* said: 'The public's fears may be exaggerated, but even the pros are flying scared these days,' and quoted Captain Ron Cole about the flight of a seriously deficient 727: 'They're just not fixing the damned aircraft. They're pushing it to the limit of the law.'

Similar allegations are heard in Europe, mainly about keenly priced charter operations. Although all airlines claim that the pilot has the ultimate say to go or not to go, the pressure, unspoken though it may be, is always there. 'When a pilot is down the line with a planeload of passengers and a *minor* problem crops up, he's going to think jolly hard as to whether the problem is one he can live with or not, in which case he may have to put passengers into an hotel, cause a great deal of inconvenience all round, and a lot of expense to his company,' said John Chaplin of the CAA. 'Regulations give people the muscle to resist pressures, but we mustn't overdo it or else we make the whole business uneconomical. If we require something Spain does not, we can put our charter operators to a serious disadvantage without a corresponding advantage to the public who may simply transfer to the cheaper flight. Our rules are sometimes more stringent than the Americans' because the FAA, under industry pressure, cannot maintain much of a margin above minimum requirements. That's why when an older American aircraft is bought here, we often demand some extra modifications to comply with British standards.'

A certain erosion of the above-minimum safety cushion is inevitable because the competition is increasing the pressure on the best, who'll then find it harder to keep taking the extra

* *The Times* 29.12.1987.

precautions on which some airlines and manufacturers like to pride themselves.

Several dozen caring experts on both sides of the Atlantic and the Pacific echoed these thoughts to me. They all seemed to agree that the safety cushion on which we sit so comfortably had become so substantial over the years that it could take diligent whittling away for a surprisingly long time before the full effect would be felt. But even if, fortunately, their fears are not yet translated into accidents which can be fully attributable to the deflation of the safety cushion, Dr Alfred Kahn, who used to chair the U.S. Civil Aeronautics Board and was known as the father of deregulation, acknowledged in the *Los Angeles Times* (5.4.1987) that 'it may well be that the margin of safety has narrowed'.

The level of aircraft maintenance and the standard of pilots are probably the two areas where one could pinpoint most readily the clearest manifestation of what Geoffrey Wilkinson described to me as 'the risk that the legally required minimum may become the desirable maximum'.

As aircraft grew bigger and more sophisticated, they could carry more duplication and triplication of fail-safe systems. Our reassuring air safety statistics were the chief beneficiaries of these built-in features.

Then the temptation crept in: pushing the gap between overhauls to the limit and deferring the maintenance of non-essential and redundant items could save a fortune to none-too-fussy airlines. That led to debatable interpretation of what should be on the lists of maximum allowable deficiencies or minimum serviceable equipment. How long can faults be carried without breaking the law? When cabin comfort items, like reading lights, didn't work, only passengers complained. (In 1988, on a Continental flight to New York, I counted seventeen broken or non-functioning small items.) But by deferring more serious maintenance, 'we're attacking the very thing that made everything so safe,' Wayne Williams, president of the U.S. National Transportation Safety Association once said. When Eastern was fined almost ten million dollars in 1987, it was guilty of more than 78,000 maintenance violations. Many of those might have been *minor* items but it would be a brave man who claimed that they would always be superfluous.

Everybody I talked to had a story about the MEL (Minimum Equipment List).

This was Captain McClure's favourite: 'It was raining heavily at Panama City when a Tristar made a perfectly good, soft landing. As it touched down, the right main gear collapsed without any warning. Running at high speed, the captain struggled to keep the aircraft on the runway because, if it veered off into the mud, a wing might be ripped off and a tank might get fractured, starting a blaze. The natural behaviour of the aircraft would be to turn and pull towards the damaged right side. To balance that, the pilot used asymmetrical reverse thrust on the other, the left side, for directional control. He succeeded. The Tristar stayed on the runway. Only some minor injuries occurred during evacuation. Good technique, one might say. But it was more than that: it was sheer luck. The gear collapse was caused by stress corrosion. It could have happened to the left gear instead, in which case the right-hand engine would have been needed for directional control. Except that it wouldn't have worked. The right-hand reverse thruster was inoperative and could have been left so indefinitely. It is not an MEL item because reverse thrust is not a part of the certified braking process. It's an extra. Except that in this case, it was needed not for braking but directional control to avert a catastrophe.'

Brian Richardson of the NTSB chose this classic example. When an aircraft that belonged to Simmons, a commuter airline, crashed short of its destination runway, it emerged that the crew had got lost. Whatever the reason was, it certainly did not help them that they did not know how far they had travelled on their journey because their Distance Measuring Equipment was out of order. 'We found a sticker attached to the faulty DME showing that only the previous day it had been designated as a deferred maintenance item. Perfectly legal, but when the crew got lost, the DME would have helped them in their fuel computations, and might well have saved their lives.'

The risk of running out of fuel may seem to be ridiculously remote, but it is not. It has happened many times. In 1980, a Viscount had to crash-land in a field when its tanks ran dry. Did the pilots fail to look at their fuel gauge? It was immaterial whether they did or did not. The gauge was known to be 'chronically unserviceable' and had been the subject of twelve complaints in the previous twelve months, yet nobody seemed to regard it as an issue of urgency because other methods *could* be used to ascertain the amount of fuel on board. A Bulgarian TU-154 suffered total fuel exhaustion and crash-landed with pilgrims in the desert near Benghazi. The pilot 'couldn't even imagine' how it happened. Aboard a Republic Airlines DC9, allowable deficiencies helped to distract the pilots from fuel management, and they found themselves having to make an emergency landing with barely four gallons of fuel remaining.

'Unfortunately, pressure by their peers plays a part when pilots accept imperfect aircraft for flight,' said Captain McBride of the Canadian ALPA. 'Aircraft are getting bigger, manuals are getting smaller. We fly aircraft with more snags than ever because pilots are sometimes made to feel sort of challenged: if another captain could fly it all the way to here, why can't you take the next leg?'

That kind of macho might have played a part on July 23, 1983. An Air Canada 767 left Ottawa with inoperative fuel gauges and 22,600 *pounds* of fuel in the tanks. The pilots thought they had 22,600 *kilos*. Less than halfway to their destination, they suffered a to-them mysterious total power loss.

A single stroke of luck came to the rescue: Captain Pearson happened to be an experienced gliding enthusiast. To use a big jet as a glider is no mean feat – to land it on an abandoned airstrip at Gimli, Manitoba, would need outright brilliance. Pearson put it down smoothly with no room for mistakes or overshooting for a second attempt. A most laudable achievement; but how could it come about in the first place? If anybody asks now about the *Gimli glider* case, people at Air Canada duly blush, though at least they have learned from the incident.

There are three fuel gauges on the MEL lists of Boeing and the FAA: they are essential requirements for flying 757s and 767s. If only one of the gauges is unserviceable, the aircraft can, however, be despatched as long as dipsticks and a special fuelling control are applied to check how much is loaded. On the *Gimli glider* the entire system was out of order, including the visual warning device to alert the pilots if fuel ran low. The captain ought to have rejected the aircraft. 'That was the second error in the chain,' said Laird Stovel, Operations Engineering Manager. 'One pilot meets another in a parking lot . . . hears that the fuel gauges aren't working . . . communications are not a hundred per cent between them . . . Captain thinks what the heck, if the other guy could fly it into Ottawa, I could fly it on as it is. Then more rules and set procedures are broken, one after another. The man who checked the fuel on the 767, our first metric airplane, didn't appreciate that a litre of fuel weighs 0.8 kilo. He multiplied litres by 1.77, expecting to get the weight in kilos when it gave him, in fact, pounds.

'There were several stages where the mistake should have been discovered. It wasn't. Mechanics and the brilliant pilot have been punished by temporary suspension; training and fuelling procedures have been tightened; double-checking was introduced, and we enforce the strict application of the MEL. Unless several people go mad simultaneously, the incident can never happen again, and I hope that everybody else will have learned from it.'

One could add that, had there been a flight engineer, the third man in cockpits of the past, it is most unlikely that the crucial mistake could have gone unchecked. Flight engineers, however, are being phased out to be replaced by automation that is even more reliable – when it works.

'We now review MEL deviations every morning,' Jeremy Haines, the airline's Maintenance Quality Director told me. 'On a fleet of a hundred aircraft, we have up to thirty such items a day. Most of them are just minor ones, like a broken passenger seat, which we ourselves choose to treat as an MEL item. Some may be carried for a while, but after a maximum of five days, questions are asked no matter how trivial the deviations are.'

In principle, the regulatory agencies' MELs grow more rigorous all the time, but, to quote Barry Sweedler, NTSB Safety Programs Director: 'Our investigators have noted that mechanics are under increasing pressure, everybody tries to reduce the time an aircraft *wastes* at the gate, everybody wants to cut maintenance time. A big battle at corporate level has just been won by the marketing people – and that was at one of our better airlines.' Smaller and poorer airlines are said to find themselves in the shoes of the guy who has just made a big outlay on his first Cadillac only to discover with a shock that a set of new tyres will cost him three times as much as he paid when driving around in that old banger. So he may delay the new tyres and the service, and leave a few repairs until he gets that pay rise . . . Deferred maintenance is also like using plastic money: living on credit costs more and, after a while, servicing the interest alone may grow beyond the borrower's means.

The cumulative effect of deferring the repair of seemingly insignificant deficiencies is an area of major concern, and leading agencies like the FAA and the CAA have introduced limits on the length of time any faulty item can be carried. But under subtle pressure to meet schedules, particularly in countries where controls are slack, the mood could induce loyal staff not actually to cut but to round off costly corners.

One hears of cases when new aircraft had deficiencies during the delivery flight – and several months later they were still not put right. British Airways keeps track of all Allowable Deferred Defects, so now it is becoming quite common that their aircraft come in with no ADD to report. It is also a commercial consideration. An engine may be fine to fly with one of the fire loop warning systems inoperative. But if it is down in Bengal or Qatar when the other loop goes, the aircraft is grounded until mechanics and parts are flown in. So where are the savings then? Of course an ADD can be defined in different ways by various airlines. Some like to maintain a safety cushion, others do not. 'The CAA created a matrix of defects that might be acceptable in combination,' said an airline safety specialist. 'But the variety is too large, and to my mind, whatever is in the book, it's the captain who has the final say. If there's undue pressure on him, you get nothing but trouble. Not worth it.'

A fine principle, but would it work everywhere? A Continental flight was ready to leave Denver when a mechanic discovered that the public address system was faulty. While he went to get some parts, the crew grew anxious about the delay and took off. Only the mechanic knew that he had left an access door to the fuselage hanging open. He alerted the control tower, and the plane had to return right away. An American commuter line's Twin Otter crashed killing eighteen people in a wood near Rockland, Maine. The investigation revealed that the company owner and president had pressurized pilots to take no reserve fuel, to fly in almost any weather, and to carry more than standard loads. 'Anyone who refused to toe the line was ridiculed by his colleagues or even fired.'*

* *Safety by Stress Management,* Swissair Flight Safety film.

WHAT'S IN AN ACRONYM?

The fight to define strict MELs and enforce even stricter adherence to them has become a matter of life or death since we have learned to live with EROPS. The acronym stands for Extended Range Operations which is a euphemistic replacement of ETOPS – Extended-range Twin-engine Operations. The new word has been introduced to emphasize that there is no reduction of safety when flying nonstop over oceans and deserts in whatever sort of aircraft, but perhaps also to make us forget that, whenever we mention EROPS, we still mean *two* engines as opposed to four.

Currently, hundreds of EROPS flights are completed safely every day, their record is unblemished so far, and the manufacturers' order books for new generation twinjets are bursting. But the doubters suggest that *Engines Running Or Passengers Swimming* is the true meaning of the acronym.

Aviation developments have always been pushing the limits. Better equipment in better aircraft has made the navigator and the engineer redundant, reducing the number of people needed in the computerized cockpit to two (with back-up when a 'heavy crew' is required to give everybody a chance to rest on long trips). Now more reliable technology has also reduced the minimum of engines from four to two for flying nonstop far away from any suitable haven.

EROPS are good for business. It is more economical and convenient to fly frequent and fully loaded twins than half-empty Jumbos. Medium size towns can now get direct services to faraway destinations, and the savings for the airlines are huge considering that a 747 burns some 13,000 pounds of fuel per hour while a 767 consumes 5,300 pounds. If a 747 is sixty per cent full, with 270 passengers, it makes no profit. If 170 seats are full on a twinjet, the revenue is most impressive.

Opponents of EROPS claim that manufacturers, airlines and even the regulators use 'salami tactics': from 1984 onwards, they have allowed the new twinjets, like the Airbus 320, the MD80 and the Boeing 767, to venture further and further away from the nearest airport where they could land in the event of 'losing' one of the two engines. The limit for single engine diversion went up to 75, 90, 120, and now 180 minutes. The regulation, that in any emergency one engine is deemed safe for 180 minutes' flying time (about 1200 miles), opens up almost all routes for the twins. Once

EROPS excluded winter flights. Now they do not. Bad weather is still a consideration, but less and less so. At first only certain airlines got the right to fly to the new limit, with certain engines and extra precautions. But many senior pilots fear the dawn of blanket licensing, at least in some, maybe the least capable, countries. Pilots continue to identify a barrage of deadly technical problems that can arise. They like to quote Lord Hives, a great engineer and ex-chairman of Rolls Royce aero engines, who was once asked by a journalist:

'What's a really safe aeroplane, sir?'

'Well, hard to say, but if, above the Atlantic, the engineer reported to the pilot: "Excuse me, captain, I think we've lost No. 29 engine," and if the captain answered: "Thank you, Mister Engineer . . . No. 29, you said? Which side?" – I'd be happier.'

Although these days 'No. 29' is genuinely superfluous, pilots worry not just about engine reliability* but about the ways some airlines may treat the entire propulsion system. Ed Smart, the Montreal representative of IFALPA told me that 'ICAO standards are too vague and minimal. They say, for instance, that you must have *navigation equipment appropriate to the route*, i.e. each state must decide what's appropriate.'

It is a fact that EROPS regulations demand higher safety standards, more rigorous maintenance, more exhaustive MELs, and special crew training, but as Captain Steve Last, Aircraft Design and Operations Chairman of IFALPA said: 'The salami tactics have now begun to work on MEL and the total package, and, if too much is sliced away, what we have left may not be a salami at all. And when the authorities quote the truly outstanding safety record of EROPS, they're talking about the excellent airlines that were first to fly these twinjets. El Al, Air New Zealand, American Airlines, to mention but a few, all have the resources to comply with the extra requirements; but what will happen when the practice spreads everywhere? An African airline has disposed of its twinjet fleet because it realized it had insufficient expertise to keep it flying. Others may never recognize such facts until it's too late. A small Austrian line is flying 737s

* For EROPS certification, manufacturers have demonstrated extremely low engine failure rates, such as 0.01 times per hundred thousand flying hours.

from Vienna to Rio, I hear, with a refuelling stop on some tiny island. The current ICAO regulations don't prevent them from doing it, but do they know what they're doing? Okay, they use a *new* 737. But what if some clown decides to stick a couple of wingtip tanks on a clapped-out Mark One DC9 and proposes to fly it from Vienna to Miami?'

Air Canada pilot Gary Wagner, who has fought long battles on behalf of IFALPA and the Canadian union for the safety of EROPS, shares Last's concern: 'Even the best airlines dread the proliferation of the system. Some day somebody will ditch a 767, and the effect will be felt everywhere. I heard that a South American country had started to fly a domestic version of the 767 across the ocean to Europe. They have no business to do that. But in some countries it's easier to get licences for all sorts of things if you happen to know someone like a general of the armed forces who runs the civil aviation authority.

'My own airline takes a whole range of extra precautions, but will everybody do so? The pressure is on everywhere to shorten those MELs because the reports and so the statistics on engine shut-downs are very incomplete. Some operators come clean, others don't. And there are other loopholes. Even our own rules state somewhere that if you have a major system failure, a flight test must be done unless you're positively certain that the repair you've done has fixed the fault. Now that's the most ludicrous statement I've ever heard. What mechanic would be satisfied with repairs if he did not think that it had fixed the problem? Mostly he'd just *know* what the problem was and how to fix it. Or else he'd seek advice. Yet we've had a case when an EROPS flight came in with an engine failure that was fixed and ground tested before being dispatched – only to have a recurrence of the same problem on the next EROPS flight! Our argument is that no aircraft should be allowed to carry passengers on an EROPS flight after a major system repair without a flight test or domestic sector being flown. As there's hardly any room for error, and the balance between safety and economics is becoming very precarious, we must balance on the side of safety.'

Some European airlines still hold out against EROPS or are conducting computer experiments with it, but they are ordering the types suitable for it, and once the aircraft is available, the temptation will be irresistible to take full advantage of its capability. So EROPS are here to stay. In 1988, a 767-200 flew non-stop from Halifax, Nova Scotia, to Mauritius – 8,727 miles in sixteen hours and twenty-seven minutes – setting a distance record for commercial twinjets. Yet the British Air Safety Group, having been assured by the CAA about the safety requirements for EROPS, fears that success may breed complacency, the airlines will use the weapon of favourable statistics to press for the relaxation of the rules, and 'the CAA may not be able to maintain its present high level of monitoring indefinitely.'

The introduction and superb performance of the widebodied aircraft with its quadriga of mighty jets eliminated ditchings almost completely, and airlines would dearly love to get rid of the costly maintenance of life jackets. Under their pressure, the CAA seems willing to let them, even though the *low* failure rate of engines is still an admission of potential failures. No wonder that I have come across numerous people who think that the first twinjet ditching in an ocean may become a blessing in disguise. They pray that when the inevitable happens there will be no fatalities, hope that it will tighten up regulations (including those for life boats and vests), and trust that it will serve as a timely reminder of the once dreaded hazard.

'Maintenance needs firm enforcement particularly in a deregulated environment,' said Jim Burnett, 'but at least we have a fully codified MEL system. Operations and flight training are more worrying because the regulations are far less clear cut, the pressures are even greater, and there's a shortage of experienced pilots.'

THE SECOND CASUALTY

'Milwaukee, Midex 105 ready on 19R.'

The Tower acknowledged the call, and twenty-eighty seconds past 15.20, on September 6, 1985, Flight 105 of Midwest Express was cleared for take-off. Fifty-eight seconds into the flight, when the DC9 was 450 feet above the ground, a loud, clunking noise was heard in the cockpit. Then came a sound that was similar to an engine spooling down. The following is a simplified extract from the cockpit voice recording:

15.21:26.7	(Captain:) What the # was that? (No response from first officer.)
15.21:29.5	(Captain:) What da we got here Bill?
15.21:33	(Captain:) Here . . .
15.21:34	(FO does not answer but responds to the local controller:) Midex one oh five roger ah we've got an emergency here.
15.21:36	(Sound similar to stickshaker starts and continues until end of tape).
15.21:38	(Captain) Oh #.
15.21:39	(FO:) * heads down. (* signified the sound of power interruption to the CVR for 0.1 second)
15.21:40	(FO:) Heads down.
15.21:41	(FO:) Heads down.
15.21:41.7	Whoo- (Sound of the beginning of first "Whoop . . . Whoop . . ." of the ground proximity warning system.)

Two tenths of a second later the recording stops, coinciding with the crash that ended the 101-second flight and the fifteen-second emergency. The aircraft was destroyed by impact and post-crash fire. All thirty-one people on board were killed.

There were a hundred witnesses. They saw flames and/or smoke coming from the right engine, heard loud reports like gunshots, and noted the weird manoeuvres of the crippled aircraft at the apex of its brief climb. Witness accounts were supported by the wreckage examination as well as the DFDR and CVR read-outs. Everything seemed to point towards probable right engine failure. The crew's proper response ought to have coped with that. Yet on the voice recording, neither pilot had called for 'Max Power' or 'Ignition Override – Check Fuel System' as required by the airline's emergency procedure following engine failure after reaching V_1 speed.

Did the engine failure create conditions that were uncontrollable? The answer was 'no'. Was the control system affected in a way that would render the captain helpless by, for instance, the fracture of the rudder pedal just when left rudder was essential to control the aircraft after right engine failure? No. Did disintegrating engine parts hit the aircraft in flight? Yes, but the damage was of no consequence. When the right engine failed, was there a total loss of power? No. (There was ample proof of that: both recorders worked until impact; though the right wingtip navigation bulb was smashed by ground contact, its filament was found unbroken and in an elongated shape which revealed that the filament was *hot*, i.e. powered, working, and malleable when it was subjected to impact forces.) Was the small fatigue crack that caused the engine failure serious enough to make the crash inevitable? No. (Recommendations for engine improvement and better maintenance checks were issued nevertheless.)

The investigator's mind followed the branches of the logic tree – *what? when? how? why?* This arduous process of elimination left only the crew in the spotlight of suspicion. So why had they neglected the proper responses? Some adverse medical condition? No. Excessive fatigue, habitual negligence or suicidal tendencies? No, no, and no.

The next step was to see how exactly a DC9 would behave when an engine failed, and how the pilot could cope with the problem on the basis of the information that would actually be available. A test flight was arranged. An engine failure was simulated. The pilot and the observers were in no danger – and a cursory scan of the instruments gave them all the information they needed to initiate corrective action. It led to the conclusion: *'The failure of the first officer to respond to the captain's questions and the failure of the captain to maintain control of the airplane suggests that there was a breakdown in instrument scan by both pilots in the critical seconds which followed the right engine failure.'*

But the investigators refused to stop and take the easy way out by blaming human error. Admittedly, had there been no engine failure, there would have been no emergency. Had the pilots scanned the instruments properly, they would have known the nature of the emergency. Had they followed the book for such emergencies, there would have been no accident. So why didn't they?

The examination of the two-pilots' professional history revealed that both men were rather inexperienced on DC9s. The captain was thirty-one years old. He had been a corporate pilot of Beech 90 turboprops, and had a total of 104 hours' instrument flying experience when joining Midwest Express as a DC9 first officer. 600 flying hours and 369 days later he was upgraded to captain.

Another 500 hours and seven months later he crashed and died at Milwaukee. The accident report revealed that 'advancement to captain can occur much sooner' at a relatively small carrier like Midwest than at 'more established airlines' where a new captain would have about ten years' seniority with 10,000 flying hours, three-quarters of which would have been on turbojets as first officer. Not suggesting that all that experience was essential or legal requirement for upgrade to captain, the NTSB emphasized that the 'extra experience does provide a greater margin of safety to the travelling public'.

And that was not all. The investigation found that the dead pilots' training was inferior to what larger airlines would provide *above* the required minima, particularly for dealing with engine-out emergencies in critical phases of flight like at Milwaukee. The NTSB was also concerned 'about Midwest Express utilizing a "silent cockpit" philosophy which was not outlined in its approved training and operations manuals'. This might have contributed to the less than ideal communications and coordination in the cockpit. (The captain's 'What the # was that?' might have been rhetorical, but his next question was a cry for assistance he never received possibly because his FO was influenced by Midwest's unwritten guidelines against 'unnecessary callouts or even verbalizing the nature of an emergency' in the cockpit, a concept that was 'not approved by the FAA and was in conflict with approved emergency procedures'.)

Finally, another two crucial questions were raised. How come that the potentially disastrous cracks in the engine had not been picked out? And why had the relatively young airline's training and philosophy for emergencies not been queried? The accident report criticized the FAA whose 'surveillance of Air Carrier Engine Service (AeroTrust) was deficient' when failing to detect the shortcoming (e.g. faulty test equipment calibration) of the servicing company's inspection practices. The report also pointed out that, during the vitally important first two years of the airline's operations, the FAA 'oversight of Midwest Express procedures and training . . . was less than optimum and probably suffered as a direct result of the inexperience of the POI'. (The Principal Operations Inspector was not fully qualified to inspect DC9s, had to rely on other inspectors, and could devote only a fifth of her time to Midwest, the first and only scheduled passenger carrier in her care. As a result she dealt mostly with administrative matters, and she accepted verbal assurances about the completion of pilot training without checking the records.)

The significance of this case is much greater than the tragedy itself, and carries grave warnings for those who fail to learn from the American experience that staff problems are a major casualty of the system. For the numerous post-deregulation reports of the NTSB offer the free lessons to anyone who cares to read them that, in the era of increased competition and fare-slashing, traffic grows faster than the availability of genuinely qualified and experienced airline pilots, and that deregulation must mean *extra* regulations and *stricter* inspections in the field of safety.

Fast expansion and the mushrooming of airlines have created a voracious appetite for aircraft and pilots. In 1985, for instance, U.S. airlines 'hired 7,840 pilots, an increase of forty-three per cent over 1984'.* In 1986, another 6,341 pilots were hired. In those two years, American alone employed almost 2,000 new pilots. Where did they all come from? The regional lines. The cream of their people had been siphoned off. *Aviation Week and Space Technology* reported that in a single two months' period, some commuters lost a quarter of their staff, who would then be harnessed for national airline service within forty-eight hours. A Tennessee commuter line once lost sixty-two of its 232 pilots in a forty-one-day recruiting raid by the majors, and had to lay up aircraft with no one left to fly them. On average, the smaller lines lose half their pilots every year.

Demand was created for new entrants. No wonder that the training levels, and particularly the allocated maturing time, dropped. In 1983, years after deregulation, the average new pilot of a major airline had 2,300 hours of jet flying time. In 1984, the average dropped to 1,600 hours, and that was halved a year later. 'This will be a safety problem for several years,' cautioned Jerome Lederer.

The Air Florida crash into the Potomac, already discussed, was one of the classic examples of rapid upgrading with no time to season pilots (in that case for winter operations). In 1987, a DC9 of Continental crashed on take-off at Denver killing twenty-eight

* Article by Jalmer D. Johnson of ALPA, in the *Airline Pilot*, September 1987.

people. After a three-year strike, the pilot had been type rated for DC9s and accumulated 182 hours, then upgraded to captain, and gained barely thirty-three hours more in that capacity before the crash. His co-pilot was very experienced – but on turboprop and straight-wing commuter aircraft; on DC9s, he had logged a grand total of thirty-six flying hours. Some accidents revealed that the pilots involved had struggled to qualify at all for the left-hand seat, and crashed long before they could mature operationally. In a speech in May, 1988, Jim Burnett commented: 'In the current environment, pilot training requirements simply are not adequate. The FAA's requirements for pilots are based on historic and archaic practices. For years, it was the industry's practice to give far more than the minimum training required by the FAA.' Now many operate at the minimum level and, worse, 'often request – and receive – exemptions from regulations'.

'Even without statistical proof as yet, logic tells us there's going to be a price to be paid for that,' said Barry Sweedler, Director of the NTSB Safety Programs Bureau. 'We already see cases where we know that a crew with greater experience could have flown out of a bad situation.' His colleagues reeled off examples – all affecting major airlines: The pilot who lost three engines due to fuel starvation, and made an emergency landing on one engine at Tokyo, had simply mismanaged his fuel because he had too little experience on 747s. Aircraft strayed dangerously from their allocated paths because the pilots had fed the wrong data into the computers. Frequent landings on the wrong runways; cutting off the fuel flow inadvertently; taking off without clearance. Not just mistakes, but signs of poor decision-making in the cockpit.

A senior pilot of Eastern told me in 1988, before the big strike and the sale of the airline: 'I get some very bright young co-pilots. Only one thing is lacking: proper apprenticeship. Yet they move on in a couple of years to become captains somewhere else. When I got hired, we had the luxury – didn't see it as luxury at the time – of flying as FOs for nine years with all sorts of captains, learning from them and their mistakes, understanding that good command includes the ability to listen. Sure enough, simulators now save money and reduce training time, but the savings ought to be ploughed back into training because, whatever you do, you can't speed up the process of gaining experience. Besides, training now tends to be decentralized. As a result, standards can vary a lot.

Pilots may meet for the first time in the cockpit, and find out only when running at V_1 that they're going about the job in different ways. Even the emergency routines they're using may differ substantially.* And the problem is exacerbated by the inherent risk of the huge pilot turnover. It's sixty per cent a year at some commuters. But even the major carriers are hit by it. At Eastern we've lost 900 *senior* pilots in fifteen months. That's partly because they're lured to other airlines, and partly because our operating environment is so bad, that they express their views with their feet even if they lose seniority or some of their pensions.'

At a time of world-wide air traffic explosion, the shortage of skilled pilots is a world-wide phenomenon. Some airlines will virtually retrain pilots to fit in. Others may have to drop them into the left-hand seat right away if they want to stay in business. European feeder airlines may soon be forced to employ whomever they can get at short notice.

There is serious concern about the haphazard British way of training professionals for the industry: it affects not only pilots, but also flight managers, instructors, training captains, everybody, that the airlines have their pecking order, that the big ones cream the smaller operators. It's the system that's wrong and unplanned. For years, young pilots had to support themselves throughout their studies. Now at least sponsorship is coming back to help them through approved, independent flying schools. But private pilots are allowed to act as instructors, they can teach and gain assistant instructor rating at flying clubs while they amass the required 700 hours enabling them to take exams for a commercial pilot licence.

* A Boeing study of crew-caused accidents found clear proof that the management of the safest airlines in their sample recognized 'the need for aircrews performing in a standardized way', using standardized checklists and operations manuals, and working in cockpits with the configuration 'as nearly alike as possible'. It was an eye-opener to go through preflight preparations with a BA crew. Captain, pilots, engineers – including the relief crew for the long haul to Tokyo – gathered in a small no-frills room, reminiscent of a Battle of Britain fighter station except for the weather radar screen and the rickety, postwar antiquity of a coffee machine. The first thing they did was to introduce themselves. They had never flown together before, but thirty seconds later they spoke exactly the same language, and followed identical routines, knowing, without any need to spell out, what action each could expect from the others.

Historically, the system has relied on a free supply and sometimes overproduction of military trained pilots, but that source is now drying up, and we are also losing the traditional *learning* seat in the cockpit. Sooner or later a purposeful, perhaps an all-embracing European concept will have to be developed to follow maybe the French pattern which is much better. There the state subsidizes and controls a nationwide system which aims at developing pilots for modern, sophisticated aircraft from day one of their training. It would help the entire industry, starting with the smaller companies that badly lack pilots as well as managerial skills. Sometimes the CAA inspectors have to become their 'consultants', writing instead of checking their ops manuals. Their job is to certify not manage, but what can they do? Small companies, even if they are purely holiday agencies, see that the money is there for the taking, so they go into the aviation business. Soon they go from flying Aztecs to twin turboprops, then to jets, 1-11s – or from Piper Navajos to the big jets in a single leap. They don't even need to buy, they can just lease a fleet, and start up without mature managerial back-up. A small taxi company's operations manager may be told: get ready, we're getting 727s and 737s. Where will they get experienced pilots, engineers and managers? Nobody knows.

Maybe there is still time for a warning before the small lines of the world take-off in an even bigger (the American) way to shuttle millions of people without much of a thought for what happened in America. Would it be too cruel to suggest that the crashes of commuters or feeders get less attention because they kill only dozens, as opposed to hundreds, in one go?

COMMUTERS ON THE TIGHTROPE

Air Illinois Inc. ran regularly scheduled passenger flights. Its small fleet included a twin turboprop HS 748-2A which suffered numerous generator problems in the last weeks of September and the beginning of October, 1983. That the generator had to be shut down repeatedly did not amount to emergencies (in fact, the manufacturers and the authorities knew of eighty-one such cases world-wide, and every time the aircraft had landed safely), but it was obvious that the malfunction would need urgent attention. Despite FAA and company regulations, nothing was recorded in the log of the aircraft because pilots were in the habit of making verbal reports to the ground engineers. (Unfortunately, neither the company nor the FAA inspectors had ever spotted this serious irregularity.) When the maintenance personnel failed again and again to cure the problem, they sought the manufacturer's view, and eight telexes were exchanged with British Aerospace. The last of these messages from Britain, dated October 11, 1983, pinpointed the parts that were the potential culprits.

Meanwhile, the aircraft had carried on flying. On October 11, the weather was bad, and Air Illinois Flight 710 to Carbondale, Illinois, was running late when it departed from Springfield. It had three crew members and seven passengers on board. Ninety seconds after take-off, it reported having a 'slight electrical problem'. It might well have been slight to start with, but it was growing by the second. The generator was playing up yet again. The first of many gates between safety and disaster was open.

At that stage, it was not obvious that many such gates were opening. Just to mention a few: the faulty operations manual and emergency checklist; the company's inadequate recurrent training programme for double generator breakdown; the FAA's inspector's failure to put an abrupt end to this unsatisfactory state of affairs. And finally a whole batch of safety gates could be blown away by the captain's known, though not always illegal, flying habits and personal preferences. He was self-confident enough to take small risks. He was a stickler for schedules, and suffered a fair amount of 'press-on-itis' because he 'liked to be a good old boy' who brought in the flight on time against all odds, even if he would need to disconnect the overspeed warning horn whenever he exceeded the descent speed limit. He would also fly very close to or under thunderstorms if circumventing them would cause delays. Reluctant to listen to co-pilots, he could become angry and uncommunicative if suggestions were offered. And above all, he hated to be stranded overnight in Springfield.

Now, less than two minutes out of Springfield, the left generator suffered a complete mechanical failure. The first officer mistakenly shut down the *right* generator, and all attempts to bring it back to life failed. Although valuable minutes were wasted, there was no acute danger because the electrical system was so designed that, within limits, the 748 could be flown safely on batteries alone after the total loss of power from the generators.

Since they still hoped to bring the right generator back on line, power had to be preserved. The captain, however, was disinclined to follow routine procedures to shed every non-essential consumer of the vital supply or even to alarm unnecessarily his passengers by turning off their reading lights. The first officer warned that the battery power was dropping 'pretty fast'. Time was beginning to run out, but basic 748 training had taught pilots that they could expect thirty minutes' safe flying on fully charged batteries, so everything was still under control. But the captain would not turn back. Still only six minutes' out of Springfield, he continued towards Carbondale which was thirty-nine minutes away. Various 'factors would have influenced the captain's choice: training and experience, psychological and environmental stress (e.g. get-home-itis), and cost and safety considerations'.*

The batteries worked for thirty-one minutes.

At 20.53, the descending Flight 710 lost all its power, turned blindly 180°, and crashed in a rural area near Pinckneyville, Illinois. There were no survivors.

While the NTSB was investigating the crash, a special FAA inspection team descended upon Air Illinois. In just six days it 'identified several major safety deficiencies affecting the overall operations' of the company. It amounted to a tacit admission of inadequate FAA surveillance prior to Pinckneyville. Four days later, on December 14, 1983, knowing that a revocation order was in the pipeline, the company voluntarily surrendered its key operating certificates. Though strangely enough, the certificates were reissued and some operations were resumed within the following two months, Air Illinois filed for bankruptcy, and ended scheduled services in April, 1984. The NTSB report made numerous critical remarks about the company and the FAA, offered recommendations, and blamed the pilot.

It is interesting to note that despite the clear analysis of the report, a European specialist disagrees with the 'cause'. He said: 'These days, we'd say that the pilot lost control in bad weather following the loss of electrical power which deprived him of his flight instruments. Who can tell that the pilot was wrong trying to continue? Perhaps another, more capable pilot would have made it.' Yes, one could see the good intention to avoid blaming the

* NASA study of decision-making, 1975.

pilot, the usual *easy way out*, but in this case the bare listing of the catastrophic sequence without naming *the cause* could have implied that it was impossible to cope with the emergency, and would have kept the spotlight off the real culprits behind the cause: management and FAA controls.

Commuters, feeders, the so-called *third level carriers*, come between air-taxi operators and the major national/international airlines. The aircraft they use are up-to-one-hundred-seaters, turboprops rather than jets, and mostly small ones that may haul a dozen people over short distances, often from one inaccessible place to another. In America, the dividing line comes with aircraft that seat a maximum of nineteen passengers: if they carry twenty people, there is a big leap in the costs involved, for suddenly there are much higher requirements for airports they can fly to, for increased crashworthiness, for the carriage of flight recorders and cabin crew. Although U.S. authorities refute allegations that commuter safety is below the national standard, and although they tend to claim that 'the difference is statistically insignificant', an FAA task force found training problems across the board, and the frequency of operator certificate revocations has been ominous in the last eight years.

Charles Fluet, manager of the FAA Safety Analysis division, who must have lived in the shadow of crashes all his life because his father was an outstanding, now retired, NTSB sleuth, has the unenviable job to try spotting incident trends long before they cause accidents. 'We knew there was an increase in the number of commuter accidents,' he told me, 'but this was meaningless until analysed in the light of the general traffic exposure to see if it was statistically significant. We found it was, and was not just a chance occurrence.'

In a five-year period, the number of people transported by American commuters doubled to almost thirty million a year, still growing fast. Meanwhile, their apparently varied accidents began to reveal common factors. In some cases, like Ryan's, already discussed, Bar Harbor's, Simmons' and other crashes, pilot training and inexperience were blamed again and again. The pilots of a Henson Airline Beech-99 had not yet completed two months in their respective positions when, in 1985, they got lost in bad weather, and crashed into high ground killing fourteen people. In some cases, the warning signs of previous incidents were ignored. In some cases, other factors, such as the slow dissemination of weather information, were blamed. What figured with most painful repetitiveness were the lack of good management and the lack of FAA surveillance.

One commuter airline had already had a fatal accident but considered certain regulations affecting pilots an 'unwarranted economic hardship' which could be circumvented. The company ran both scheduled and so-called 'on demand' charters using small twin-engined aircraft with fixed landing gear. Flight 901 was to be a scheduled run carrying eight passengers, baggage, and cargo. The available pilot who was twenty-one years old was qualified to take an 'on demand' flight but *not* a scheduled one. So, using the excuse that 'too many passengers had booked in for one aircraft', an *additional* 'on demand' Flight 901A was created – on paper. In fact, the original Flight 901 was cancelled and substituted by the same plane with different documentation.

The rest of the drama was an inevitable sequence of doom.

The flight was twelve per cent over the permitted gross weight. The pilot taxied to the gas pump, and put in an extra thirty gallons of fuel. He failed to drain the water from the fuel system. Soon after take-off, one engine failed due to its water contaminated fuel. In the developing emergency, the pilot then failed to follow the proper procedures which would have saved his life.

At the root of all these mistakes and malpractices was the fact that the FAA had never put an end to the management's habit of cutting corners. The operator's licence was now rescinded, but nine people had already been killed.

'FROM THE HEAD STINKS THE FISH'

Perhaps the Hungarian fishermen's old proverb could help to guide modern crash investigations. If management is under severe pressure, investigators must look deliberately at management itself. That is why there could be a good case for recruiting a management specialist for the traditional GO-team, because the bigger the carrier the more managerial and paper veils may disguise the potential causes behind the seemingly obvious one.

Management has remained, however, perhaps the most underrated causal factor of accident investigation. Safety specialist

Chuck Miller is one of the few who have studied this subject in depth.* He noted innocent looking yet crucial policy decisions behind many hazards, but found neither any 'management' sections in the international manuals, nor any suitable headings in the regular accident records. Olof Fritsch can claim proudly that his ICAO Accident/Incident Reporting Manual has a fine list of management factors (as explanations rather than causes) to look for, but this document is rated only as advisory, not compulsory, and so it has not been adopted universally even by the leading nations of aviation. And only at one authority, the Canadian Aviation Safety Board, did I come across serious thoughts about establishing a post for an investigator with relevant business background:

'There'll be resistance when we start looking beyond traditional areas of risk, but it'll have to be done, just as we had to expand our spectrum when we added medical investigations to the technical and operational aspects,' said Ken Johnson, Executive Director of the Board. 'Even if it's harder to investigate and judge managerial than, say, metal failures that are more definite, scientific, and repeatable in the lab, management structure is essential to safety, and I'm convinced that airline finances, for instance, will become a legitimate target to look for clues to accidents.

'We're not just groping about in the dark. We've had plenty of warnings. If management can affect the safety of the biggest carriers, no wonder it affects more frequently the small ones. Take our crash at Port Franklin in the Northwest Territories. In bad weather that was unsuitable for the flight, this Twin Otter struck a radio tower and killed seven prople. On the surface, it was aircraft handling *[a euphemism for crew errors, favoured in Canada]*, but there was more to it. We found that the approved management structure was not effective. It was not just some mom-and-pop operation. Nahanni Air Services had about a dozen aircraft. It had a proper organization chart with all key positions marked – but there were only a few managers, each man doing several jobs, and it led to some questionable judgments on the assessment of operational risks.' (The accident report listed numerous Cause-Related Findings including the lack of reliable news

* Management Factor Investigation Following Civil Aviation Mishaps, (manuscript), 1988.

about local weather conditions. It found that 'there exists in the more remote areas of Canada a different attitude with regard to the assessment of risk in flight operations; it is likely that this attitude influenced the decisions and actions of the pilot'. It also disclosed that psychological stress and recent events in the pilot's life – a major accident, the loss of a job, the start of new employment – might have added to the prevalent 'bush-pilot' attitudes. But this was not fully appreciated by management, who simply delegated decision-making about take-offs to pilots.)

'The management problem was even more pronounced in our Wapiti Aviation case, a Piper Navajo that crashed at High Prairie, in Alberta. The pilot had clearance to descend to 7,000 feet, but although the published Minimum Obstruction Clearance Altitude was 5,600, he continued to descend in cloud, and struck a tree-covered hill at some 3,000 feet. Now why would he do that? Why would he knowingly contravene written company rules?'

Six people were killed in the crash, but the pilot survived. The investigators found that the company's working environment might have influenced his decisions. Contrary to regulations, he had taken off with only one of his radio direction finders working, and was slightly lost at the time of the crash. He had been behind schedule. He had already had three confrontations with the management that day alone. He had been reprimanded for spending too much time in the weather office. He had asked for engine covers, and had been told off for that. He knew what a high pilot turnover the company had, and he was anxious to prolong his employment. After all, his descent below minimum safety height was no worse than what he had had to do as a short-cut during his route check with a senior company pilot barely six weeks earlier. Irregular or not, he thought this was acceptable company procedure.

Such a violation of rules was not unique at Wapiti. Aircraft unserviceabilities were not entered in the journey logs. Preflight weight and balance calculations were not an enforced routine. For two years (!) before the accident, ministry inspectors were 'concerned' about the company. Following complaints by several pilots, a special surveillance plan was to be initiated. Just a fortnight before the crash, an airworthiness audit had found enough maintenance deficiencies to ground eight of fourteen Wapiti aircraft. And finally, that old potato – management structure. All positions were held by two people. One was the operations manager. The other, who played the roles of chief pilot, chief engineer and chief flying instructor, also worked as a line pilot, training pilot and flight test examiner. Occasionally, one of them took over all these duties. Even worse, Transport Canada inspectors were aware that this practice would never allow sufficient time for any of the key jobs.

'The trouble is that the words "small airline" do not always mean the operation of a couple of Twin Otters or something like that,' said Allan Winn of *Flight International*. 'But what they have in common is that usually they're strapped for cash, lack human resources, and will employ whomever they can get because of the great scarcity of pilots and engineers or, for that matter, experienced managers. They'll find it impossible to maintain their fleet which tends to be second-hand or worse, or leased without ever feeling the owner's responsibility for long-term care. At least third world carriers will try to hire specialists from the more advanced parts of the world. But perhaps the biggest risk is in the West where setting up a "small" airline is often seen purely as a money-making proposition rather than a service. A financier or package holiday specialist may buy or lease one or two aircraft, concentrate on the business side, and regard operations as just an inescapable extra cost. Maintenance will, of course, be contracted out, usually to the cheapest bidder who may be different, in a different part of the world, every time and who will never get familiar with the specific problems which an aircraft may exhibit repeatedly. There will be no continuity of recurrent pilot training either. And if there's an accident, investigators will find it hard to pinpoint where things have truly begun to go wrong – in management.'

On April 22, 1989, *Flight* carried an article by Charles Tyler about Vayudoot, India's big little regional. In eight years, its network had spread at great speed to serve vast, remote areas, and in 1988 it carried almost a million passengers. The man who runs it, 'the ambitious thirty-three-year-old assistant to the chairman and managing director of Air India, was catapulted into the job of general manager'. Most of its fleet, some two dozen elderly Avro 748s, Fokker F-27s and Dornier 228s, have been leased from or handed down for a token fee by its elder brothers in the industry. Some of its telephones have been cut off for non-payment of the bills, credit for spares is running out, parts are made to play musical chairs within the fleet (the word 'cannibalization' is apparently objected to). Growth is pursued vigorously – though Vayudoot suffered eight accidents in the previous twelve months.

Prevention through management, risk management or even managerial acceptance of responsibility for mishaps are more the exceptions than the rules. Following air accidents, Japanese managers sometimes apologize to the public or even resign to take all the blame themselves. But virtually everywhere else, a chief executive's dismissal for his company's disasters would require something stunning.

Iberia's president was fired after two years, during which he had cut losses but labour troubles had multiplied, his airline's safety record had been shattered, and some preventive measures had been neglected. For example, only two of Iberia's thirty-five 727s were fitted with Ground Proximity Warning Systems. Yet it remained unclear whether he was sacked after a series of mishaps and disasters or because of power struggles with the Spanish aviation authorities.

Usually, when a top manager is eased out of his job, it happens discreetly. Often he has the chance quickly to remount the merry-go-round of the industry, as if to act out the old joke: 'All right, General Dayan, you keep winning all those wars, but what if you lost the next one?' – 'I'd start another one under my wife's name'.

In America, people are aware of numerous odd coincidences: airline executives may go board-hopping – only to be followed by slipping safety standards. When air operators' certificates are withdrawn or suspended, the authorities sometimes specify that a small airline may start up again only if certain executives, usually the owner and his family, renounce all their operational control. According to a DoT spokesman: 'There's no actual black list of managers, but some people may become known for running a less than tight ship, and a wary eye will be cast upon any new application they may care to submit.'

Yet the idea of licensing qualified airline executives – like they license pilots, engineers and air traffic controllers – is anathema to regulators and the industry everywhere. The argument is that a manager can manage anything, be it a pawn shop or a public service. It is only the financial track record that seems to matter. 'If an airline is closed down, those who own and run it lose their investment while mechanics lose only their jobs,' said a don't-you-quote-me American financier. 'This country has been made great by people who had an idea and put their money into it. The market place deals harshly with failures, and in aviation, failures cost lives, so there's a moral and ethical obligation to safety, but how can you set the standard by which to measure managers? Admittedly, if their safety track record is bad . . . well . . . ' he hesitated but soon regained his composure, 'well, you'd just have to prove it was his policy that crashed.'

A TIME TO RE-REGULATE?

As well as unbridled competition and doubts about safety, American deregulation brought outcries against the abominable deterioration of services. By the late 1980s, there was much talk about re-regulation. Would the public tolerate the return of higher air fares in return for more comfort and punctuality? The heady days of Texas-Air-style cartel-busting and union-bashing were almost over. It became the age of take-overs and mergers. Mega-carriers already dominated the market, and had the power to increase fares. It had also become obvious that, ultimately, deregulation itself did not encourage the lowering of standards, but it created opportunities for inexperienced or sharp management to get away with murder. What was needed was not a deluge of *new* regulations, but stricter enforcement of the existing ones.

However, in line with the Reagan administration's zealous determination 'to get government off the people's back', the budget for controlling the industry was cut repeatedly, even though aviation taxes, designed to promote safety, had accumulated a 'surplus' of nine billion dollars by 1987. While air traffic, the number of airlines, and flagrant flouting of the rules were increasing at breakneck speed, the FAA experienced an actual drop in its inspection force, losing almost a fifth of its field personnel. In 1983, it had 1,331 people to inspect all the manufacturers, airlines, thousands of aircraft, pilots, engineers. It endured a great deal of criticism from every direction – and rightly so. It just could not cope with the new workload. Rampant violations were found across the board. To circumvent the antiquated thousand-dollar ceiling for fines, set in 1938, cumulative penalties were enforced but, until the early 1980s, the FAA's average revenue from fines barely exceeded 200,000 dollars a year – less than an airline's profit on flying just one fully booked Jumbo from coast to coast.*

In the mid-eighties, FAA field personnel began to increase at last. There were 2,000 inspectors in 1988, and the plan was to reach the 3,000 mark by the early 1990s. Previously inspectors had been expected to make snap judgments and break new ground every day; now the field staff's work was made a little easier by the introduction of new rules† – but, to avoid even a hint at re-regulation, many of them are referred to as standardized requirements. Though more licences were revoked, and the fines grew bigger and more numerous, enforcement could only scratch the surface. Maintenance and a great variety of operational violations on a massive scale resulted in record penalties exceeding $100,000: nine in 1985 (including one and a half million against American Airlines), eleven in 1986 (the nine and a half million against Eastern, and three point nine million against PanAm), eleven in 1987, and almost as many in 1988, with Eastern and Continental, its sister company in the Texas Air conglomerate, figuring most frequently on the lists of shame.

Owned and ruthlessly steered by Frank Lorenzo, Texas had grown amazingly fast from meagre beginnings into the Western world's biggest air carrier in half a decade. Its story encapsulates most of the ills of deregulation, and could be copied in Europe in the 1990s. EC Commissioner Sir Leon Brittan insists that merger mania and other pitfalls revealed by the American experience will be avoided. But have we *ever* learned from other people's mistakes?

Demanding, and getting, 'voluntary' wage reductions, Lorenzo fought acrimonious battles with pilots and mechanics. His airlines gained the poorest reputation for service. A further million-dollar fine for safety violations (such as 511 flights of a non-airworthy Jumbo, and 160 flights of an Airbus with an eight-inch crack in an access door) was but another fly-speck on the books. The

* 'Air Travel: Safety Through The Market' by Dr Andrew Chalk, ISASI forum, November, 1987.

† The imposition of strict limits on how long the repair of any MEL item could be deferred was an example.

carrier and its management came under serious suspicion in 1988. A huge federal investigation was mounted. Was there a danger of financial instability? Had there been any sham deals or dangerous asset stripping? Was the management disposed to comply with safety rules? The U.S. Secretary of Transportation warned that if any American airline contemplated a strategy to put financial pressure above safety, it would have to take the Texas investigation 'as a shot across the bow'.

Two months' intensive scrutiny by a large team of specialists gave Texas a clean bill of health – but could not allay some lingering doubts. Although the inspection had been pre-announced, giving the airlines time to put their houses in order, a third of Eastern's and fourteen per cent of Continental's aircraft were grounded temporarily because of maintenance deficiencies; Eastern had to promise to improve quality control; the war between labour and management was attributed to a dangerous lack of communication ('In a company so divided, the risk is increased that the labour-management discord will, at some time, either through inattention or design, have an adverse impact on the public safety'); a mediator was appointed to deal with this problem though nobody could 'predict a successful conclusion'.

In fact, a few months later, the biggest ever strike erupted. The fleet was grounded; losses ran at a million dollars a day; the company went into voluntary bankruptcy; major assets like profitable routes were sold off; other assets were 'shuffled' (Eastern sold its reservation computer system to Texas, its parent company, for $100 million, then pledged to pay $130 million a year for the use of its own baby); and now the greatest asset, Eastern itself, may have to be sold slice by slice. That could end the great deregulation 'success story' that broke the comfy cartel of the old major carriers, created new ones, and brought slashed fares to the public . . . at a price.

Pre-announced inspections are to be stopped, but extra safety regulations and the imposition of heavy fines are set to continue, yet some people doubt the usefulness of such penalties which 'made airlines more reluctant to warn the industry if something went wrong', said Earl Weener of Boeing. 'When one U.S. airline ran into some problem with door seals, they warned everybody at once, and the first thing that happened was . . . a several-hundred-thousand-dollar fine from the FAA. Some airline people's immediate reaction was "see if we tell you guys anything about maintenance problems any more". I can only hope they didn't mean what they said, but others might.'

COMPETE ON SAFETY

If surveillance (with sometimes crippling fines) is still not an entirely sufficient and certainly not a very sophisticated way of controlling the conflict between profits and safety, what other pressure could be brought upon the mischievous and the corner-cutters? What about allowing the public to catch a glimpse of the invisible – each airline's, each manufacturer's record and attitude to taking risks?

The proposition *to compete on safety* is The Great Unmentionable. And yet some safety indicator could compete with cheap fares for passengers' loyalty. Alluring smiles, flashes of thighs, cocktails, legroom, steaming titillation of tastebuds – such inducements would lose their appeal by comparison with safety. Even some vague categorization – *Very good, Not so good, Good, Goodish* (no line that is allowed to fly can be *Bad*, can it?) would help passengers to make the *best* choice as opposed to the *cheapest* one.

The nearest anybody came to advertising implied safety was the slogan that 'British Airways take care of you'. The nearest anybody came to breaking ranks was Robert Baker, a senior vice-president of American Airlines: the 'FAA should encourage and reward airlines for excellence as well as reprimand sloppiness . . . If the public believes every carrier to be equally safe, where's the financial incentive for a carrier to do better than minimum standards?' And Professor Ian Savage of Northwestern University summed it up suggesting that, in an ideal world, consumers in a free market can select the level of safety they are willing to pay for.*

Arguments against are numerous, and some of them are certainly valid. The feeblest is that there are so few accidents that 'a single mishap could screw up an airline's statistics with a *hundred per cent increase* – from one to two'. Over, say, a five-year period

* Quoted by J. W. Steenblik, 'Debate Without Dialogue' in *Airline Pilot*, September, 1987.

such arithmetic would lose its significance. More debatable is the theory that 'safety doesn't sell tickets'. This has never been tested. ('Large powder extinguishers are not required by law,' said Fritz Hofmann of Swissair, 'but, unlike others, our planes have carried them for ten years. It's a matter of pride, but has it ever sold a single seat?')

To compete on safety? An executive of an international organization mulled over my question. 'You want me to answer frankly or to represent our members and give my name to the quote?' Having settled for the former, he said: 'We as an industry are proud of our safety record. You could fly for ninety-nine years nonstop without being even in an incident. But a lot of information isn't made public because of the vultures like lawyers and some of the media who'd play it up to their advantage. And then there's the question of comparing oranges to lemons. Qantas is always quoted as the safest. Well, maybe. But it would take Qantas a quarter of a century to fly as much as United does in a year! UAL had about one accident every four years – that's one accident per every four and a half million departures. On that basis, to match the UAL record, Qantas would have one fatal accident every thirty-two years, and would have to fly something like 128 years to equal UAL's exposure to date.'

One of the unspoken arguments is rooted in the traditional attitude of airlines: many of them won't talk about safety, let alone accidents. For it was not all that long ago that newspapers with disaster headlines were barred from being carried aboard any aircraft, when Galbraith's *Great Crash* was not displayed on airport book stands, and when airlines would not recommend flying with safety belts on* in case it worried passengers. And anybody tempted to sneer at the *old* days, when after a crash company men would race to the site first with buckets and ladders to paint out the name of the airline, needs only to look at the photograph of the Eastern Airlines jet that cracked into two at Pensacola: the name was painted over before photos would be permitted. The date? 1987.

Specialists who promote the immense benefits of confidential incident reporting, express serious concern about any form of

* Airlines now ask us to keep seatbelts on when seated, but many still try to blur the shadow of menace by claiming that it is for our comfort *and* safety.

competition in safety. Like Weener, they are afraid that the flow of information may dry up. Do they have so little faith in the airlines' integrity, morality and dedication to safety? Could they be right? Roger Green of the RAF Institute of Aviation Medicine who runs CHIRP, the pilots' confessional: 'Airlines may start keeping the best safety ideas to themselves.' Peter Gray, head of the CAA Safety Data and Analysis Unit: 'Utilizing occurrence report statistics alone, i.e. numbers of reports raised, as a basis for comparing the safety performance of airlines could be both unfair and counter-productive, in fact, a retrograde step. The more professional airlines will continue to report all that is required with no restrictions, resulting in apparently large numbers of occurrences. The less professional ones may try and cover things up or may not have the will and adequate staff to submit effective reports, resulting in lower, more flattering statistics. So how could one say which is the safest?'

Undoubtedly, to achieve both accuracy and fairness in a 'chart of safety', vast difficulties would have to be overcome, but the already mentioned Boeing study did it (albeit without names), the *Sunday Times* did it once on a limited scale, publications of *accidents per airline per hours or cycles flown* do it regularly, at least by implication – and implication can be more damaging than a format carefully devised for this specific purpose.

The American Department of Transportation has started to issue *report cards* of complaints about service, delays,* overbooking and baggage problems. Continental was shown to be the worst offender. The initial effect was stunning: 'on time' departures, i.e. with no more than fifteen minutes' delay, increased from sixty-six to eighty-two per cent, because travellers could check the large carriers' timekeeping† through newspapers and travel agents but the improvement does not seem permanent. Now the CAA also publishes its 'league of shame' for airlines, airports and routes. Many airlines, particularly charter operators, argue that delays are mostly due to strikes and airspace overcrowding, but the fact is that all-too-often they themselves are responsible through unrealistic scheduling and lack of back-up aircraft.

* In 1986, at the busiest U.S. airports there were 367,000 delays – some due to weather, a third due to overcrowding, bad scheduling, etc.

† In 1988, Northwest came bottom with seventy-three point four per cent.

Public exposure may exert extra pressure, particularly on small and charter operators, to take-off on time *no matter what*. But preventing ill-advised take-off is a task for the regulatory agencies.

One in five European scheduled flights took off late – not in some peak summer month, but February, 1988. (The figure was twice as good a year earlier.) In Britain, a sampling survey for travel agents Thomas Cook revealed that a million Britons, two thirds of the passengers to Greece, experienced an average of six point three hours' delay, and four million people, two thirds of those flying to Spain, were delayed an average of three point eight hours in the summer of 1988. If it cost more to avoid the most likely delays, passengers would at least have the choice – to pay less and wait, or to pay more and be more punctual.

Numerous experts believe that punctuality can be a measure of an airline's attitude to safety, too. Peter Spooner has long included 'on time departures' in his safety audit when reviewing a company: 'The pattern shows how much they care, how well they maintain and prepare their aircraft, how well they plan their operations.'

And now the FAA may take another step forward – if not quite take the plunge. Former Administrator Allan McArtor, who was keen to dispel 'conjecture about declining safety standards' and regain public trust in aviation, initiated a safety programme and the development of a *Dow Jones* indicator of air safety.

Associate Administrator Anthony Broderick told me: 'We all would like to develop some means of comparison to show which airline is safer or rather which is more exposed to risk. So far this has been an admirable but elusive goal despite several studies in cooperation with other agencies. The difficulty is how to measure safety as opposed to areas like financial stability. Accidents are only a measure of past performance, and can be a matter of bad luck. People in the U.S. General Accounting Office think that the result of FAA inspections may be the best measure of a particular carrier's safety standard. This is a format we're about to automate to see who does better in what. We do an annual quarter of a million individual inspections ranging from major surveys that last for weeks, to fifteen-minute checks on pilot qualifications. That can provide an impressive data base, and may yield a useful comparative index.'

It may also turn out to be a great incentive to the industry to attract customers by means other than fare-slashing alone. CAA

Chairman Christopher Tugendhat was quoted by *The Times* (14.3.89) saying that passengers might have to decide whether they were prepared to pay more for extra precautions by certain airlines. To do so, they would have to *know* about those extras. Airlines will never advertise safety, of course, but will they *inform* us that both their new and old aircraft conform to the latest, some would say superfluous, safety standards? Will BA find some fine euphemism to make it a publicized 'selling point' if and when they carry smokehoods and widen the bulkheads opening? The extra improvements would cost millions. They ought to pay for themselves by being used as 'marketing tools'. To counter that, will other airlines be ready to go on record with adverts that *they* see smokehoods as an extra risk or at least a non-viable idea?

If that happens, the meaning of the expression *freeforall* would certainly attain a new and perhaps healthier dimension.

– 10 –

GERIATRICS CROSSING – BEWARE

Nobody lives on the windy ridge of Mount Ogura in Japan. Yet a winding path up the southern slopes is well maintained. It ends at a small hut in the pine forest. A swath through the trees also leads to that desolate spot. The crowns of the pines were razor cut, and the slim, swaying trunks were slashed by a 350-ton hurtling, screaming, thunderous mass of flesh and metal, when 520 people aboard a Jumbo perished on August 12, a hot summer day of 1985. Since then, young green shoots have begun to grow, filling the gaps.

Japan Air Line's staff have always accepted collective responsibility for the disaster. During weekends, volunteers man the hut as a form of penance. They offer simple comforts to visitors. Year in, year out, hundreds used to go up there on a weekly pilgrimage. Their number has lately begun to dwindle, but once a month, always on the twelfth, a coach is still available for a free ride from Tokyo to an open-air memorial shrine in the clearing near the ridge on Mount Ogura. Some who come are surviving relatives, others just join them in a prayer for the victims. Mourners bring gifts for their dead, a toy for a child, a can of beer for a friend, an LP for Kyu Sakamoto, the pop singer.

In Kyoto, the old capital, I was introduced to an old man, one of the last few working cobblers. He lost a friend in the crash. 'I mourn him, but I never go to Tokyo, not any more,' he said. 'My daughter says the old must take care of themselves to live longer. Like the old Imperial Palace. With care it can stay alive. Forever? Who knows? I used to read everything about the accident. Not any more. People just blame people, they say the aeroplane was old and tired. Maybe. But it's strange. A young thing like a machine being old. My own tools are very old but will remain young even when I'm dead. It's just that there'll be no one to use them.'

Until a couple of decades ago, few airplanes had a chance to grow old. Technological advances made them obsolete long before age, fatigue and corrosion forced them into retirement.

Then hugely grown demands prolonged their usefulness. Improved technology offered them longevity, like medical advances extended people's lifespan. Except that, like old people, the pressurized, high-flying jets began to suffer hurt in parts they never knew they had.

On May 14, 1977, one of Dan-Air's 707s approached Lusaka. It was carrying freight, with a crew of five and one passenger on board. At 09.28, the co-pilot had the airfield in sight. The descent was uneventful. Four minutes later, the flight was cleared to finals. Then witnesses saw a large portion of the aircraft separate in flight. From 800 feet up, the 707 dived into the ground. There were no survivors. The right-hand horizontal stabilizer and elevator assembly was found 200 metres back along the flight path. That fact spoke for itself, and confirmed the witness accounts of a break-up in the air. But why did it happen? Did the pilots overstress the structure by some crazy manoeuvre? What else could induce such a sudden, massive failure?

The Zambian authorities called for expert help from Britain, and decided to hand over the entire investigation to Geoffrey Wilkinson of the AIB. Within a few hours of his arrival, he telexed disturbing news to London: he had found signs of metal fatigue in the rear spar structure, an area that was designed and thought to be fail-safe. It was a Saturday night, but AIB chief Bill Tench was called right away. He, in turn, telephoned John Chaplin of the CAA who was also notified by the operator, and, by Sunday morning, international lines were buzzing. Although nobody knew exactly what had happened, the risk was obvious: hundreds of the vast 707 fleet in service could fall out of the air without warning. (An urgent world-wide alert and special inspections found identical defects in another thirty-eight aircraft.)

But the greatest concern was caused by the fact that nobody knew exactly how it could have happened. Fail-safe design meant that redundant elements should have been capable of carrying flight loads even if a primary load member failed. That was why the regulatory authorities did not require fatigue tests, and relied on the philosophy that the aircraft could virtually live forever provided that mandatory inspections were duly carried out. Notional 'lives' were assigned to various models and structures, but those were set so high that nobody needed to bother much about them. After all, the 707s had been very successful workhorses for twenty years, not old at all compared to turboprops, such as Viscounts, still flying in large numbers at the time. (A full decade later, thirty-eight years after entering service, fifty Viscounts were declared fit to fly for yet another fifteen years or 75,000 landings.) If the fail-safe design was less reliable than its name suggested, the 737s, DC8s, DC9s and all jets designed in the 1950s and early 1960s would also be at risk.

The plane that crashed at Lusaka was fourteen years old, the first off the production line of a convertible model: it used to carry passengers in America, spent a short while in mothballs, then it was bought by Dan-Air for conversion to transport cargo. Its logs revealed no serious incident that could have weakened the structure. The fact that another three dozen 707s were found with their rear spars in an identical fatigued condition precluded the easy way out of treating the event as just a one-off mishap. So was it a matter of age after all? Lusaka became a major landmark that brought into focus the *geriatric jets* syndrome.

Wilkinson's outstanding accident report threw new light on the ageing process of jets. It found that the fatigue crack had begun some 7,000 flights earlier, growing steadily all the time.

Boeing now devised special checks, and the horizonatal stabilizers of the entire 707 fleet 'were beefed up'. Two-thirds of them showed signs that it was high time, too. After further research and consultations with the manufacturers to identify potential, invisible trouble spots and reassess their fatigue/damage tolerance, the CAA introduced a revolutionary concept within a year. All aircraft would be given ever-shortening 'lives'. At the end of each life, each aircraft would have to undergo a rigorous Structural Integrity Audit if the owners wanted to get it certified for yet another life. The safe-life system was costly, but other countries quickly followed the British example. The problem seemed to be solved, and the new philosophy of 'design for damage tolerance' would protect new aircraft. (In 1988, an Israeli commando operation depended on coordination from the air. The aircraft serving as a command post was a very old 707. It must have been fully trusted for reliability.)

The certification of an aircraft's airworthiness is both the duty and the privilege of the regulatory authorities. It is, however, an inescapable fact that the FAA or CAA could not duplicate every single test a manufacturer has done – not unless thousands more engineers were employed by them and the certification was allowed to last for several years more. And even then the process could not be foolproof. Two men who ran the engineering teams building the first commercial jets, one for Boeing and one for Lockheed, once got together and declared – so the saga goes – 'No matter how many tests you do, there's no substitute for the big wind tunnel in the sky so, at the end of the day, you've got to fly the damn' thing and see what.'

In 1980, concern about a DC10 mishap at Chicago (a pylon cracked, an engine flew off taking with it the hydraulic lines) led to a little known gathering of a dozen American backroom boffins. Under the aegis of the National Academy of Science, and calling themselves the Blue Ribbon Panel, they examined the development of the type. They found that of the 40,000 required tests that had been made, only 4,000 had been shown to the FAA, and only 500 of those were even date stamped.

'We tried hard to suggest just one thing that could make a real difference to airworthiness certificates,' said aviation lawyer Don Madole, leading counsel in numerous major disaster litigations, who served on that panel. 'In the end we came up with the important proposal that each manufacturer should designate hazardous parts and areas where uncontained damage could occur, i.e. a failure causing serious damages to some other parts.

'Our rule was proposed in 1982. Manufacturers, the Air Transport Association, everybody fought it. On August 5, 1985, the proposal was scrapped: the industry had won. A week later, a 747 took off from Tokyo . . . Pathetic, isn't it? And not only that. The same Blue Ribbon Panel made a prediction in 1980 that the top skin of the 737 could be suspect, and it might just peel off. When? Maybe in 1987 or 1988, we thought.'

Those somewhat preposterous predictions were not taken too seriously though, in the 1980s, growing attention was paid to the fast ageing fleet of the world. It became known that *cycles* (take-off – pressurization – flight – depressurization – landing) were more important to the ageing process than the date of birth or the number of hours flown, but manufacturers were unshaken in their belief that aircraft could live 'forever' if maintenance requirements were strictly adhered to. Boeing was specially exposed to the problem partly because of its immense success and domination of the market, partly because its aircraft continued to give, reliable service.

The story of the 737 began in 1965, when the first of those stubby little twin-engine craft was delivered. It was ideal for hard work and fast turn-rounds on short routes. Everybody wanted 737s. United bought some four dozen including Nos. 150 and 151 which rolled off the production line in February 1969. No. 152 was bought by Aloha Airlines for their Pacific island-hopping routes. Many of them would change hands several times, but, by the late 1980s, 1,500 737s would still be in service.

Because of the frequency of pressure cycles and prolonged exposure to saltwater atomosphere, there was intensive corrosion that was hard to detect visually under the skin and inside joints. Inspectors would notice cracks only around the smoking area of the cabin where nicotine stains woud appear on the outside. After the production of No. 291, Boeing strengthened the fuselage skin, and introduced improved metal bonding.

Meanwhile, in 1976, Far Eastern Air, operating from Taiwan, bought No. 151 from United. The aircraft had a long history of corrosion and skin cracking, but pressure to complete turn-rounds in a hurry, the essence of false financial prudence, must have limited the time available for inspections. If not before, the damage ought to have been discovered in July, 1981, when the FAA issued a worldwide Airworthiness Directive for extra skin inspections. On August 22, 1981, No. 151 had to cut short a flight because of cabin pressurization failure. Some minor repairs seemed to have cured the nuisance, and, the following day, the aircraft took off again. Just after it had reached 22,000 feet, the fuselage burst open. No. 151 began to disintegrate, scattering wreckage and bodies over a vast area; 110 people were killed.

Being only seven years old, that aircraft was no geriatric, but geriatricity comes in many forms, whatever the age. Its youth was shown by the number of cycles it had completed: a mere 33,313 when Boeing fatigue tests showed that 130,000 typical flights should be safe. The patchy Taiwanese accident report revealed severe corrosion and cracks. It found that when the skin blew out, sudden depressurization led to the collapse of the cabin floor which sheared all the control runs underneath. There was speculation that saltwater from raw fish cargoes might have eaten away the underbelly, but the airline insisted that fish was always in leakproof steel containers. Boeing claimed that many airlines would wait and spend money on replacing badly rusted parts rather than conducting preventive maintenance. According to a Boeing internal memo, Far Eastern had been supplied with a skin replacement kit in 1980, but sixteen months later, at the time of the crash, the kit was still in the airline's spare-part inventory.*

In January 1982, the FAA issued an another AD: the forward and aft cargo compartments of all 737s would have to be checked regularly for corrosion. That is how Aloha, for instance, found in the bottom of No. 152 substantial rust and cracks which needed several immediate skin replacements. 737s as well as DC8s, DC9s and other jets continued to suffer geriatric symptoms, but most cases were viewed as one-off incidents. In 1986, Jim Burnett of the NTSB testified to the U.S. Senate on the ageing problem. He called special attention to the environment: flying in salt air or near sulphur-producing industry would develop rust problems long before an aircraft would reach its fatigue safe-life limits. He was unhappy with the level of specific rust monitoring which seemed to be merely incidental in the course of fatigue inspections.

Two years later, on April 27, 1988, a passenger boarding No. 152 noticed the crinkles atop the fuselage skin. Having mentioned it to a stewardess, he was probably regarded as a fusspot. The day after that, the nineteen-year-old jet took off from Hilo for Honolulu. It had already logged some 90,000, the second highest number of cycles completed by any 737 to that date: 'Aloha flies what may be some of the hardest-working planes in the U.S. fleet. Because the airline's inter-island flights in Hawaii average just twenty minutes, compared with about an hour for flights of other U.S. carriers, the planes are subject to more frequent

* *The Wall Street Journal*, 31.5.1988.

pressurization and landings. And because its planes fly mostly over saltwater, corrosion is always at work, too.'*

As the flight levelled out at 24,000 feet, both pilots heard a loud 'clap' and a 'whooshing' sound which were followed by a whirlwind from behind. First Officer 'Mimi' Tompkins felt her head jerk back. Various debris-like pieces of grey insulation filled the cockpit. Its door to the cabin had simply disappeared so, when Captain Bob Schornstheimer glanced back, he saw 'blue sky' where the ceiling of the first-class section used to be.

Unknown to the pilots, an eighteen-foot chunk of the fuselage top had been blown away.

A flight attendant was sucked through the gap to vanish literally in thin air. The other attendants grabbed anything within reach to hang on for dear life. The inrush of air began to inflict windburns on the eighty-five petrified passengers. The aircraft stretched into a bow shape and, with the floor bent down by three feet, aft passengers saw only open skies as they stared ahead in shock and disbelief. They thought the whole cockpit had gone. But the pilots were fighting for survival. They donned oxygen masks, and began an emergency descent, diverting towards the nearest airport, Maui.

The noise on the flight deck was so deafening that the pilots had to communicate by hand signals – anathema to all training principles. Controls felt loose, the jet was rolling left and right. Despite the use of speed brakes, at one point the rate of descent reached 4,100 feet per minute. Mimi Tompkins radioed Maui Tower to declare an emergency but, as she tried to outshout thunderous air, exchanges kept deteriorating into slapstick double acts dreamt up by some sick comedian. *('Is that Aloha two forty four on the emergency?' 'Aloha two forty three' 'Ah, two forty six' 'Aloha two four three . . . ')* The crew's simulator training to deal with multiple failures helped them to evaluate conflicting indications, and to select their priorities. They lowered the landing gear, but the green light to indicate that the nosegear was 'down and locked' would not illuminate despite repeated attempts. No time to use the viewer device – they would have to land whether the nosegear was down or not. They asked for all the emergency services to await them beside the runway if they ever got down there in one piece.

* *U.S. News & World Report*, 16.5.1988.

Time alone was on the crew's side: they were only ten minutes from Maui. Reacting to problems, compensating for each symptom of instability, they flew the jet with the 'sunshine top' like test pilots, nursing it along.

The captain manoeuvred around a few clouds to keep the Maui runway in sight. He would have to come in over the Pacific, at the northern end of a valley sandwiched between mountainous terrain. But as he slowed to below 170 knots, the jet became less controllable. He used speed to steady it. But then the No. 1 engine began to cough, and packed up. Neither pilot could recall shutting it down. On the final approach, the aircraft began 'shaking a little, rocking slightly, and felt springy'. An observer watched it through binoculars from below. He saw some unidentifiable debris emitting from the fuselage. His good news was that the nosegear appeared to be down.

There was a perfect touchdown.

The gaping fuselage seemed to wiggle and buckle – but refused to break up. Slides shot out, all passengers evacuated smoothly. It was sheer luck that there was no urgent need of an ambulance: the captain had asked for one, but some bureaucratic application of the rules prevented its despatch until five minutes too late.

The magnificent airmanship of the crew was never in any doubt. (IFALPA would eventually honour them with its *Polaris* award.) It was thought that the work done on the bottom of No. 152 might have helped to protect the integrity of the structure. But general maintenance by Aloha and the condition of that particular aircraft raised numerous grave questions. Accusations and counter-accusations erupted. Boeing claimed to have warned that the airline had only seventeen technicians per aircraft while other carriers had twenty-seven. It had once suggested that Aloha should hire a structural engineer to monitor the safety of its older planes which kept flying in a 'deteriorated condition' under pressure to keep to schedules. The airline rebutted the accusations, but the manufacturer released its confidential correspondence: the roof of No. 152 peeled away 'nearly six months after Boeing first warned Aloha that all ten of its jets were seriously corroded and its oldest four were in need of major maintenance work'.*

* *Washington Post*, 27.6.1988.

The NTSB accident report criticized the airline maintenance programme, the FAA inspections and, to some extent, Boeing's tests which might not have been fully realistic for actual fleet experience. Board member Joseph Nall took the long-term view: 'There's a critical issue here. We've put all our eggs into the inspection and maintenance basket. We rely on the awareness, training and equipment of our mechanics. That's really in the focus of this accident. What kind of maintenance regulations should we have? What follow-up should there be by the manufacturers? Although the strength and crashworthiness of the aircraft proved to be amazing, and although the death of a flight attendant is one too many, this time we dodged the bullet, avoiding loss of lives on a catastrophic scale. But this was not just some solitary maintenance slip-up. If it can happen to an economically viable airline, we must have an industry-wide problem that mustn't depend on luck alone'.*

With the exception of Transport Minister Yamashita, the Japanese had no such luck with Flight 123. (Once, the Minister experienced a hairy landing on the wrong runway at Bombay; another time he made a last-minute cancellation of his seat aboard the flight that ended in Tokyo Bay; and on August 12, 1985, he flew from his Fukuoka constituency and disembarked in Tokyo *before* the Jumbo flew into Mount Ogura.) The pilots' struggle to circle Mount Fuji and land a crippled aircraft lasted for more than thirty-two horrific minutes. Parts of the fin broke away. Having lost the rudder and vertical stabilizer, and soon all four hydraulic pressure systems, too, the jet could be steered by the use of the throttles alone. At some point the Tower was told that the pilot intended to descend, but the radar trace showed the jet was still climbing. Amid fright and general confusion, the well-trained captain kept his head. But his was a losing battle. Trying to make yet another turn towards an air force runway, the jet made a complete circle; the CVR recorded the chilling computer voice of the ground proximity warning: *'Pull up, pull up, pull up . . . '* Nine seconds later, the sharp noises of ground contact. Two seconds later, the crash. The jet hit a ridge, bounced, hit another ridge, bounced again – then fire.

* Soon after this interview, Joseph Nall was killed in a private aircraft accident, while on official business in Venezuela.

The accident happened at dusk. John Purvis, Boeing's Air Safety Investigation manager, was woken up by the telephone at three a.m., Seattle time. 'At least I was at home,' he told me, 'otherwise my wife would have to alert people on the list that is always by her bedside. I ran downstairs to my desk where all my *things to do* are kept. It's best to go by the checklists. I'd contact some people at once but spare others, if possible, from calls at such an ungodly hour. At that point, the 747 was not yet found. No news how many were on board. By the sound of it, there would be no survivors . . . Waiting for more news, I started to pack my essentials – heavy boots, socks, foul weather gear, camera, measuring tape, small handbooks with diagrams of the aircraft.'

Of the 618 747s in service, only fourteen had been lost and, if the first telexes were correct, JAL Flight 123 would be the first crash due to some structural failure. So it was essential to take the right people out there to help. The team was assembled, and they were in Tokyo twenty-two hours after the first call. 'Japan was in a state of shock. People seemed to take personal responsibility for the deaths.'

1,500 police and a thousand volunteers from JAL had swarmed all over the precipitous mountainside. Yamashita as well as high-ranking JAL executives would soon apologize to the nation and resign. That was expected of them: hadn't the Emperor offered to take the blame for World War Two? Hadn't the shoguns of lost battles committed harakiri hoping to spare their soldiers' lives? An airline mechanic, who had nothing to do with 747s, and a ministry inspector, who was merely questioned by the police, would hang themselves. JAL employees wearing uniforms in public would be abused and spat upon. The national trait of hysteria-cum-vengeance manifested itself long before anything was known about the cause of the crash with the highest toll on record.

'I prefer not to go to the site until the bodies are removed – the search for survivors has priority,' said Purvis, 'but this time it was imperative to get there early, and preserve every bit of possible evidence. Two endless chains of Japanese soldiers were passing wrapped bodies down the mountain while we began our search. Eventually, it was our structures man who picked up a piece of wreckage that looked odd to him. Nobody would have noticed it so quickly without his thirty-five years' experience and thorough knowledge of the 400,000 *types* of parts in a 747. It was a spliced plate from the rear bulkhead. There ought to be two rows of rivets here, our man said, so where's the second row? If at all, a spliced plate would normally crack along the row of rivets furthest away from the edge. Some foolish failure to insert that second row of rivets could explain the signs of premature fatigue'.

Yet if that was bad news, another, even bigger blow was in store for the Boeing man. The logs revealed that, in 1978, that Jumbo had made a heavy landing at Osaka, and dragged its tail on the ground. An AOG (Aircraft On Ground) report would have alerted Boeing, and an assessor team would have flown out to survey the damage and bid for the job. JAL assigned Boeing to do the repairs. Somehow, only one row of rivets was applied to fix that plate, a part of the rear bulkhead web, reducing fatigue resistance by some seventy per cent. From then on, disaster was on the cards. At some point, the bulkhead would not be able to endure the cabin pressure in flight and, when it blew out, the vertical fin structure and the controls would be destroyed. But how come that the improper repairs were never noticed? In the seven years to the crash, six 'C-checks', one every 3,000 flying hours, had been conducted, but the visual inspections had found nothing partly because the damage was in an inaccessible section, and partly because nobody would *expect* to find any damage there: that's why no access door was provided by the aircraft designers who assumed that premature fatigue could not and would not occur in any part of a properly maintained bulkhead. Although Boeing still believed that the accident was unrepeatable, access doors were now added belatedly for safety.*

* True to itself, Tokyo is still teeming with suspicions and rumours of cover-up. Whatever the conclusions of the Accident Investigation Commission, the local police, acting as final arbiter, want to bring criminal charges against several people, including ministerial inspectors. There are rampant allegations that Boeing accepted responsibility for the faulty repairs, the causal root of the crash, only to protect the honour and commerical viability of JAL, their greatest customer. Hideo Fujita of the JAL Flight Crew Union sees more holes than

Let us use the 747s as an example. They are among the safest, most popular airliners ever built, but they are 'getting on'. Whenever there is an accident, even if sabotage seems probable, the structure is suspected. Yet a PanAm veteran, the first off the production line twenty years ago, still flies safely . . . after a virtual rebuild for twenty-one million dollars. The Jumbo that died in Japan had completed only 18,000 cycles. It was, therefore, not exactly a geriatric – but faulty repairs had made it age prematurely. That accident is regarded as a freak, but those crucial lines of control could again be severed† and such freak conditions are more likely to arise with any ageing aircraft as the demands on maintenance increase.

In the late 1980s, the Jumbo suffered a spate of mishaps that seemed to be freakish only until they started to occur repeatedly, and were viewed in conjunction.

* texture in the report: he criticizes the CVR and FDR data analysis; alleges that some clues in the wreckage, pointing at rudder trouble preceding the bulkhead failure, were ignored; and asks that if there was *rapid* decompression, as claimed by the Commission, why didn't the pilots don oxygen masks? And, if they didn't, why didn't they suffer excruciating earache and respiratory complications or even loss of consciousness while flying above 20,000 feet altitude for some twenty minutes? The report admits, in fact, that hypoxia might have reduced the pilots' capabilities.

† After the Japanese crash, the bulkhead was strengthened to survive even attempted terrorism, and *fuses* were inserted in the hydraulic system to prevent the total loss of hydraulic fluid if the lines were severed. Yet the British, German, French and Dutch airworthiness certification for the new 747-400 series ran into problems. Though there are already more than 200 orders for the type which will probably be the mainstay of long-distance air transport for decades, the Europeans demand the introduction of the latest safety standards, and claim that the FAA has just rubber-stamped the design, accepting it as only a derivative of the old one. Boeing and the FAA suspect that the criticism is merely a sign of anti-Americanism in support of European products, like the Airbus. Nevertheless, delayed acceptance of the new 747s could play havoc with the plans of the industry, so Boeing has immediately yielded to the pressure and agreed to some modifications. The upper deck floor will be strengthened, and wiring and control runs will be re-routed through separate channels instead of a single duct to increase the chances of survival if any one area were seriously damaged. After crashes like the 747 at Tokyo, a Tristar at New York, and a DC10 at Sioux City, the FAA has belatedly set up a task force in 1989 to improve aircraft survivability in cases of major in-flight structural damage.

1986: a chance discovery almost led to the grounding of the entire fleet. Boeing's Vice President Ben Cosgrove, who heads the Engineering Division, told me: 'We happened to be looking at an aircraft in Wichita, and found three cracked frames in Section 41 up front. I couldn't believe it. I mean it wasn't just skins, but frames! We had to issue an alert to all lines: we want you to check all your Section 41s within twenty-four hours. We got a lot of whining – oh, it's Friday afternoon, can't it wait? – but they did it, and found several of the older 747s with such cracks. Visual inspections can spot the skin cracks, but the really serious problem could be seen only when those skins near the cockpit are removed. In retrospect, we know that we could have waited, it wasn't vital to do the checks overnight because the aircraft was still safe, but we couldn't take the risk. [In fact, some 747s were flying with fatigued ribs, so the structure was held together only by the one-eighth inch thick skin.] The FAA issued new ADs to tighten the inspection schedules, and that was that. It's a pity that every time we issue a warning, we get hysterical demands in the press to ground everything in sight, but that can't be helped.'

1987: a PanAm Jumbo was climbing though 20,000 feet when the pressure blew a cargo door open, damaging the skin around its overhead hinges. Luckily, it became just an incident. Investigations found the latch mechanism faulty, and the FAA ordered all users to modify it – by 1990!

1988: reports of fuel leaks in the main wing tanks of at least five 747-200s led to another immediate alert to check all Jumbos.

1988: a British Airways captain heard a loud *thump* and his aircraft veered violently as it landed at Dubai. The cause turned out to be a flap failure, the second that year, calling for ultrasonic fleet inspections right away.

February, 1989: United Airlines Flight 811 from Honolulu to Auckland, New Zealand, had 336 passengers and eighteen crew on board. Some twenty minutes after take-off, a loud *thump* was heard: once again, a cargo door had burst open, but this time it ripped a massive hole in the fuselage, and three rows of seats with nine passengers were ejected by the escaping pressure 22,000 feet above the Pacific. The aircraft yawed to port, No. 3 engine failed, No. 4 engine was on fire, but the captain managed to jettison fuel before landing back at Honolulu – a tribute to his skill and the overall strength of the structure. Then it emerged that UAL inspected the latch mechanisms from time to time, but was still taking advantage of the three-year leeway granted by the FAA in 1987, and the modification had not been carried out. (The jet was nineteen years old with a relatively low number of cycles logged.)

March, 1989: it became known that Northwest Airlines, one of the American giants, had been flying two dozen 747s on more than 8,000 trips without carrying out a directive to check suspect engine mountings.* Three weeks later a UAL 747 lost a large part of its wing over the Pacific but returned safely to Manila.

April, 1989: another Flight 811 by another UAL Jumbo was forced by engine trouble to return to Auckland – as it landed, its undercarriage collapsed due to metal failure.

A jinx? A series of coincidences? Or examples of the menace of ageing? Should they junk the geriatrics or is it enough to watch more closely those 'golden oldies'? Besides, what constitutes a geriatric jet? The claim that modern aircraft and, particularly, engines live forever is misleading: in responsible hands, by the time '*forever*' comes, so many parts are replaced that nothing but the production number remains the old one. Good design adds to longevity, but also to the geriatric risks that come with it.

The traditional representation of hazards in the life of a model takes the form of a *bath tub* (see drawing): the high level of early

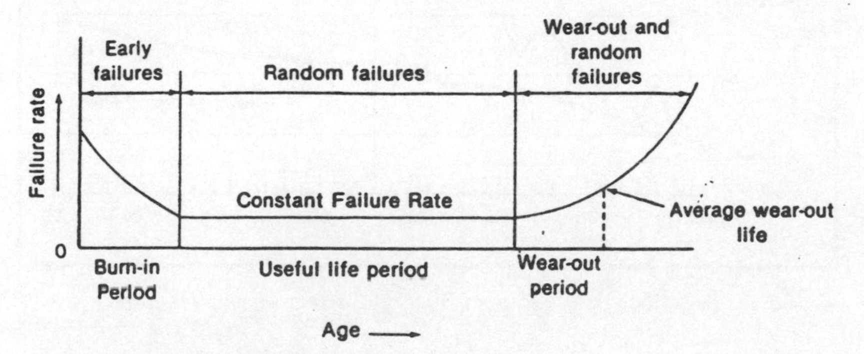

Source: ICAO Accident Prevention Manual

learning failures used to drop sharply, but now most of those potential teething troubles are eliminated by improved tests of a design before a new type ever takes to the air; the long, second phase, when the aircraft is in the prime of its useful life, shows a low level of mostly random failures; finally, as the aircraft approaches its pensionable age at the far end of the tub, the likelihood of mishaps rises more and more steeply – this is the dangerous wear-out period of an increasingly geriatric machine.

Due to the feasibility of longer life, and to the *fin-de-siècle*

* *The Sunday Times*, 5.3.1989.

traffic explosion with voracious appetite for airliners, it seems reasonable to squeeze the safe maximum of service out of all jets, and postpone their retirement as long as possible. Despite the rejuvenating effect of the boom in new aircraft, the world's fleet is losing its youthful vigour. The average jet was five years old in 1971. In 1988 the average was eleven and a half years. Now a third of the 7,424 jets in service are over the not-so-sweet sixteen mark, and almost a thousand are in the twenty to twenty-nine bracket. Consequently, the gap between *average* and *retirement* has widened considerably.

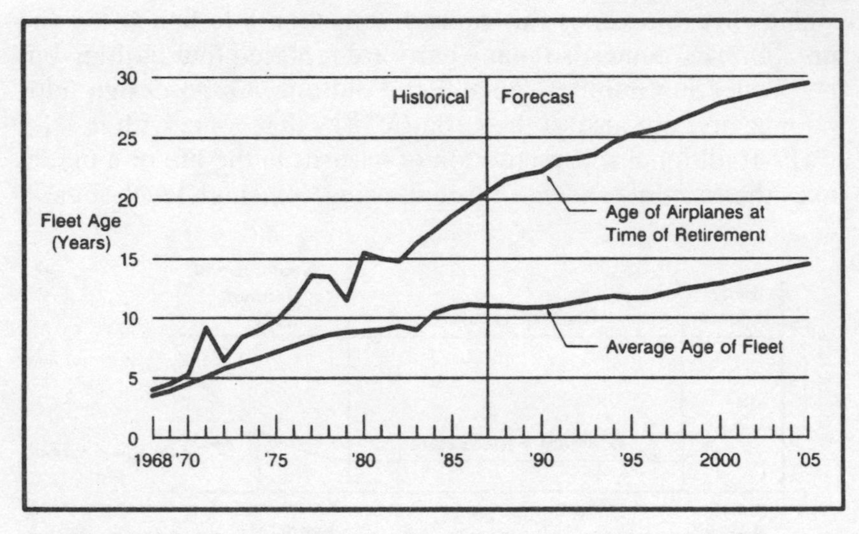

World airline jet fleet age. (Source: Boeing' *Current Market Outlook*.)

While most aircraft are now retired at twenty-two (having flown more than twice the hours and cycles of their original design lives), the widebodies and new technology derivatives are expected to live for up to thirty-two years on average.

Such graphs conceal, of course, huge variations in the fleet medial ages. Singapore Airlines: fifty-two months; Swissair and KLM: nine years; Delta: eight point five years; BA: ten years; Thai International: six years; PanAm, TWA, Northwest: fifteen years. (In 1990, PanAm announced that the average age of its transatlantic fleet was under seven and a half years because half of its Jumbos, some of them twenty years old, had been rebuilt so extensively that they were now effectively *new* ones! Other airlines rushed to ask for FAA and CAA rulings whether such claims were justifiable. At the British Parliamentary inquiry into cabin safety in 1990, CAA chairman Tugendhat classified the claim as a PR exercise.) The Chinese national carrier bought thirty-five old Tridents in the early 1970s – and still flies them. The U.S. commuters' turboprop fleet averages ten years – including active relics like Convair 580s that are at least thirty-five years old.

In October, 1988, a Uganda Airlines 707 tried to land at Rome's Fiumicino airport, and crashed, killing thirty-one of the people on board. The Italian authorities quickly blamed the pilot for a series of blunders. The Italian Civil Pilots' Association was equally quick to point out, however, that the runway was very poorly equipped, and that the 707 was a 'twenty years old museum piece'. It had been nursed by a British maintenance company until a year earlier when the contract was scrapped by the airline to save money. Ever since then, the aircraft had carried its own travelling mechanic to check and patch up all systems before each flight.

The proper assessment, update and complete structural overhaul – as well as stricter noise control – of the world's geriatric aircraft will cost well in excess of one and a half billion dollars. A recent study recommended seventy-four major modifications for the 727, fifty-six for the 737, and thirty-one for the 747. The task force examining the old and battered McDD jets has demanded many similarly expensive modifications. According to the U.S. Air Transport Association, the Boeing fleet alone includes 1,200 jets flying in a truly geriatric condition, and their overhaul would cost $800 million. The equivalent cost to rejuvenate the old McDD fleet will be $563 million. Regulatory authorities are reluctant to enforce such vast expenditures in a hurry, for they would cripple the financially least secure airlines, but they have no choice. In the early 1980s, airframes were blamed for less than five per cent of all crashes, in 1984 they were held responsible for more than twenty per cent. But once again, it may become a question of *cushions:* the best and richest operators will exceed the basic legal requirements – others will try to get by on the minimum or less, exploiting every possible excuse to delay modifications. All of them are troubled by the problem – more and more – of determining the moment when rocketing maintenance bills become uneconomical and so unacceptable, even if a new 'stretched' 747 may cost $130 million, and the latest 767 half as much.

Most airlines have, therefore, begun to phase out and replace older types. American Airlines, for instance, flies 164 seventeen-year-old 727s, but plans to retire, i.e. sell off, most of them and reduce to five years the average age of its fleet by 1993. Others, too, queue up to place multi-billion orders. (In a ninety-day stretch in 1989, Airbus Industrie sold almost sixteen billion dollars worth of aircraft – more than in the whole of 1988. UAL broke all records for a single order when it signed for new Boeings costing fifteen point seven-four billion dollars.) The manufacturers' order books are bursting, and they expect to deliver 8,417 jets by the fifth year of the next millennium. Although McDonnell Douglas had to delay the roll-out of its new MD-11, orders for that long-range trijet have almost reached the 300 mark.

Increasing geriatry is a compliment to fine design for long life. But it has also raised serious questions about airworthiness standards. Manufacturers have stepped up non-destructive testing of old models and, for the first time, the FAA is setting limits to the life of critical structures in aging jets. But, while Europe tests entire hulls for fatigue, the Americans test sections. The European JAA is pressing the FAA to agree on standards: no type certificate should be extended automatically beyond ten years. After that, even if coming fresh off the same old production line, it should be considered a *new* design, and made to comply with the *latest* standards of airworthiness.*

THE OVERTIME-BOMB

The planemakers' successes breed new pitfalls. As production cannot keep up with sales and demand, delivery delays stretch

* The differences in certification standards cause much chagrin to European planemakers. 'A prime example is the higher standard in take-off/stop criteria which the A320' must achieve but the 737-300 need not, said a report in *FLIGHT International* on June 17, 1989. That extra requirement can either reduce the allowable payload in bad conditions or impose the weight penalty of heavier brakes for the A320. Airbus vice-president Bernard Ziegler finds this discrimination so appalling that if harmonization of rules cannot be brought about speedily, he will propose to *advertise* the higher safety standards of the A320. Could that lead to a break-through in attitudes to competing on safety?

longer, and slip-ups occur on the assembly line. The FAA instituted Operation Snapshot for stricter inspections of the manufacturers, who also tightened their quality control.

Just like non-fatal airline *incidents*, even minor complaints against manufacturers must be taken seriously as potential forerunners of accidents. This is particularly true when reputable airlines feel compelled to take the plunge of going public with their dissatisfaction. Three years after the disastrous experience with a job of shoddy repairs, Japan Air Lines President Susumu Yamaji complained to Boeing about lapses in 'production quality' such as incorrect installation of a fire extinguishing system: '. . . These production errors, small though they may seem, should never have been overlooked.' Within twenty-four hours, Chairman Schrontz of Boeing apologized to his 'largest 747 customer', and described what corrective actions had been initiated. Soon after that, BA and others made similar complaints about badly fitted wings as well as cross-wired instruments and fire extinguishers in the cargo hold. Dangerous faults in new jets were found by Australian airlines, too, and the FAA ordered immediate checks.

A report in *The Times* (21.1.1989) quoted a Boeing fitter saying that 'the pay is great . . . But they work us hard and there's people out there like zombies. Twelve-hour days for weeks in a row without weekends.' Boeing answered that overtime had been compulsory only for a while and only on 747s, and that engineer training and controls were now reorganized.

It is frequently overlooked that traffic explosion brings about a world-wide shortage in well-trained aircraft mechanics as well as pilots. It affects manufacturers as much as airlines – and the smaller the airline the greater the problem because the majors skim the manpower market. Some important countries like America choose to differ from the ICAO 'recommended practice' for licensing maintenance engineers, others like France, do not even license such peoplc, who hold lives in their hands. In most countries, including Britain, the training facilities are inadequate, and freelances, who have licences but no company loyalty or specific experience, are employed. British Airways plans to train 250 engineers a year. Most airlines find it hard to attract apprentices who could eventually expect to earn up to £25,000 a year. Airlines which cannot afford to run their own major maintenance base begin to come up against manpower and capacity shortage, and experience difficulties with contracting out jobs. Monarch, for instance, has recently been forced to turn down profitable work for foreign airlines because they could not man all their shifts.

Modern aircraft also add to the growing demand made on engineers who need to specialize. Small companies cannot afford the manpower to follow the example of airlines like JAL, which assigns each aircraft to a team who learn to know their charge intimately, and act as godfathers to it. A member of the team meets the aircraft on its return from each flight, in order to obtain first-hand information of any new problems.

At the sprawling Air Canada base in Montreal, Senior Director Roger Morawski told me: 'We used to expect an avionics mechanic to be competent on all the six types of aircraft we operate. He'd fix a DC9 in the morning, work on a 747, then do a 1011 after lunch. A radio technician was a radio technician, he could fix any radio on any aircraft. Not any more. We give mechanics type ratings – three types maximum. With specialization we also try to protect them from long-term fatigue. There's a limit to what the mind and memory can take.'

Metal and human fatigue in the air have become recognized as a menace. But concern about the human variety is all too often limited to-pilots, despite many ground engineers' conviction that a monkey can be taught to fly a spaceship but it takes a mechanic to fix it. Fatigue in engineering remains one of the hidden hazards. *Overtime-bombs* are planted every night; they are allowed to tick away. Rumour has it that at a prosperous third world maintenance-for-hire base, the staff are paid 'fabulously' – for fifteen-hour shifts day in, day out.

Small outfits are specially vulnerable. Roger Morawski used to run a local operation in the Arctic: 'I saw people do pretty dumb things up there. A mechanic once misrigged an aileron by crossing cables on a DC3. About the most lethal thing he could do. Luckily, the pilot's preflight checks revealed the mistake, but how could it happen in the first place? It turned out that the guy had worked at incredibly cold Arctic temperatures for almost twenty-four hours without sleep. Bad in itself, but there was more to it. That mechanic had been the last to see the DC3 before it was offered to the pilot. There was no inspection, no supervision of the work done! Which is a serious management problem – and not unique even today.'

When, in February, 1983, a de Havilland DHC-6 of Sierra Pacific Airlines crash-landed, injuring seriously all aboard, the cause was found to be that the wrong elevator linkage bolt had been installed – and it was not even secured properly. Yes, it was the mechanic's fault. But the NTSB discovered that the rot went further: 'the carrier's maintenance director had supervised both the maintenance work and the inspection, mingling the two functions' so that there was no one to spot mistakes. Trying to do two men's job can lead to fatigue-induced negligence, just as long hours can.

Roy Lomas, head of the BA Air Safety Branch, had an experience that taught him a lesson for life: 'When working as a young engineer, I was sometimes on duty all night. One evening our airport closed down because of bad weather. Two aircraft had to stay with us overnight. One had already been refuelled, but could not take-off because of the weather, the other wasn't refuelled for we knew it wasn't going anywhere until the next day. At that time, all the documents were taken from the cockpit to the engineers' office. In the morning, everybody was in a hurry to go. I was still on duty – twelve hours after starting my shift. I picked up a log in the office, went up one aircraft, read the fuel gauge – 450 gallons – and checked the log: it didn't show any refuelling! So I rechecked the gauge and tank: yes, 450 it was. It seemed obvious that the engineer who refuelled the aircraft forgot to endorse the log which I now corrected. The captain checked the gauge against the log – and he was off. Meanwhile, my colleague, also on long hours, checked the other aircraft. He saw his log showing 'refuelled with 450 gallons' and a captain's signature confirming it. That was good enough for the tired man as well as for the pilot who now accepted the aircraft in a hurry. He took off – with no more than the "arrival fuel" left over from the previous flight, which luckily, was adequate for his next sector.

'What happened was that the logs had been mixed up, and the mistake went unnoticed because everybody was tired and under pressure. Seeing a log in which *a* captain had already signed for 450 gallons, the pilot of my colleague's aircraft accepted it at face value. He might have glanced at the fuel gauge but being convinced that the aircraft had been refuelled, the facts might not have registered with him. So he failed to really check the gauge – until he was airborne.

'In those days I saw engineers sleeping in the hangar. They were overworked, yet that was perfectly legal. Now, of course, we have stringent controls, but nothing, not even decent pay, can stop people moonlighting to earn extra on, for instance, repairing cars at night or during the weekends. We're very careful with pilot selection, but in engineering nobody looks at the man's personality and background, things that can affect the safety of the whole industry. Perhaps it's time to pay attention and invest more heavily in engineering staff selection and training. We need people with integrity, and we must back their decisions. At BA, for instance, our engineers won't be crucified if they pull a marginally unfit aircraft out of service.' (Unfortunately, some small operators cannot afford to be so trusting: if there's no other aircraft to roll out and take the flight, a tired engineer may sign off a dodgy aircraft just to get rid of it – and avoid any hassle with operations.)

World standards, both in the severity of laws and their enforcement vary a great deal. ICAO recommends duty limits for pilots, but not for engineers. In Germany, nobody may work more than ten hours a day – twelve with at least a two-hour break – and Lufthansa teaches its inspectors to ask colleagues to double-check their work if they feel tired. Pilots flying to Spain and Portugal complain about mechanics 'who talk and work like sleepwalkers out there. Only their red eyes indicate how long it's been since they saw a bed.'

CHIRP receives numerous confidential reports about pilots falling asleep at the controls, but not about mechanics who are so tired that they misrig elevators that will make an aircraft descend when the pilot tries to climb. It is not easy, however, to legislate for fatigue. 'Sadly, fatigue has . . . no satisfactory definition,' wrote Roger Green in *The Log* (October, 1987). Regulations limit duty hours because there is no known way of measuring the many intangibles that induce fatigue through factors like lack of sleep, dehydration, emotional problems, stress or boredom which may well wear out engineers. American and Canadian accident investigators are most unhappy about the elusiveness of 'shop-floor fatigue': they claim with justification that it is not good enough that unions should be the ultimate safeguard against excessive overtime, and even more unsatisfactory that brave union-busting mavericks of the aviation industry can hire people with doubtful qualifications who will accept lower wages but can supplement their income by long hours and moonlighting.

Fatigue can also threaten air safety from such unexpected quarters as the duty time and morale of *airside drivers*. 'Many of them are worked very hard. Perhaps they should be kept to stricter duty hours and, like pilots, they could be type-rated in line with vehicle size,' said aviation insurance risk management specialist Peter Spooner. 'They work under great pressure, their vehicles sometimes swarm over the apron to provide services, and sometimes aircraft are damaged. The drivers may not realize the economic implications of the physical damage that can result in vast hidden or consequential costs such as meals and hotel bills, staff rescheduling man-hours, crew re-rostering, leasing of parts for repair . . . all have to be added up. In Japan, airside drivers at major airports have to take an annual test, and some other countries have now followed their example.'

Concern about spreading drug habits crept into numerous interviews, particularly in America, where some people saw narcotics as the shop-floor answer to fatigue and boredom. 'Add to this the means, I mean the affluence of the industry, and we have a problem,' said an NTSB investigator. 'We know the problem is there, for there's no reason to suppose that airplane mechanics are any different from the rest of the population, but what can you do? If you demand that regular health checks and kinda snooping into engineers' private lives should be conducted, the guardians of privacy will be up in arms. Yet something will have to be done about it. If someone is on the bottle but can hold his liquor, at least you can smell the alcohol. But a junkie can go on for a long time, and cause a lot of damage in a sensitive job like airplane maintenance or the nuclear industry, before he's caught out.'

THE WRONG END OF THE TUB

The aircraft retired by major airlines is not pensioned off for good: it remains big business in the hands of 'Second-hand Rose and Associates'.

The 'average' retirement age of twenty-two years conceals the fact that many aircraft remain active for thirty birthdays, may soon clock up a hundred thousand hours in the air, and will vastly exceed the lifespan for which they had been designed and tested. They are sold and resold repeatedly until they end up in the hands of poverty-stricken, inexperienced or outright unscrupulous operators in countries where the authorities cannot afford to be too fussy about certification. There are widespread rumours about some clapped out, neglected Caravelles, Tridents and other old jets being sold 'for scrap' – only to carry on flying without much ado. A most creditable source said: 'Everybody knows that in Africa and South America, many such aircraft are in service. The sellers knew what they were bought for, but it was good enough for them that the buyers were willing to sign that the aircraft were purchased for scrap.'

The *U.S. News & World Report* (16.5.1988) quoted 'airline watchdog groups' saying that 'frequent leases and ownership changes often cloud an older jet's maintenance history'. It mentioned the story of a DC9-14 which began service with Delta in 1966, was sold to Southern Airways in 1973, leased to the Latin American AVENSA in 1975, then returned to the United States to be flown by Republic Airlines. In 1988 it was with Northwest Airlines, but it was still in the Republic colours and the silhouette of the previous owner's logo could still be seen on the tail. This is, however, far from being the most 'tangled tale' of aircraft – for, at least on its return to U.S. service, the different standards of maintenance would have come to light, and one hopes that essential repairs would have been carried out according to FAA requirements.

There is, in fact, an unwritten *dodgy aircraft register* in quiet world-wide circulation, and people 'in the know' are familiar with it. Peter Onions, Marketing Manager of FFV Aerotech Manchester Ltd., the largest independent maintenance company in Europe, said: 'When, for instance, a visitor to some small country sees an old plane take-off on just two of its three engines yet carry on regardless towards its destination, he'll spread the word about the aircraft and its operator. You can't do more, because the country of registration may not be too hot on the type – perhaps the only one based there – or just can't be bothered to demand better maintenance because the operator could not afford it anyway. When resales are agreed, they are subject to survey and the checking of records, but, depending on who the seller is, there is no guarantee that all the recurring faults of the aircraft have actually been logged.

'When somebody buys a jet from a bankrupt company in, say, South America, he may be duped into believing that the local certificate of airworthiness is worth the paper it's written on. He brings it to us saying, "It's only a bit out of hours" – i.e. maintenance is long overdue – and wants us to do the barest essentials on it. But our quality manager, with his hard-won licences riding on it, will specify all that must be done before he will release the aircraft.' (Some specialists may also tip off the CAA who will run a computer check, dig up some unpaid landing fees, and ground the aircraft until debts are paid and all the repairs are carried out.)

Sometimes even such precautions fail. The owner may say: 'Just do the essentials, we'll do the rest ourselves.' He will then have to sign papers to that effect, but everybody knows what will happen: the maintenance of a 707, for instance, may need a work force of some seventy people, plus costly stores of spares – surely uneconomical for one or two aircraft – so the owner will go to smaller, cheaper outfits which charge less and offer no embarrassing advice. In third world countries, the job may be done at half price. (While specialist labour may cost up to U.S. sixty dollars an hour in Sweden or France, and forty-five dollars in Britain or America, a mechanic with allegedly the same skills would be paid twenty dollars or so in the Far East or Africa.) Several South American companies will do a cheap job 'according to printed schedules' but no more: no overview, no evaluation, no advice. Even some dodgy American firms worked like that. They have now been closed down, but they were in business long enough to earn Miami the name '*corrosion corner*', and it is suspected that most of them have already set up shop somewhere else. They thrive on offering something scarce – maintenance for those who wish to afford only the absolute legal minimum or a semblance of it as long as papers are signed and stamped by someone. Anyone.

Instead of enjoying extra vigilance, checking and double-checking, the increasingly geriatric aircraft may cross several frontiers as it passes from owner to owner, losing some of its identity. Its *curriculum vitae* may change faster than the history of the Soviet Communist Party, and blemishes of its past will be omitted just as conveniently. It is often the investigation of an accident (incident if the owner is lucky) that sheds the first light on some shady affairs.

The Vanguard accident at Basel in 1973 was a classic example of calling attention to some 'side issues'. It was snowing heavily, visibility was poor, and the approach beacons did not comply with ICAO standards when the crash occurred after two unsuccessful approaches. The Swiss authorities blamed, in effect, the dead pilot for 108 deaths and thirty-six injuries, naming 'loss of orientation . . . under instrument flight conditions' as the main probable cause. But some contributory causes were more revealing: maintenance, instrument and procedural flying techniques were substandard; some licensing requirements were questionable; the approach navigation was inadequate – perhaps because deficient navigational aids had sown confusion.

'We found serious faults in the ADF (Automatic Direction Finding) and ILS equipment,' said John Owen who led the British team helping the Swiss on behalf of the AIB, the operator and the aviation industry. 'Some of the soldering must have been done by amateurs, and a resistor was glued with Araldite! The aircraft technical log gave no indication where the botch-up jobs had been carried out. Some faults had been reported by various crews, yet the aircraft was recertified repeatedly without anybody bothering to find out what was wrong.'

Invicta, a British airline, owned the aircraft. Most of the maintenance was done for them by ATEL, a firm licensed by the CAA. Some repairs were done by Invicta's own unlicensed workshop under ATEL's supervision. Britannia Airways did Invicta's radio rectifications by contract. But, after the crash, a letter from Invicta argued that some faulty work might have been done by the previous Canadian owner of the aircraft. Whoever was responsible, somewhere down the line a trap was laid for the pilot – and nobody spotted it.

Whenever aircraft change hands, it is vital to examine the maintenance logs, particularly if certain third world countries were among the previous owners, because maintenance requirements vary widely, and record-keeping can be patchy. Many national authorities, the manufacturers and large leasing companies are increasingly concerned about second-hand deals in aircraft. They are trying to standardize maintenance records, but universally enforceable standards for technical log entries may never come. (JAA, the Joint Aviation Authorities of West European member states, is on the verge of unifying maintenance requirements, and its 'ultimate goal is an even closer unity with a single autonomous body to manage safety regulations for all European countries',* but to achieve that, each country would have to give up some of its precious sovereignty in this field. Certain authorities, specialists and pilots' organizations view the prospect with some apprehension because they are worried about the possibility that in the process the lowest common denominator will prevail.)

Caravelles suffered probably the highest loss rate among long-range aircraft, but that said nothing about the design – many later models are bound to suffer the same fate – because as Captain Caesar of Lufthansa pointed out: 'such aircraft are often traded by brokers like old cars, changing hands six, seven times, owned occasionally by an airline for no more than a few months, going to *bush* fliers with inexperienced crews and mechanics, then returning to civilization, carrying the ominous marks inflicted by age, poor technical back-up and hostile environment that ought to make accidents predictable.'

If more reminders were needed, they arrived on the wings of a Jumbo in August, 1987. The aircraft, operated by Flying Tigers, was on final approach, descending to land at Hongkong. The pilots had just selected thirty degree flaps when a loud bang was heard. The aircraft began to yaw and roll to the left. Flaps were reduced to twenty degrees, and so at least partial control was regained, but by then over-populated skyscrapers were looming on the crew's horizon. The choice was between ditching into the bay and trying for the runway with no room for error because, if they missed it and hit the buildings, there would be a massacre. They went for the runway, and landed the partially uncontrollable aircraft safely. The left inboard trailing flap was then found to be not only broken, but actually hanging by its safety straps. The investigation quickly found that both flap spindles were broken. The fractures were identified as the products of stress corrosion, something that does not just happen overnight.

* *Aviation Safety Authorities in Europe*, paper by Gerald Le Houx, Director of JAA Secretariat.

The life story of that 747 left nothing to guesswork: potential disaster had long been on the cards. In the 1970s, the aircraft was owned by Seaboard which contracted out its maintenance to PanAm. In 1978, the airline took charge of its own maintenance, using an American Airlines programme. In 1980, Seaboard sold the 747 and several other aircraft to Flying Tigers. 'They received all prior maintenance records – in a box,' said Jim Burnett of the NTSB in a speech to the ATA Engineering and Maintenance Forum. 'Can you imagine – and maybe you can – vital records being turned over just like another piece of office furniture?' The new owners then put the 747 into service 'without determining if critical parts had been inspected. The plane at that time had 20,000 hours on it. Flying Tigers normally does its (major) D-checks on its 747s every 28,000 hours. So what did they do when they received this aircraft? They scheduled its first D-check 28,000 hours later.' In other words, they treated it as a new plane – with zero hours.

Further investigation led to more startling discoveries. Boeing recommended visual checks but it was impossible 'to visually detect the fracture location of these two spindles'. The NTSB then noted 'an interesting fact': all but one of the major U.S. airlines operating 747s had a special inspection programme because they had already found spindle corrosion that could not be seen because it was hidden by a sleeve. The one exception? PanAm. PanAm adhered to Boeing's basic recommendation. 'When airplanes change hands, it is not enough to determine if they have complied with all the ADs and service bulletins. You must focus on their maintenance program and compare it with your own.' This was not done when Seaboard took over its maintenance from PanAm or when Flying Tigers bought the Jumbo. But not only that: why did all the other majors fail to alert everybody else that they had found a hidden hazard?

Similarly it is often overlooked that operators somewhere down the resale chain seek out the cheapest source. They buy from all sorts of airlines in any part of the world. Their aircraft, thus, differ from each other in many ways. It is too expensive for them to introduce modifications to achieve standardization for the last years of elderly aircraft. So they end up with a hotchpotch fleet. Every time a crew take charge of an aircraft, they have to get acquainted with the cockpit layout. The process deprives them of knowing instinctively what is what and where to find it; so eventually they may lose vital seconds when they have the least to spare – in an emergency.

'It is extremely tempting because it can be financially rewarding to keep old aircraft flying,' Ferdinand Füller, a Lufthansa Deputy Manager, told me. 'If spares are no problem, and it's the right plane for the route where you can achieve eighty or even ninety per cent seat utilization, it's well worth it because the plane is all paid for, it has already earned its keep, and now most of its earning is pure profit on your original investment. Most, not all, I must emphasize, because to keep an old aircraft in top condition is costly and requires a lot of expertise. Some of our 727s and DC10s are more than fifteen years old, but give us good service.

'And here is probably the greatest anomaly: it is usually the small, the get-rich-quick, the poverty-stricken or the vanity airline that claims it can afford only second-hand aircraft. Sure, it's cheaper to *buy*. But can they afford to *maintain*? If I were the airworthiness authority of the world, I'd allow only the big, prosperous airlines to operate so-called geriatrics.'

So what is it that top-class airlines, as opposed to others, can do?

'When it comes to the evaluation of older aircraft, age alone will tell you very little,' said Ben Cosgrove of Boeing. 'You just can't take a simplistic view. Cycles could also be meaningless: on average, a 737 would do four times as many landings as a 747 in the same period. Besides, the fuselage is ground-air cycle critical because of pressurization, the wing is partially cycle critical, the fin on the other hand must withstand special flight loads in the air.

'Fatigue? We give our test aircraft a hell of a bashing, but that alone won't help either, because you can't *test* for corrosion or for damages inflicted when a truck runs into the fuselage, when the airplane is battered by hailstones and hard landings repeatedly, yet all of these may speed up the ageing process. So all you can do is to watch that machine and stop the rot wherever it occurs, and before it makes your airplane truly geriatric. When operators boast to me: "Look how wonderfully our aircraft are doing having completed X thousand flights," I worry about those people, for I'm not really sure that they know what they're talking about. We have the experience of forty gillion* flights world-wide in all shapes and sizes, yet even we could not claim to know everything about our products.'

Leading airlines have, indeed, the financial means and sophisticated techniques – if not always the will and care – to monitor the geriatric progression by spotting cracks and corrosion with penetrative dyes, ultrasonic and X-ray examinations and, lately, thermal imaging which was pioneered by British Airways.† Probably the greatest step in safety, cost-cutting and efficiency has been the introduction of human and machine performance monitoring that has opened the way to 'on condition maintenance', i.e. doing the job when necessary rather than just according to time-based schedules. With the aid of dozens of extra flight recorder parameters, those operators monitor whole systems, including engines. Old Tristars, for instance, bought by the RAF from British Airways, now serve as tankers monitored by a 250-parameter recorder and ground-based analysis programmes.

'Some twenty years ago, TWA was among the first to start monitoring, but spent vast sums unwisely, and ended up with paper mountains of recorded data – but no support service to analyse it,' said Eddie White of The Flight Data Company at Heathrow. 'That misadventure was one of the reasons for two decades of American reluctance to embrace the technique. On the other hand, leading European airlines – and more lately, Singapore, Japan and Australia – were keen to take advantage of monitoring.'

Monitors are one form of listening to what the hardware is saying, but the various techniques determine who will be able to hear the message and at what point. In February, 1986, a walkaround inspection of an Eastern 727 led to the discovery of a hole in the No. 1 engine. A combustion chamber was then found cracked and severely burnt. The other combustion chambers were also cracked and badly damaged. The 727 was operated under

* U.S. billion, equivalent of 1000 millions, 10^9 in Britain.

† The technique had long been available, but BA applied it to aircraft as a means of detecting moisture in composite materials. Media coverage of this innovation was a good case to show why airlines can be jittery about *any* publicity concerning safety. When a TV report hailed 'this break-through in non-destructive testing', it also tried to question why BA was doing it at all: what was wrong with their fleet? They must have a serious problem!

an engine monitoring programme. It either failed to detect the problem – or its message was never listened to. The combustion chambers of that engine should have been removed long before the incident.*

Some airlines use QARs (Quick Access Recorders) which are removed after each flight, and will alert the ground-based data analyst to 'special events' – incidents that might have gone unnoticed. Others have airborne computers to store and analyse data, with print-outs available to the crew during the flight – the monitor can thus do some of the now redundant flight engineer's job.

Beyond the increased safety of detecting trends, some savings can be spectacular. 'The traditional *hard life* of an engine on a commercial aircraft is about 2,000 to 3,000 hours,' said Eddie White. 'Then it must come off, be stripped, checked, and so on. *On condition* maintenance means that within certain *fatigue life* limits, the monitored engine can stay on as long as it's performing well and has no incipient faults.

'Here's a simple example. The RB211 engine's so-called Variable Inlet Guide Vane controller is a complex component that must be adjusted regularly to avoid the engine spool speed exceeding prescribed limits or the engine surging. If it goes outside the limit, the pilot may have to abandon take-off, and that alone is very costly. Therefore every 150 hours or so the aircraft must be taken out of service to run the engine, take readings, see if limits are exceeded, open the cowlings, adjust the ViGV controller, close the cowling, run the engine, check the spool speed – and possibly, repeat the job two or three times to get it right. In all it may take two or three hours of a fitter's, engineer's, supervisor's, fireman's and equipment time that could easily add up to a couple of thousand pounds per adjustment, not to mention lost revenue. With the monitor, the information is on tap, a single adjustment is done when necessary. The record is so good that BA, Gulf Air, Air Portugal and the RAF follow this routine which is safer and saves thousands every time.'

Some crew unions object to monitoring, which they see as yet another *spy in the sky*. Nobody but pilots work under such continuous, relentless scrutiny, with every word they utter, every

* J. Burnett, speech referred to earlier.

duty they perform being recorded. But monitoring can save lives through early warning of bad habits and any tendency to deviate from approved procedures. 'Take hard landings,' said Colin Murfet, development engineer in BA's Flight Data Recording group. 'A pilot's perception is subjective, and the feel of a landing can be strongly affected by pitching movements. Analysis may sometimes show that an apparently not very hard *[i.e. non-reportable]* landing was firm enough to require precautionary checks on the undercarriage and related structure as well as sensitive parts like wing root mountings.* Or it can be the other way round: a reported "very firm" landing has caused an aircraft to be removed from service while the QAR was checked – only to show that it was well within acceptable limits'. In one case, a pilot got worried about a landing at a particular airport, but when the monitoring unit compared his problem with other records, it transpired that the runway was known to have a slight hump, and so 'an otherwise excellent landing at that precise point could produce an acceleration level high enough to disconcert the pilot who didn't know of the effect'. (All pilots were then warned about the peculiarity of that runway.)

With the advent of ACARS (Aircraft Communication Addressing and Reporting System), real-time monitoring becomes possible, and airlines like BA invest millions in it. Hundreds of sensors will monitor the systems, and transmit directly to a global network of ground control stations not only to spot and solve problems in the air, but also to predict potential troubles long before any serious fault could develop *during* the flight. Pilots can now radio symptoms, like engine overheating or vibration in flight, but the engineer on the ground can only guess the cause without real data. With ACARS, a genuine diagnosis will often be possible. Further damage could then be prevented, the cure of some malfunctions could be suggested to the pilot, and, if necessary, maintenance could be alerted to reserve parts and capacity for certain 'on condition' repairs to be carried out as soon as the aircraft has landed.

Air Canada has such data-link to their 767s on EROP routes.

* Lockheed plans to introduce airborne computers to monitor structural stress, and alert users whenever design tolerances are exceeded. It could help to prevent structural failures, diagnose signs of premature ageing, and so extend the useful life of the aircraft.

'That's a hell of an extra reassurance to us pilots,' a Canadian ALPA representative told me. 'But what worried me is that a lot of small and third world vanity airlines want to join the club – without the resources or the will to invest in sophisticated hardware.'

The same concern was echoed by people talking about the operation of old aircraft. 'Monitoring can help to slow down or at least warn about premature ageing,' said Eddie White. 'Overly cost-conscious airlines would draw a huge benefit from it, but the irony is that it's the most safety-minded airlines that invest heavily in the techique, while those who'd gain most from it show the least interest. It's easy to set up a new airline with old aircraft almost overnight. How long they will last – I wouldn't like to predict. The regulatory authorities ought to be more concerned about the safety of those shoe-string outfits.'

A few years ago, during the big recession in the industry, the Arizona desert was becoming a massive aircraft cemetery where the dry air helped to embalm the machines that waited for resurrection. A Jumbo could be bought, *as seen* or *as is*, for peanuts. These days, aircraft are still parked there, but the buyers are eager, and an old 747 may fetch thirty million dollars – almost a quarter of the price tag on the latest model. If the aircraft has a good pedigree, i.e. it comes from a reputable stable, the price will reflect that. A Lufthansa reject was sold for thirty-five per cent more than the average market price. An old TU-134 from Aeroflot may be cheap, but it uses more fuel, for aerodynamically it is not as clean as a 737, and it needs an engine overhaul every 1,500 hours – more than four times as frequently as the 737. To compensate for that, new owners will have to find cheap maintenance. It will probably be bought by a third world country. The Ilyushin 62 which resembles the old VC10 flew first in 1962. Its history is full of technical problems, and its accident record is bad even though only a few cases are known to ICAO. (A Russian joke suggests *Fly by aeroplane* to be its slogan.) When one engine fails, other engines and the tail also tend to get damaged. Yet 228 IL62s keep flying, form the backbone of the Soviet long-haul fleet, and are sold – exclusively to Cuba, China and East Europe. (Aeroflot has begun to re-equip A310s – maintained by Lufthansa.)

FLIGHT International (27.5.89) quoted Kenneth Taylor of International Lease Finance Corporation saying that structural reports on older jet airliners can be 'absolutely frightening'. He warned that 'we are heading for thirty-year and 100,000-hour-old aircraft', and it must soon be defined what standards the public should accept before passengers start asking at the ticket counter: 'How old is the aircraft?'

The purchase of relatively dilapidated geriatric aircraft can still be attractive because rejuvenation – unlike that of humans – can really work. But if safety is not to be compromised, the cost is so high that it will be recouped only if the owner knows his business inside out, and can forecast precisely the profits he can expect. And the undoubtedly ever-growing demand for seats alone is no guarantee for that.

'Recently we've rejuvenated a seventeen-year-old DC8 for a German owner,' said Peter Onions of FFV Aerotech. 'We performed a D-check, used a vast amount of material, skin and floor panel replacements, you name it, apart from 40,000 man-hours – almost as long as it takes to *build* a 737. But they now have a virtually new machine to fly safely for many years.

'At the age of twenty, an early 707 was laid up as a write-off in the late 1970s never to fly again. It was owned by a bank, and totally neglected for years. But when the 1980s renewed demands for 707s to work as freighters, more than two million pounds were invested in it. It needed *hush-kitting* by TRACOR, an American specialist, to comply with current noise regulations.* It had to have several major parts replaced, many areas had to be changed or beefed up, it needed a new crown skin [top of the fuselage], wing surface, wing root support and scores of other structures, but a virtual double D-check and 80,000 man-hours later it could be sold with profit. The question now is: will it be properly cared for under new truly stringent regulations? If not, it could be a menace wherever it flies to.'

Many Jumbos are twenty years old, and corrosion takes its toll even if they sit mostly in a hangar in the Gulf area where the royal families who own them may fly them no more than 500 hours a year. But one of the gravest dangers is that whenever an aircraft is pressurized for flight, it gets inflated and then deflated like a

* Latin American and African countries, supported by East Europeans and the U.S.S.R., object to the phasing out of the noisy jets because the bulk of their fleets belong to the noisy generations of aircraft, including BAC 1-11s, early McDDs and Boeings as well as Russian types.

balloon. The higher it flies the greater the pressure difference. A long-haul jet, like the 747, will experience permanent stretching over the years. Though this growth is invisibly small, the stress is considerable, so it is essential to replace not only the cracked and corroded skin, but also some crucial parts of the structure in the life extension programme for 747s. Nose Section 41 is a good example. After 8,000 cycles, small areas need inspection. As, through various checkpoints, the aircraft reaches 18,000 cycles, every inch but the windows and doors in Section 41 must come under scrutiny. This alone takes 3,612 man-hours, keeping the Jumbo out of service for twenty-four days – excluding any repairs.

Should small and poor airlines, the keenest customers for old aircraft, be allowed to fly geriatrics at all?* Even if they are reasonably regulated, try to do the minimum, restrict overtime and don't skimp on costs, where will they find good enough engineers and supervisors to defeat *Murphy?*

Murphy's Law is that whatever can go wrong, *will* go wrong, whatever can be assembled the wrong way, will be. That over-worked mechanic misrigging ailerons in Arctic conditions was not unique. Murphy may strike in even the best conditions. When Alexander Onassis, son of the shipping magnate, died flying his father's Piaggio, the investigators examined the wreckage, and found that the ailerons had been reversed. The overhaul had been faulty, the supervisors had not spotted it, and magnate junior had failed to do his preflight checks properly. When he took off, the crosswind caused some drifting to the right which needed correction by the pilot. As he tried to turn slightly to the left to keep straight on the runway, the reversed ailerons actually turned him even further to the right. The more control he exercised, the worse the situation became. Time ran out. At slow speed, still close to the ground, he never had a chance. The Greek tragedy ran its full circle: the overhaul was done by Olympic Airways, his father's own airline.

When investigators, working endless hours on a seemingly

* An old Dakota, chartered from the South African United Air, burst into flames in the air, and crashed near Bloemfontein in April 1988, killing all aboard, twenty-one jockeys and the crew. Though the South African investigators found 'overtones of slipshod practices', they saw no evidence of improper maintenance, and responsbility for the accident could not be ascertained. The victims' families cannot sue the airline, for it is in liquidation.

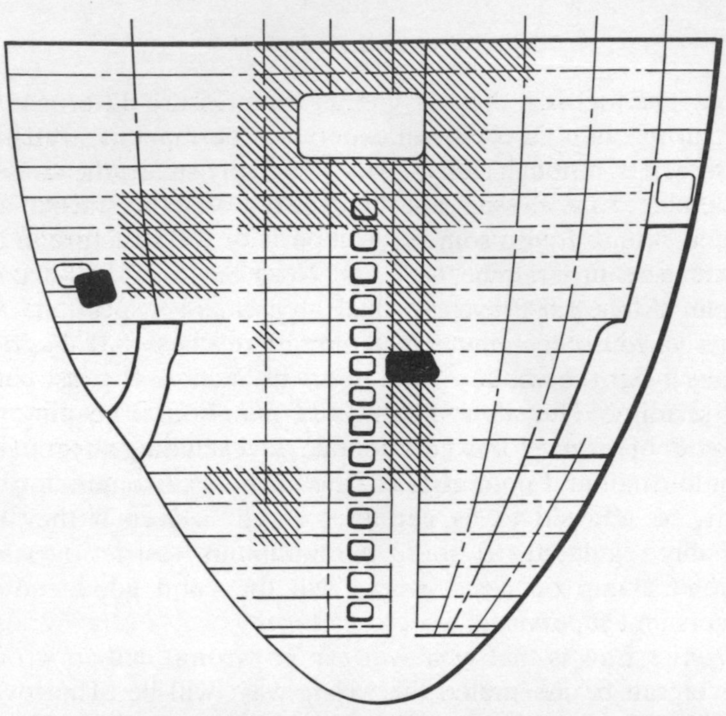

Nose section 41 of a Jumbo: inspection areas at the 8,000 flight threshold (above) and at 18,000 (below).

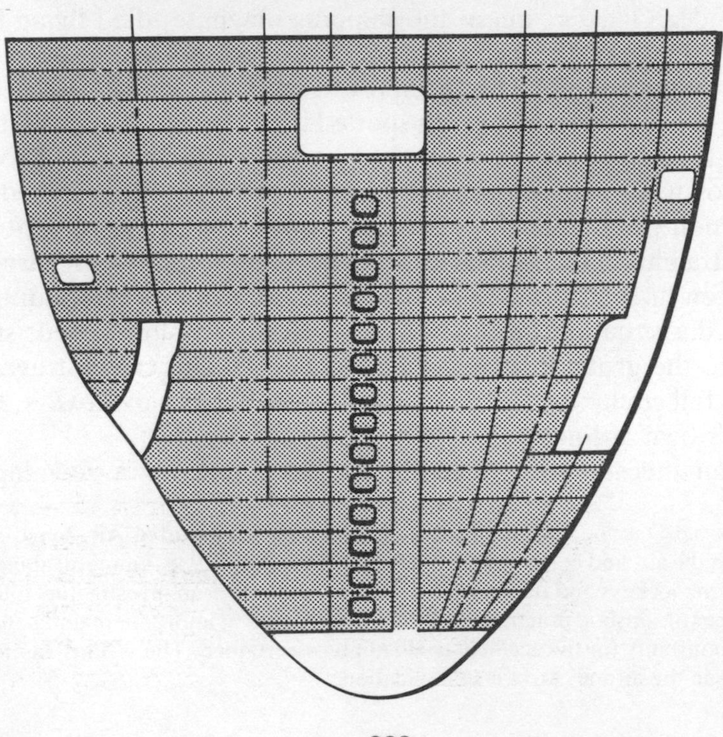

hopeless case, gather for a late-night drink, they often recall a horror story about a four-engine military transport aircraft. It encountered a 'minor' problem soon after take-off: fuel was seen by a passenger streaming from the vicinity of No. 1 engine. The engine was shut down to avoid fire, but the gauge showed that No. 2 of the four portside tanks had lost 3,000 pounds of fuel in ten minutes. The pilot chose to return to base. His second portside engine packed up. Speed was dropping, but the captain fought to maintain height because there was a village along his path. Powered only by the starboard engines, the aircraft went into an uncontrollable turn to port, hit an electrical cable, swung round, decapitated some trees, and crashed into the village. Several houses were demolished. Three people on board and two in the village were seriously injured, sixteen aboard and two on the ground were killed.

The wreckage was spread out over a large area. The AIB investigator in charge listened to a survivor's account of the fuel leak, then read the maintenance log, noting that some minor service had been carried out on the fuel system to stop a leak from the portside collector tank. Could it be that what the passenger saw was a recurrence of the problem? Was it a leak that led to the loss of fuel and the death of No. 2 engine? As the investigator knew nothing about the aircraft, he studied the fuel system during the night, and learned that each wing had four tanks which were connected to a collector tank which in turn fed the two engines on that wing. Governed by various cocks, the two collector tanks were connected so that engines on both sides could be cross-fed. In the morning, walking around the wreckage, he picked up a small tank fitted with non-return valves and bits of broken pipes still attached to it. He labelled it, marked the spot of his find on the map, and put it into his car. He told his search party that special attention should be paid to any piece of wreckage that was part of the fuel system.

By late that evening three fuel cocks had been found. Their position indicated the configuration that would permit the flow of fuel to No. 2 engine. So it could not have been fuel starvation that killed it. Or could it? Could the collector tank stem the flow? And where was the leak? Near engine No. 1 – close to the tank overflow vents. Why would the tank overflow? The investigator's second lonely night exercise then conjured up a mental note he had taken: there was a little arrow on each non-return valve fitted to the collector tank showing the right direction of fuel flow . . . wasn't one of those arrows pointing the wrong way?

At dawn he examined the tank he had stowed in his car. The valve was indeed reversed. The implications were shocking. No. 2 tank must have been feeding the collector tank which, in turn, fed two engines *and* tank No. 1. No wonder it was emptied so fast, while No. 1 was overfilling until the vents duly began to discharge the surplus. So it would be venting, not a leak, that the passenger saw. When No. 2 ran dry, the cock was shut, and tank No. 1 would supposedly start to feed the remaining engine on the portside. But from that tank not a drop would come through: the non-return valve did its job – *the wrong way round*. It soon emerged that the threaded portions at both ends of that valve were identical. Though the arrow was there to warn the engineer, it was *possible* to fit the valve the wrong way round. The supervisor also must have overlooked the position of the arrow. Then there was a test flight. It was too short to cause problems but, when it ended, the log showed there was more fuel in No. 1 than before take-off. That bit of silly discrepancy must have been ignored or attributed to a mistake in the figures. From then on the crash was inevitable.*

Manufacturers learned their lesson: the two ends of that non-return valve were changed to different size and different thread; thousands of other parts were specifically altered to make it impossible to fit them the wrong way unless something truly drastic was done to them. (And even that happened occasionally. A 'peculiarly acting elevator' caused a crash when it was already designed in a way to make it impossible inadvertently to fit and reverse the controls. There was an 'interference pin' which would prevent the mistake every time – until some bright spark found the pin too long, and sawed it off.)

Murphy's Law has savaged many aircraft over the years. When, in 1988, the engines of a certain DC9 were started up, sparks were seen coming from the APU (Auxiliary Power Unit) compartment. The captain reached for the APU fire switch in the cockpit, and discharged the extinguishing agent. The sparking ceased – but the extinguishant had flooded the No. 2 engine itself. Had this

* *Aircrash Detective* p.104.

happened in the air, there could have been problems. The cause? Simple. The discharge plugs were incorrectly connected to the fire extinguisher. The type had been in service for several years. Now DC9 users were alerted, and a new system was introduced to make those plugs more positively identifiable.

Another tale takes us back to the recent complaints about cross-wired fire extinguishers in the cargo hold of new Boeings. When, after the British Midland 737 crash in 1989, 1,600 Boeings were checked, ninety-five cases of Murphy-style instrument misconnections were discovered, eight of which could have resulted in the display of misleading information to the pilot. Design changes will now be introduced to prevent wiring and plumbing misconnections. As well as labels and colour codes, varying sizes and fittings will now ensure that, when wires and tubes get together, only suitable partners can be mated.

Different aircraft, different manufacturers, but very similar problems. Traffic explosion exacerbated the pilot shortage, and stretched manufacturers' resources to the limit and beyond; airborne geriatry exposed the menacing gaps in the ranks of truly skilled aircraft and maintenance engineers. We need a return to structured appenticeship, strict licensing of mechanics, and stronger inspections at every level.

Neither experience nor technology will provide excuses when geriatric aircraft go slumming, and begin to grow senile or actually to kill in earnest. But unfortunately, in the third world, excuses may not even be needed every time: the public may never know about events of 'limited local interest,' and of slaughter on a small scale. Had the spectacular 'sunshine top' incident of Aloha occurred to some African or South American bush charter, nobody outside a circle of experts might ever have heard of it.

– 11 –

DESTINATION OR DESTINY?

If airlines ever start to advertise their special ways of taking extra care of their passengers, would airports follow suit? Glittering bars, shopping malls, high ceilings and low prices may be great attractions, but how about the lure of competitive safety?

Once again, the idea will be resisted on all sides. The bad ones have nothing to advertise, the good ones are disinclined to call attention to potential hazards passengers have never thought of. After all, if in a moment of greatest need hydrants can run dry even at a busy and above average airport like Manchester, Robert Louis Stevenson's 'To travel hopefully is a better thing than to arrive' could attain a new, prophetic meaning for end-of-millennium air travel.

The world is peppered with airports known euphemistically as *marginal*. We put up with poor standards because there is a vast shortage of airports and suitable runways.

'People complain about crowded skies and lack of airspace when the biggest problem is airport capacity, the rate at which we can get airplanes down on the concrete,' Keith Mack said when he was CAA Group Director and controller of Britain's air traffic services. 'Heathrow is bursting with up to eighty landings and take-offs handled in an hour. Gatwick has grown enormously but it has no right to build a second runway, and can utilize its taxiway only for emergencies.' In Britain, the airlines share out runway capacity by consensus, but in America the cut-throat gate ownership and hub system, twin daughters of deregulation, create massive traffic jams at peak times when people are flown in along spokes to switch planes at the hub. At Atlanta's Hartsfield International, the airlines may schedule thirty-two arrivals in fifteen minutes – and hope for the best.

Ground facilities have failed to keep up with the traffic explosion. Growth is restricted not only by lack of resources and pressures of environmental politics, but also by the NIMBY factor: what? more noise? more pollution? more roads and trains? Not In My Back Yard!

It can now take at least a decade to obtain agreement even to planning a new major airport or just a parallel runway. Choosing the site for a third airport near London is a perennial debate that begins to bore even the cartoonists. Western airports have no space for additional runways and terminals because the surrounding areas have become populated right up to the perimeter fence. A proposal for a third runway at Sydney's busy airport sparked off a vast political row. In America, no new airport has been built in fifteen years. 'New York, Chicago, Atlanta, Washington, Dallas, Los Angeles and many other American towns badly need new airports,' said Dave Kelley of the NTSB. 'As a result of endless battles, only hard-pressed Denver has at long last a new airport – on the drawing-board. It will be the world's largest, but it will take up to ten years to fight out land acquisition, zoning, and environmental quarrels before we could even think of actual flying there. Silent aircraft could make use of night capacity, but that isn't when people want to fly.'

Airport safety is dependent on the attitude of regulatory authorities. In Australia, the airports have the unusual right to close down for weather or operational reasons, and pilots then have to divert rather then consider taking a chance. On the other hand, many airports, even in the most advanced countries, show cavalier disregard for ICAO minima, and keep going with minimal approach facilities no matter how marginal the conditions may be. Pilots talk about the 40° parallel as the *gateway to the unknown*, beyond which 'in parts of Southern Europe and Africa it's anybody's guess what awaits you on arrival'. Sydney Lane, former assistant Technical Secretary of BALPA, singled out Africa as 'a case all of its own as one of the world's most deficient areas in the provision of air navigation services, airports, lighting, communications, etcetera. A vast continent with many emergent nations whose financial resources are strained even at the best of times when not engaged on "prestige projects". Money sometimes tends to go on gilded, elegant terminal buildings rather than on the provision of ILS or on the repair of faulty aids. It's only because the weather tends to be relatively good, and because there are fewer aircraft movements, that accident statistics don't show up the facts. But their airports often figure prominently in the IFALPA *bad boys* guide.'

'It's no accident that the most defective ports are usually in countries where the political climate is also troublesome,' said Captain Steve Last, who chairs the Aircraft Design and Operations group of the International Federation of Airline Pilots' Association. 'Africa and South America are of special concern in this respect because instability is combined there with lack of resources and expertise.' IFALPA, in fact, has an *Annex 19* with a register of airports that are awarded 'black stars' for dangerous deficiencies. The list is confidential, partly because it deals with sensitive areas of national pride, partly because it is felt that criticism is more effective behind the veils of secrecy, and partly because much of the information comes from pilots and their sometimes vulnerable organizations which could be embarrassed. But such considerate confidentiality, not only denies the public's right to be warned, but also discards the powerful weapon of public consternation which would be firmly on the pilots' side.

The following list is not a truly representative sample, and by the time of publication cannot be fully up to date. It illustrates some of the airport problems that have been highlighted by accidents, incidents, special studies, and pilots' confidential comments in the 1980s.

ACCRA: IATA mission helped to eliminate frequent flooding, but after rains aids and runway lights tend to be unserviceable.

ADDIS ABABA: no room to extend for decent runway overrun area.

ATHENS: a 'black star' port for years; ditches at both ends of runway, three-metre drop where overrun area ought to be; limited radar only; radio congestion: flights must dodge heavy traffic often without getting through to Athens control. Greek CAA 'law to themselves irrespective of pilot and ATC strikes'. Runway resurfaced after Swissair accident, but the material was the same as that used in streets; its limestone content creates microstructure that is slippery even in humid, let alone wet, conditions; extra length ought to compensate for that; an added hazard is too infrequent cleaning of burnt tyre contamination.

ATLANTA: runways were lengthened to accommodate fully laden non-stop flights to Frankfurt and Hawaii, but now reach out to drops over interstate highways at both ends.

BANGALORE, INDIA: Air India A320 Airbus crash in 1990 exposed appalling conditions. Airport licence expired in 1961; no emergency plan; rescue vehicles get bogged down on road through marsh; firemen had no key to rusty gate in security fence; no radio contact between tower and emergency service; shortage of foam.

BILBAO: neither the 3,266-feet Mount Oiz, nor the 178-feet TV mast atop it, was marked on approach charts when Iberia 727 (with no ground proximity warning) clipped the mast and crashed in 1985.

BOGOTA: 'tricky location, yet, like most Colombian safety equipment, fire and rescue services not quite up to "rubbish standard" '; alternates are feared by pilots.

BOSTON: undershoot and overrun problems: runway terminates at the water's edge, but that might have saved 208 lives when the airport had 'poor' or 'nil' braking conditions, and a DC10 skidded off through snow-covered bank of rocks.

CAGLIARI, SARDINIA: furious Mediterranean storms may cut power supplies; navaids may be left out of action for long periods.

CAIRO: despite high ground on approach, poor landing aids; 'dodgy firefighting equipment'.

CHARLESTON, VIRGINIA: probably worst U.S. overrun safety area.

CHARLOTTE, N. CAROLINA: steep embankment, no grooving on runway which ends 440 feet away from railway track. (In 1986, when a Piedmont Airlines 737 crashed due to-pilot errors, 'poor frictional quality' of the runway and 'the obstructions presented by a concrete culvert', localizer antenna and a chain link fence contributed to the severity of the accident.)

COLUMBUS, GEORGIA: forty-foot overrun area when now the FAA demands at least a thousand feet for new airports.

CORFU: pilots say airlines go in only because others do it, too.

DENVER: 'real crummy port to approach in a snowstorm when runway and taxiway look deceivingly alike; when big airplane landed on taxiway by mistake, only luck averted disaster.' The best diversion airport, Grand Junction, Colorado, decommissioned its standby power system to save costs, so if power fails in a storm, this alternate has no lights or aids and must close down – a grave risk if the diverting plane is low on fuel.

DOMINICA: narrow runways, fire services notoriously poor at both international airports, IATA regional office alerted.

FUNCHAL, MADEIRA: An 'aircraft carrier made of stone' where safety depends on the pilot's competence. A serious landing accident in 1987 raised questions about the airport's suitability for use by large aircraft.

GREECE: strikes by air traffic controllers for safer equipment and better pay; after serious double airmiss by British aircraft over the Aegean in 1988, government ordered civil mobilization to thwart industrial action, but the desperate controllers began a hunger-and-stay-awake strike for safety to get disqualified from duty on medical grounds.

HOUSTON: confusing markings of runways and taxiways.

IZMIR: surrounded by hills, no lead-in lights, buzzing with fighter aircraft on low-flying exercises.

IDAHO FALLS: ALPA complained – taxiway markings hardly visible.

JAKARTA: short runways with bad surface; poor ATC and landing aids; glide slope unreliable even when ILS is working; inadequate search and rescue facilities. (At some stage, Australian pilots boycotted it.)

KABUL: difficult terrain yet poor air traffic control.

KANO, NIGERIA: risk to aircraft because villagers use the unprotected tarmac as a short-cut; a woman pedestrian was hit and killed on the runway in 1989 by BA DC10 about to lift off.

KATHMANDU: aircraft must virtually 'drop in' between mountains, good route training for pilots essential.

KETCHIKAN, ALASKA: a ravine only 700 feet beyond runway threshold, aircraft in trouble cannot be chased by fire and rescue vehicles to accident site. (Problem was highlighted by 727 crash in 1976!)

LAGOS: congested approach frequency on 1243 causes delays and confusion; poor quality transmissions, language problems; frequent power failures and ILS unserviceability. (Similar deficiencies at **KANO** and **PORT HARCOURT**, the other Nigerian airports.)

LA PAZ: after an Eastern Airlines flight crashed into a mountain in 1985* 'investigators discovered that navaids, VOR, ILS, had last been

* See 'Blood Response', Chapter 2.

checked when installed, never again. Bolivian authorities had no equipment to do it, checks would be done at long intervals by the Argentinian air force or the FAA.

LIVERPOOL: airport ordered by CAA to review and coordinate emergency plans; runway close to Mersey, yet insufficient equipment to rescue passengers from the river.

LOS ANGELES: non-standard manoeuvres are necessary because of special noise abatement requirements.

LUTON: thirty-foot escarpment 200 feet beyond eastern runway. In 1983 it transpired that local firemen had refused to use breathing apparatus for eight years.

MADRID: 1983 was the year when serious problems with world-wide implications were exposed; a Colombian Jumbo crashed on approach to Barajas airport killing 181 people. Ten days later, Avianca and Iberia had a ground collision in fog, with a death toll of ninety-three. The Avianca was to turn forty-five degrees right into a taxiway, but made a ninety degree turn into a runway where the Iberia flight was taking off. Though, allegedly, Madrid facilities complied with ICAO *Annex 14* standards, the need of Surface Guidance System, less confusing runway markings, and clearer signs for intersections was already known because of previous incidents. Calls for visual aids, e.g. stop bar of lights for runways, were made to ICAO, but the Visual Aids Panel assigns such problems to a working group whose eventual recommendations must be discussed by the Panel which sits only once every two or three years. Improved standards could not be hoped for before 1992.

MALDIVE ISLANDS: no ILS despite pressure by some airlines; 'without proper landing aids, pilot must be thoroughly familiarized to hit the right spot as he virtually drops into the ocean'. To *enhance* its fire services, a prosperous Far Eastern airline shipped out a good fire tender as a gift but, allegedly, they can't maintain it properly for it goes frequently on the blink.

MANILA: very poor ATC though dangerously heavy military air traffic. Radar service and electricity supplies unreliable; runways often full of litter; emergency service patchy.

MIDDLE EAST: allegedly, nepotism rules the selection of applicants for many key jobs; some sellers of equipment mark up their prices to allow for kickbacks. 'If you must have an accident, avoid siesta and hospital visiting hours'. Military use of civilian airfields results in tragicomic incidents of secrecy. (A CASB study recalls the case of a DC10 taxiing to the runway of a large airport on a foggy night. As there were no signs to determine the position of the aircraft, the pilots had to rely on the landing chart that showed that the third turn-off ninety degrees to the right was the runway. The pilots counted the turn-offs carefully, but, on take-off, they were horrified to see that they almost ran out of concrete as they got airborne at the very end of a supposedly long runway. The mystery deepened when the pilot revisited the airport in fine visibility, and found that the landing chart was wrong: there were four, not three, turn-offs. The airport authority then revealed 'since one of the run-offs was for military use only, prohibited for civil aviation, it had not been depicted on the landing chart'.

MOSCOW: 'if in trouble in Soviet airspace, particularly over Siberia, you must *ask* where to go because alternate airports are a military secret.'

NEW DELHI: several crashes and serious incidents have been attributed to inadequate landing aids and services; at Delhi, Bombay and Calcutta, cows and cyclists use and cross runways.

NEW YORK, LA GUARDIA: built into East River on Manhattan's doorstep; wet surface hazard because a twenty-foot drop into the river is the overrun area; in 1989, a 737 ran into the water, hit a wooden pier, broke up, killing two, injuring forty-five.

NEW YORK, J. F. KENNEDY: 'approach more hairy than at Hong-kong'; 'useless ditch at end of Runway 4RT; could be filled in completely but New York Port Authority claims "there's a four-lane highway on the other side of the river, and we don't want airplanes to cross that highway". Does it mean they'd rather stop airplanes in the ditch? An SAS flight already ended up in there with badly damaged wings.'

ORLANDO, FLORIDA: problems with edge and centreline runway lights may induce departing aircraft to stray on to taxiway.

PENANG: aids have not kept up with growing popularity of the island; people and animals may wander across runway.

PHUKET, THAILAND: installation of modern radar was *considered* after Thai Airways 737 had crashed into the sea, killing eighty-three, in 1987. (Controllers messed up approach of two aircraft.)

RHODES: the improvement needed because of location and dangerous wind conditions was thought to be impossible for technical reasons; construction of new airport recommended by IFALPA in 1970s; since then airport improved only in popularity.

RIO DE JANEIRO: runway liable to flood.

RIYADH: since Tristar fire tragedy, 'airport is now over-equipped, and has twenty times more extinguishing agents than necessary – maybe because of kickbacks to some buyers – but maintenance is inadequate, fire tenders are dented and rusty, and crew training is still poor.'

SEOUL: frequent complaints about heavy traffic by fighter aircraft without any positive control.

ST LOUIS: difficult weather conditions with windshear; one of the airports of huge traffic explosion since becoming a major TWA base, with sixteen million passengers per annum; on busy day, up to fifty aircraft in holding pattern that almost overlaps with busy airports nearby, so in a twenty-five-mile radius, a hundred aircraft may be circling, yet the old-fashioned ATC relies on just six controllers – three for each end of the airfield.

SPAIN: airports there have enjoyed a certain notoriety, but with growing traffic the situation can get worse; in 1987, fourteen new companies sprang up, each flying one or two second-hand 737s or DC9s, and using any available airfield; air traffic controllers often strike because the government fails to honour written agreements to install modern airport equipment.

TAIPEI: much military air traffic under separate control – lack of coordination with civilian ATC.

TAMPA: at north end of Runway 36 Left, a fourteen-foot wall was built at angle towards airport for noise abatement purposes, creating serious risk if aircraft needs to overshoot.

TANGIER: when an Air Maroc flight suffered a gear collapse in 1986, luck alone saved passengers from potential disaster because the fire service took more than five minutes to arrive.

TEGUCIGALPA, HONDURAS: pilot familiarization essential to negotiate weaving route in between mountain tops.

TEHERAN: 'unreliable beacons lead into U-turn over hills to runway'; 'human inefficiency is the greatest risk'.

TENERIFE: criticized on several counts in accident reports regarding facilities, aids, communications, etc.

THAILAND: 'airports, including Bangkok, are as dirty as the cities, but debris on a runway can be a killer'.

TORONTO: pilots fight for 'runway end delethalization'; after a DC9 crash into the ravine there, only the embankment of the ditch was lowered for planes to fly over. Captain McBride of the Canadian ALPA is unhappy about it: 'I'd prefer to fill it right up or build a bridge over it. They say it would cost too much. But we think some people are afraid that proper action would be an admission of a hazard they've long found acceptable.'

TRIPOLI: ILS and other landing aids frequently inoperative. (When South Korean DC10 attempted to land in thick fog, crashed and killed seventy-eight people on board and on the ground in 1989, the ILS had been out of order for several months. Libyan authorities claimed that the information was available to airlines.)

U.S. AIRPORTS: FAA Administrator McArtor once said: 'Hubs and congestion are not just an airport issue. I am much more worried that America's aviation competitiveness is at stake because of an ageing, outdated, and limited number of airports'.* After the midair collision over Cerritos, Calif., an FAA task force made forty recommendations to foster safety at the twenty-three busiest hubs.

WASHINGTON D.C., NATIONAL: pilot experiencing problem at V_1 must take-off, rather than abort, to avoid splashing into ditch, known as Roaches Run, at north end of Runway 36; it could be filled to eliminate hazard and create 3,000-foot overrun area, but nothing has been done. (The Air Florida icing crash illustrated some other tragic airport problems in Washington.)

Airlines can exert pressure through the airport users' committees. But if they do not by consultation enforce the necessary corrective measures – and apparently they can't – why do they not stop flying to some substandard destinations, at least temporarily? 'Because of the loss of trade to the competition. Some airlines are less fussy, and particularly the cheap package tour operators would gladly fill the gap, no questions asked.'

They say they alert their station managers. They say they keep urging ICAO and IATA to press for improvements. They say that their pilots get frustrated and report the deficiencies. But a good airline can learn to live with problems. 'There is no ILS? okay, we'll adjust our landing minima, and go in only when, say, the cloud base is not less than 700 feet; or we might tell our marketing people to warn the state concerned that if they don't put in ILS, many flights will divert, and they lose out.' A Boeing study of crew-caused accidents† showed the pragmatism of operators. One operator slotted all airports it used into three categories. Bottom of the list was 'non-standard', for which stringent requirements were set down (for example, landings only by captain, who must be specially checked out and have the experience of 'minimum two

* Remarks at National Aviation Club, Crystal City, 21.1.1988.

† *Airliner*, Apr-June, 1987.

approaches, landings and take-offs within the last two months'.)

Cranfield Course Director John Owen said: 'One hears pilots' complaints incessantly about Mediterranean ports. The answer is in their hands: never mind the competition, if it's substandard, don't go there.' That, however, is beyond the initiative of most individuals.

'Without ICAO agreements, there'll never be unity,' said Captain Caesar of Lufthansa. 'Besides, we're professionals, we have a duty to provide a service. If there's a risk, we have to cater for it. People say a lot of bad things about some airports. But if EC countries threatened to stop flying there, some Americans would start up services on very profitable routes like Frankfurt to Athens. If passengers want to go somewhere, and the airport is open, they'll take them.'

Of all the operators I talked to, only Singapore Airlines came up with two examples of 'going it alone' – and even those were not all that impressive: once when at Kuala Lumpur only one fire tender was operational they stopped flying there *for one day* though others continued normal services; and once they stopped flights to Australia when the fire service there was on strike – but then it would have been illegal for the airports to operate anyway.

The ALPA Guide For Airport Standards says in its introduction: 'Throughout the history of commercial aviation the airport has been treated almost as a stepchild . . . In far too many cases, development motivation has been slanted towards the terminal area because it is a virtual extension of the community's image – as well as being a ready source of income.'

FOUR OUT OF FIVE

You do not need to be a pessimist to prepare for emergencies if you know for certain that four out of every five accidents happen on or within 1,000 metres of runways during take-off or landing.

So it is most disheartening to see the wide differences in emergency planning, as well as the provision and training of fire and rescue services, despite ICAO efforts to standardize with *Annex 14*. It is easy to say that many states are too poor to invest in emergency preparations, that dozens of African and South American airports cannot afford to stock the required amount of firefighting agents, buy machines that discharge foam at the prescribed rate, train and pay firemen and doctors in readiness, or clear runways of dangerous rubbish. But we are not talking only about the so-called third world. Incredible as it may sound, British firemen, who fought the 737 fire at Manchester, did not seem to know that the foam made escape chutes dangerously slippery, and that the powerful foam machines would knock survivors off the chutes or back right into the inferno. In the United States, the relevant 'FAA regulation, FAR Part 139.49, falls far below international standards in many respects and in its present format is not acceptable to the Accident Survival Committee of ALPA.'* And John Lodge, a former CAA fire chief, went sadly on record saying that, in his experience, 'few regulatory bodies or airport authorities have implemented in full the total package presented by ICAO'.†

Rescue and firefighting must be a unified service – but in some countries they are separate, giving extra room for errors of coordination. Firemen are usually provided with tools for forced entry into burning aircraft, but they rarely get the chance actually to practise their use. In Canada, firemen driven by humanitarian urge *may* help to rescue passengers, and *may* break down unopened exits, but it is not their *duty* to do so. Bill Tucker, CASB Director of Safety Programs, explained: 'Evacuation is seen primarily as the crew's job; the firemen's task is to provide a fire-free escape route. The philosophy is that, if there's a huge fire, nobody will be able to get inside to help.' But Captain McBride added: 'Firemen complain that their staff is cut back too far particularly at smaller airports, and that Transport Canada fails to meet international standards. The money goes on fancy tenders that discharge foam at a fantastic rate, operated by just one man who wouldn't have the chance, with the best will in the world, to deal with a door that fails to open.'

It is every airport authority's sacred duty to maintain emergency plans, but, as John Lodge says, old plans are often left to gather dust on the shelves while aircraft design and modes of operation keep changing. Though the ICAO Manual *advises* airports to hold

* The ALPA Guide to Accident Survival Factors, 1989.

† Lecture in *Airport Operations* Short Course, Loughborough University of Technology, 1989.

annual full-scale exercises, the *Annex* that is supposed to set the standards says nothing about it, and it is only now that there is a proposal for the *Annex* to take a stand. Because the actual discharge of foam is costly (ninety-nine per cent of extinguishing agents consumed by airport fire brigades goes on training and testing equipment), even that new proposal had to be watered down to biennial exercises in order to satisfy member states which plead poverty. Or take medical emergency facilities: ICAO offers some guidance only because it recognizes that, in many areas, those services would not be needed often enough to justify the cost of their round-the-year availability. The 'low benefit to cost ratio' argument frequently wins out. Others argue, in chicken-or-egg vein, that it is more effective to invest in preventing emergencies, and choose to ignore the fact that accidents will still happen.

The ICAO Accident Prevention Manual makes this devastating comment – tucked away under the inconspicuous heading 'Environment': 'Many hazards continue to exist in the environment because the people responsible do not want to become involved in change, consider that nothing can be done, or are insufficiently motivated to take the necessary actions. Obstructions near runways, malfunctioning or non-existent airport equipment, errors or omissions on aeronautical charts, faulty procedures, etc. are examples of man-made environmental hazards that can have a direct effect on aviation safety.'

There is an abundance of examples. NASA's confidential safety digest *Callback* quoted one in August, 1987: ' . . . On final I cautioned the co-pilot (flying) that it looked as if there were loose pieces of paper on the approach end of the runway. On short final I instructed him to land long to avoid [them].' It turned out that the numbers and cross-lines marking the runway 'were pieces of tape' that were 'coming loose and strips as long as six feet were blowing around with ends still taped to the runway. A very hazardous condition.'

At many airports, refuelling trucks are allowed to operate without fire extinguishers on board. The training for handling fuel tends not to be commensurate with the risk involved. Even in the United States, an NTSB study of Airport Certification and Operations* noted high turnover of personnel, and found that

* Report No. NTSB/SS-84/02.

only two of the inspected storage facilities 'administered prehire test for aptitude'. About 'seventy-five per cent of the fuel service facilities hired people "off the street" for refuelling positions', requiring only a clean driving licence, and twenty per cent were satisfied with new employees signing a statement that they had read the required regulations and manuals. The selection of firefighters may also leave much to be desired. An ALPA specialist claims that at some American airports like Miami 'positive discrimination plays havoc with hiring fire crews because the adherence to prescribed ethnic quotas may override considerations of ability to train'.

'The training of American firefighters varies far too much,' said Captain Vic Hewes of the ALPA Accident Survival Committee. 'Only Georgia and Texas train them to the standards of the National Fire Protection Association. In some states they can drive a crash truck and get the pump going, but they have no real firefighting experience since many localities prohibit hot fire drills for environmental reasons. At small airports, firemen also cut the grass, and have general maintenance duties. At some major airports, the firemen are part of the downtown structural services – with frequent rotation for airport duty. They often know little or nothing about aircraft or emergency exits, nothing, because in too many places, it's the local authority that's in charge. FAA regulations say that the men must be trained . . . but only to a minimum standard. So they can only guess how to open different exits from the outside, and whether they need a ladder or an elevated platform to do it.'

'In the wake of the Delta windshear crash at Dallas in 1985, we re-emphasized our numerous recommendations to intensify disaster drills and emergency planning,' said Terry Armentrout, Director of the NTSB Accident Investigation Bureau. 'Yet in 1987, when we had an accident at Denver, we found that better contingency planning could have prevented the extra calamity in a tragic situation. The aircraft rolled over, the wing was full of fuel, and injured people in the cabin were hanging upside down, held by their lapbelts. To get them out, rescuers had to cut away certain parts of the aircraft which therefore moved, and settled, crushing some passengers. It took rescuers a long time to free those victims. Nobody at the airport knew how to move a wing, nobody knew its weight, how to handle it or where to find the safe lift-points on it. The right kind of equipment was unavailable, and at the end they just shored up the wing with timber. To obtain correct advice took them several hours.'

Familiarization with new aircraft is a slow and much neglected process at many airports. Numerous authorities find it cheaper and more convenient to conduct only 'table-top' exercises, knowing full well that such short-cuts cannot reveal everything an actual test would highlight. In 1983, the investigation of an accident at Detroit's Metropolitan Wayne County Airport discovered some problems with the crash notification procedure which was duly outlined in the emergency plan – but the plan had not been fully practised since 1978. (The FAA only *suggested* annual live drills.) In a table-top exercise *notifications* may seem of minor importance, but the relevant item at Detroit concerned hazardous cargo: the crashed aircraft had a container of radioactive materials on board; had the container been breached, the delay in proper notifications could have caused havoc all over Detroit.

Vic Hewes says that, despite some improvement in U.S. crash preparedness, 'the FAA continues to allow the hazard known as the "remission factor" which creates a false sense of security for the public. Firefighting requirements are governed by the largest aircraft flying to any airport. If, for instance, DC9s and 727s are in use, the airport is classified as INDEX C which requires 3,000 gallons of extinguishing agent. But if there're fewer than five departures a day, say five arrivals and four departures of the same types, the airport can downgrade to INDEX B which must provide only 1,500 gallons of agent – a totally inadequate amount to put out a major fire on aircraft like those. Yet the FAA finds it acceptable to expose a thousand people a day to the increased risk at that airport. The new changes in the regulations failed to solve the problem. Some operators still complain that at small, Category A airports the meagre emergency requirements are too costly. Luckily, the FAA has refused to give way in that direction.

'Commuter aircraft – defined as seating fewer than thirty people – have lower requirements for evacuation, seat restraints, crashworthiness, and so on, and the usual airport certification doesn't apply to them at all! So they can fly to airports which are unfenced, where cows and deer and vehicles can often roam freely across runways, where there's no security, and no requirement for any emergency services.'

GROUND FOR CONCERN

Airports themselves and adverse conditions prevailing on or around runways are not named frequently among the direct causes of accidents, but they tend to figure as causal factors.

The danger zones are at taking off and landing points, as well as undershooting and overshooting. Yet *lethal runway areas* have remained unaffected by the complaints of quixotic investigators and frustrated airline pilot organizations.

'Aircraft manufacturers do a lot to make life easier for pilots, but airports do not progress the same way as aircraft,' said Captain Caesar. 'We have trouble with the lack of clear overrun areas, obstacles at the end of or around runways, trenches alongside runways that can shear off wheels, lack of devices like light stop-bars to prevent incursions that lead to horrors like the collision at Tenerife,* just to mention a few causes for great concern. You get some protective features at most airports, but it is the exception where you get them all. Non-standard markings at various airports sow confusion. Even in a relatively small country like Germany, we have no standard markings for docking. Stop signs and guide-in lights may be different wherever you go. The bad visibility lighting system can be outright misleading and, unless pilots are extremely vigilant, it can take us in the wrong direction. Why do they have to make it so hard for us to perform a simple task like taxiing in bad weather conditions? The discrepancies are becoming bigger all the time, and the traffic boom aggravates the situation because many airports will not invest in eliminating the problems.'

There is no less confusion in Canada where in the severe winter conditions pilots can get disoriented, and where plans have only now been drawn up to 'replace existing airside signs to conform with ICAO standards by 1995'.† The 'sea of blue' effect is a frequent cause of pilots' complaints – and not without good reason. Dots of blue taxiway edge-lights fuse into an ocean that creates a dangerously false sense of security, yet it would be easy and relatively inexpensive to cure by just following the ICAO Aerodrome Design Manual.

* See next chapter.

† CASB: Special Investigation into the *Risk of Collisions Involving Aircraft on or Near the Ground*, August 1987.

The inability to standardize airport aids is often quoted as a classic example of ICAO impotence but, in fact, it is the ICAO member states' fault that closer agreement is not achieved. 'In an ideal world, parking guidance systems would be simple to standardize in a way that would be self-explanatory every time,' said Sydney Lane of BALPA, who saw this as 'yet another irritant to the pilot. There's a plethora of systems. You fly from London to Glasgow, Stockholm, Amsterdam and back, and you have to cope with three or four different apron parking guidance systems. Sure, it's all in the flight documentation, but the pilot may have physically to get the book out to see what lights to follow at a particular airport.'

Runway Incursions by aircraft as well as maintenance and snow-clearing vehicles for a great variety of reasons (such as communication break-downs, errors by pilots and controllers, bad marking) lead to many near-disasters, yet most airports seem to economize on the introduction of devices that could prevent them. When, in 1986, incursions were the subject of a special NTSB study, chilling examples offered themselves from even the biggest American airports, including Chicago, Austin, Houston, Kansas City and Washington. Here is just one of them:

On March 31, 1985, a Northwest Airlines DC10, Flight 51, was cleared for take-off on runway 29L at Minneapolis St Paul International Airport. At the same time, another DC10 of the same airline, Flight 65, was cleared to follow a taxiway and cross runway 29L at the intersection 6,000 feet from the approach end. 'The controllers who issued both clearances did not recognize the hazardous situation in time to take preventive action.' Having waited for two aircraft to cross the runway, Flight 51 began its take-off roll, accelerated, but was still well below its rotation speed where it would be ready to lift the nose off the ground when the pilot noticed Flight 65 lumbering across its path. He had no choice but to take-off prematurely: hoping for the best was a better choice than to slam one planeload of flesh and bones into another. His undercarriage cleared Flight 65 by about sixty feet vertically: 501 people aboard the two DC10s could thank their lucky stars.

Were the controllers plain foolish or negligent? No, other factors had played a part. There had been a snowstorm, clearance had proceeded slowly, key taxiways had to be closed down, some experienced control staff could not get to the airport because roads were blocked, yet meanwhile there was pressure to accept delayed traffic and handle more flights. At the time of the incident, thirteen other aircraft were moving within 500 feet of the vital intersection. Braking conditions were poor (non-existent on some taxiways). The pilot of Flight 65 saw another aircraft land on 29L, so he delayed his crossing momentarily to let the new arrival clear the runway because he was afraid that, in the poor braking conditions, it might slide into his DC10. That thirty-second delay, not even worth a call to the controllers, would place him in the path of Flight 51. When his co-pilot alerted him to the fast-approaching menace, there was nothing he could do to get out of the way of that roaring mass.

Non-Standard Navigation and Landing Aids create problems. In Germany, 'despite the proliferation of small companies, the smaller airports they use remain designated as "non-controlled" ones which means that even though they may have ILS and radio beacons, those facilities must not be used as navigation aids for landings,' despaired Christian–Heinz Schuberdt, a Principal Inspector of Accidents. 'It's the law. At so-called non-controlled airports pilots must ignore the signals and land or take-off under visual rules. It's like having traffic lights all over town but telling drivers they must go as if they were on open roads, and pay no attention to red, green or amber. It's crazy to have those things if they're not for use. Pilots can get caught between two systems. If we allow them to fly in bad weather under instrument rules, we must allow them to land with the help of instruments because the present confusing conditions and switching from instrument to visual cause two or three accidents a year.

'At Lübeck, a pilot came in to land visual in very bad weather – and hit the radio beacon. Its tower was placed right in the middle of the approach line "to help" but became just an obstacle because its use was forbidden. Here at Braunschweig (the home of the Accident Investigation Bureau), we've had a radio beacon and a localizer "for testing purposes only" for many years, plus radar at nearby Hanover, but nobody coming in is allowed to use any of it. Changes in the law are overdue for we're not talking about some godforsaken bush airfields, but busy airports with good runways.'

One of the most confusing and potentially hazardous factors is the variation of visual glidepath lighting systems: VASI, T-VASI and PAPI. These aids can be particularly valuable when there is no ILS or when the ILS goes U/S (unserviceable).

VASI, the Visual Approach Slope Indicator, is an array of red and white lights shining alone the vertical approach path to tell the pilot where he is relative to the runway's ideal glide slope. The more expensive T-VASI – popular in the Far East, Australia, New Zealand, and lately at some Italian and Spanish airports – informs the pilot in somewhat ambiguous terms that he is too low already if he sees only reds, and too high or much too high if he sees more whites, let alone only whites. (Mist or water droplets can also diffuse the lights sufficiently to make them misleading. IFALPA's questions about pilots' experience with T-VASI got some highly critical reactions. Pilots flying to Alicante, Gerona, Palermo, Bologna, Venice and Rome, for example, have been alerted to the use of this system.)

The Precision Approach Path Indicator (PAPI) is more exact and helpful to much lower levels, almost to the runway surface.

'The trouble is,' said Sydney Lane, 'that each time you approach an airport you must keep in mind which system is in use there. With VASI and PAPI at least it's immediately obvious whether you're too high or too low, but with T-VASI you have to think it out, and on a dark, dirty night with poor visibility, the last thing you want is to waste time on figuring it out. You just want an immediate indication of your adherence to or deviation from the visual glidepath.'

It is not the cost. While major aids are expensive (ILS more than half a million dollars), VASI or PAPI could be installed for $20,000. But airports are reluctant to spend on changing their systems just for the sake of uniformity – even if it may prevent serious incidents. Confidential reports in CHIRP illustrate the resultant problems pilots must face. When a first officer (flying) approached a Spanish airport, the runway slope gave an unusual perspective, and there were some false warnings from the ground. He looked at the lights of the T-VASI, 'assumed it to be a PAPI and interpreted it as indicating a high position'. Fortunately, he double checked his instruments . . .

In poor visibility, another pilot made a radar-assisted visual approach to an unnamed airport with no glidepath information. He could not see the ground but spotted four *white* PAPI lights in the expected position. He now looked even harder for the runway, and suddenly four *red* PAPIs came into view. The set of 'four whites' he thought he had seen turned out to be vehicles halted by traffic lights on a public road that passes the threshold of the runway. It also transpired that, wrongly, the controller had set PAPI on low brilliance – with no runway approach lights on at all.

Runway Contamination by slush, snow, ice, water and burnt rubber deposits causes, as we have seen, totally avoidable accidents and dangerous incidents. José Santamaria, chief of the ICAO Airports Section told me: 'Until recently, ICAO always required that the surface of runways *should* be clear of debris and burnt rubber, and *should* be maintained in good condition to provide correct braking conditions. Now this has changed. We now say the runway *shall* be maintained to a good standard, and its construction *shall* be such as to provide good braking characteristics. The change of words may seem trivial, but these are big changes in ICAO terms, and the product of much work and long debates.'

The state of *wet-runway-certification* tends to aggravate the risk. Hydroplaning is a well-known hazard, yet it is still claimed to be debatable how much distance an accelerating, fully laden aircraft will need to stop safely on a wet surface if something goes wrong, and the pilot chooses to abort take-off.

'It's the accelerate/stop criteria that causes the problem,' said Gary Wagner, an Air Canada pilot and ALPA/IFALPA activist. 'If we have to abort take-off, there's often no more than a quarter of the runway left. The manufacturers are satisfied with their performance certificates but pilots have doubts about them in certain conditions. The braking tests are conducted in dry, almost ideal circumstances, with allowances made afterwards for wet, contaminated runways because they say we have reverse thrust in reserve. But what if we have an engine failure?'

'The regulatory authorities agree with the manufacturers that reverse thrust should compensate for wet surfaces,' said Laird Stovel, Air Canada's Manager, Operations Engineering, 'but if an engine fails at V_1 on a wet runway of minimum length, the chances are that the aircraft will go off the concrete anyway. Luckily, in the industry's millions of operations over forty years, we've had wet runway incidents very infrequently.'

Lucky, indeed, because the pilots' doubts about braking ability are echoed again and again by accident and incident investigations. In 1985, when a British Airways TriStar landed in a rain shower at Leeds, the aircraft was unable 'to achieve the level of braking effectiveness on which its wet runway performance is scheduled'. (The accident report by Principal Inspector Charles recommended the re-examination of the TriStar's braking performance as well as the runway surface condition of Bradford Airport.)

On July 20, 1986, a Quebecair 737 with sixty-four people on board accelerated down the runway at Wabush, Newfoundland. It was running at 114 knots when the first officer noticed a bird on the runway, and called: 'Bird!' So, at 126 knots, when the left engine lost power, the captain knew what the problem must be: the engine had ingested the bird. (It would transpire that the carcass disrupted the airflow causing an engine flame out.) Reverse thrust and maximum wheel braking failed to stop the aircraft. It veered off the wet runway, and careered on, into a bog, where it came to rest. The damage to the aircraft was substantial, but – apart from one person breaking a leg during evacuation – nobody was hurt.

The Canadian investigators concentrated on ASD, the Accelerate/Stop Distance, on slippery surfaces. Their report emphasized that although the aircraft flight manual 'recommends reducing ASD on wet runways', it is the operator who 'must select an arbitrary distance to compensate for the reduced braking action. There are no charts or figures provided to aid the operator in these calculations. This lack of published criteria has made it difficult to develop enforceable standards and regulations in this matter.' The CASB recommended that the Department of Transport should revise the relevant take-off procedures 'in order to provide a margin of safety comparable to that for dry runway operations'.*

In 1983, a CAA study of forty-six 'notifiable' abandoned jet take-offs and overruns concluded that the world-wide accident

* CASB Annual Report 1987.

rate from this cause was unacceptable and required an urgent review. Most overruns occur when the surface is wet; a third of all such accidents are fatal, like the one at Malaga in 1982. Oddly enough, the British CAA is probably the only airworthiness authority which requires wet runway accountability. Equally odd is the fate perhaps of the idea of 'an arrester bed', which is to be studied by the FAA only now – though it has been around for almost three decades. (At Wabush, the bog that brought the 737 to a halt might have saved all lives aboard. An 'arrester bed' of soft earth or crushable foam at the end of the runway was experimented with in Britain in the 1960s. British Industrial Plastics and the RAE even conducted full-scale aircraft trials. It appeared that the installation cost, not very substantial in itself, could well be recouped by savings on land for overrun areas, but eventually the effort fizzled out.

One is left with a disturbing thought: the *yuppie* who spends so much on entertainment that he cannot afford decent tyres and brakes for his Ferrari will be barred from driving; but if he cannot afford to maintain full ICAO safety standards, he may still be allowed to run an airport.

– 12 –

ALUMINIUM SHOWER

Locations where earthquakes, volcanic eruptions, tidal waves, avalanches or bush fires are likely to wreak havoc are well delineated, but aircrashes can happen anywhere. Crowded airspace and constantly sprawling suburbs have brought cities into the potential battle zone, yet the world seems to shut its collective eyes to this menace. Perhaps it is because spectators are not supposed to get hurt.

Just four days before sabotage destroyed the PanAm flight over Lockerbie, Captain de Silva told me in Singapore: 'I don't even like to contemplate what could happen if something crashed into a city. But there is a distinct possibility of it happening. Maybe one precaution could be to enforce, somehow, an international standard for issuing air operators' certificate.'

Crashes by business jets, commuters, military and light aircraft have already given us plenty of warnings. For example, in the late 1980s, a twin-engined Piper Cheyenne flew into a McDonald drive-in restaurant near Munich, and sprayed a passing bus with burning fuel, causing eight deaths and fourteen serious injuries. A French commuter jet crashed in fog near Bordeaux: it killed all sixteen people on board but, mercifully, it missed a children's home by fifty yards. In 1988, when a light aircraft ripped off the roof of a car, and crash-landed on the M62 near Manchester across the motorway, it was a miracle that several people were only injured. Shoppers in the heart of crowded Taipei were not so lucky when a helicopter ploughed into rush-hour traffic. Three

Americans died at Pembroke Pines, Florida, when a light aircraft came through the roof of a shopping mall, and another six died when two light planes demolished several buildings after a collision over a New York suburb in New Jersey.

There is frequent uproar to demand tighter flight restrictions whenever military jets damage or endanger civilian property. The risk is demonstrated by an RAF Tornado which collided with two German training jets, and exploded a hundred yards away from the village of Hinrichsfehn, and by an American A10 Thunderbolt, which razed a row of German houses at Remscheid. In November, 1989, a U.S. naval fighter plunged into a large apartment block, started an inferno and killed a dozen people in their homes at Marietta, Georgia. When a Corsair fighter-bomber attempted to make an emergency landing with an engine failure, the pilot aimed for open fields near Indianapolis, and stayed with the aircraft until the last possible moment to bail out. The jet shattered a bank and came to rest in the lobby of an hotel, killing nine people. In Karachi, a Pakistan Air Force Mirage fighter ended a dive in a crowded factory causing fifteen deaths and injuring four dozen workers.

Many lucky escapes are on record, but how long will the luck hold? When a French fighter crashed in Bavaria, the nuclear plant nearby was said to be well protected from plummeting planes, but who has ever tested the strength of those concrete domes in this way? And there was not even any protected air space around the huge liquid gas storage depot near Hamburg when a private aircraft crashed, missing it by a few yards.

Those who argue that large aircraft are safer need to look no further than the close calls by public service flights, many of which display symptoms of poor training and extended use of ageing aircraft. In 1987, a forty-year-old Stratocruiser had just taken off from Mexico City, to fly eighteen racehorses to Miami, when it ran into trouble in the mountainous Palo Alto region. It tried to land on a motorway which carried rush-hour traffic on eight lanes, smashed its way through a restaurant and four houses, cut high tension wires and set some trees on fire, then hit thirty cars, and became a skidding flamethrower that incinerated forty-two people and injured scores of drivers and passengers before exploding just forty yards away from a fuel depot that held 20,000 gallons of petrol.

In 1988, a Continental Airlines Jumbo took off from Gatwick for Miami. Filled to capacity, it had 450 people and more than a hundred tons of fuel on board. In gusting high winds (about which the pilot had not been fullly warned), one of its engines failed on take-off. The aircraft is built to cope with such mishaps, but the pilot made the mistake of trying to climb away sharply, at twice the normal angle, and almost stalled the remaining engines. Having lost his forward motion, he pointed the nose at the ground to pick up speed, and managed eventually to climb over a hill ahead – with fifty-five feet to spare. Onlookers, convinced that the aircraft had crashed into trees or the village of Russ Hill, raised the alarm. The investigators called for more thorough pilot training in simulators for just such emergencies when pilots tend to over-rotate. In January, 1989, it was Chicago that could have been at risk. Just outside the town, an engine fell off a Piedmont 737, and the aircraft, flying on one engine, had to skim skyscrapers as it limped back to O'Hare.

In March, 1989, a Transbrasil 707 on a cargo flight tried to make an emergency landing at São Paolo but, barely a mile from the runway, it crashed through a block of flats and then into a hillside shanty town, leaving some three dozen corpses in its wake. In July, it was the turn of some people on a busy Manila main road to run out of luck: the pilot of a Philippines Airlines BAC1-11 was told to go round in order not to miss the runway in stormy weather, but he landed anyway, and ploughed into rows of vehicles. The first to die was a three-year-old boy in a taxi. The initial toll was seven, but not all the eighty-two seriously injured victims survived.

At least publicly, many people in aviation pooh-pooh the 'aluminium shower' threat to cities, but a chilling episode in 1979 revealed just how seriously they would take it – if there was time. Two experienced pilots made a training flight on a ten-seater Beechcraft executive aircraft. They donned their oxygen masks, let the air escape from the entire cabin, and practised an emergency descent by diving from 31,000 to 12,000 feet where they could breathe again normally without the masks. They got the clearance to begin the exercise. What exactly went wrong will never be known. Probably the oxygen supply did not come on or was not switched on right away. With the onset of hypoxia, both men would be unable to think fast and rationally: euphoric false confidence is a symptom of oxygen starvation. They died seconds after complete depressurization.

On the ground, the loss of radio contact caused alarm. The aircraft was then seen on radar leaving Essex, flying towards the Channel. Driven southwards by strong wind, it circled endlessly as if following the route of a gigantic coil from the sky. (From its last known position, with both men dead, the plane should have descended in a straight line, but one of the pilots slumped over the controls which put it into a constant turn.) As it waltzed over the Channel Islands, the French authorities were alerted. Five Mirage fighters of the French Air Force and an RAF Nimrod were scrambled to investigate. The fighters flew close enough to the Beechcraft to see in the cockpit two figures who would not respond to radio calls or visual signals. As the circles grew tighter and lower over the French coast, it became obvious that Rennes or even Nantes could take a direct hit. Did the authorities pooh-pooh the threat on that occasion? No; they ordered the fighters to shadow the plane, and shoot it down if it looked like crashing into a populated area. It did not come to that. After five unnerving hours, having completed three dozen full circles, the phantom flight ran out of fuel, passed gliding over Nantes, and burrowed into a village vineyard.

Hongkong is the 'worst prospect scenario'. It is not an easy place for pilots, and in Kowloon and its surrounds, at the north end of the single runway, up to 4,000 people per acre huddle in highrise buildings. Usually, take-offs go south, over the sea. But in strong northerly winds, aircraft must fly towards the buildings, which leaves them with barely enough clearance. Other big cities must also live in hope that an aircraft will never crash into populated streets and houses. At London, when an aircraft comes off the stack, it is guided down, through heavy traffic, maintaining compulsory separation, aiming to be at 3,000 feet at the Tower Bridge area, pick up the ILS signal, and fly down the Thames, heading for Windsor Castle and Heathrow. The new City Airport brings aircraft right into the city. Experience shows that has been fine. But are the authorities fully prepared to deal with what they hope will never happen? Do they accept that genuine accidents follow a route that is deemed to be impossible?

In the late 1980s, Britain had more than its fair share of *impossible* disasters – in the air, out at sea, on the Thames, on the ground, and even underground. Many critics complain that the public inquiries shut the door after the horse had bolted. But was it shut at all? Bureaucracy, lack of coordination, lack of funds, lack of proper authority, lack of clear chain of command – these are just some of the reasons why the Association of Civil Defence and Emergency Planning officers went public, stating that they had 'little confidence' in the government's disaster plans, particularly for London, where delays, duplication, improvisation and confusion might lead to many unnecessary deaths.

Proper preparations for civil emergency will never be devised without guidance generated on a national level, and probably not without divorcing planning against manmade and natural disasters from civil defence against nuclear war. In a cautionary letter to *The Times*, John Holloway, London's Chief Emergency Planning Officer, attributed the inadequate state of preparedness to the government's negligence, and, as an example, singled out the menace of a midair collison between any two of the 3,000 aircraft passing over London every day.

CROWDED SKIES

Aeromexico Flight 498, a DC9, was a scheduled passenger service from Mexico City to Los Angeles on August 31, 1986. After a stop at Tijuana in Mexico, it took off with fifty-eight passengers and a crew of six at 11.20 Pacific Daylight time. The weather was to be fine all the way, but the flight would be under IFR (Instrument Flight Rules) as planned.

At 11.40, when the ten-million-dollar DC9 was cruising at 10,000 feet, a thirty-thousand-dollar Piper was cleared for take-off at Torrance, California. Thirty-six seconds later, the pilot told the tower that he was 'rolling'. With fourteen miles visibility, he would fly his light aircraft under VFR (Visual Flight Rules).

Flight 498 had an uneventful passage. At 11.44, it began its descent towards Los Angeles with the guidance of approach control. Five minutes later, the company agent radioed the DC9 its gate assignment. At 11.50, the arrival radar controller flashed a warning to the captain: 'Traffic, ten o'clock *(the direction, looking forward)*, one mile, northbound, altitude unknown'. This advisory was acknowledged, but the pilot did not follow up with a report of actual sighting that 'traffic'. The northbound blip was, in fact, a Grunman Tiger. Flying VFR, its pilot had contacted Los Angeles for guidance – that is how he was spotted on radar. It now turned out that he had no idea that he had strayed into TCA, the Terminal Control Area, without permission and at a height where there was heavy jet traffic. Unbeknown to anybody he was not the only stray in the TCA: the Piper was also there, unseen on radar, without ever making any calls to Los Angeles ATC. The Grunman pilot was duly told off by the radar controller: 'In the future you look at your TCA chart. You just had an aircraft pass right off your left above you.'

Precisely as he finished the sermon, thirty-six seconds after 11.52, the controller noticed that Flight 498 had disappeared from his radar screen.

All attempts to regain radio and radar contacts failed. Approximately twenty-seven seconds earlier, something horrific had happened in the cloudless California skies. DC9 and Piper wreckage showered down on a 600 by 200 feet area of Cerritos. Everyone on board was killed plus fifteen people on the ground. Five houses were destroyed and seven others damaged.

The burnt and broken flight recorder (an antiquated type) was not much use to the investigators, and the poor quality CVR was just as unhelpful. (At 11.51:30 the CVR recorded the captain thanking somebody for something. Then not a word until 11.52:10, one second after the estimated time of the collision, the captain is heard saying: 'Oh § this can't be' *(expletive § deleted in report)*.

If the recorders were silent, the wreckage was loquacious enough. The few autopsies that were possible at all supported the telltale marks of collision in the air. The cabin roof of the Piper had been sheared off, its pilot and both his passengers had been decapitated, and debris from its cabin roof was found embedded in the leading edge of the horizontal stabilizer of the DC9. Correspondingly, the DC9 vertical and horizontal stabilizers showed propeller slice marks and paint transfer marks from the Piper. From the wreckage it could be established with great precision where and how the light plane struck the stabilizers of the jet, and also that at impact the Piper was in the almost wings-level attitude (the shear marks occurring at the same height on both sides). 'Longitudinal control and stability (of the DC9) was lost when the horizontal stabilizer separated' and, from that point, the pilot could not fly the aircraft.

'So the unthinkable has happened,' a television commentator exclaimed, but was it also unimaginable? Had it not been the Piper, it might have been the Grunman that hit the DC9, and we shall never know how many other uncontrolled aircraft were there, unseen and untroubled, slipping in and out of that TCA zone. The investigation determined that 'the limitations of the air traffic control system to provide collision protection' was the probable cause of the accident. The Piper's unauthorized entry into the area, and the inadequacy of the 'see and avoid concept' were seen only as contributory factors.

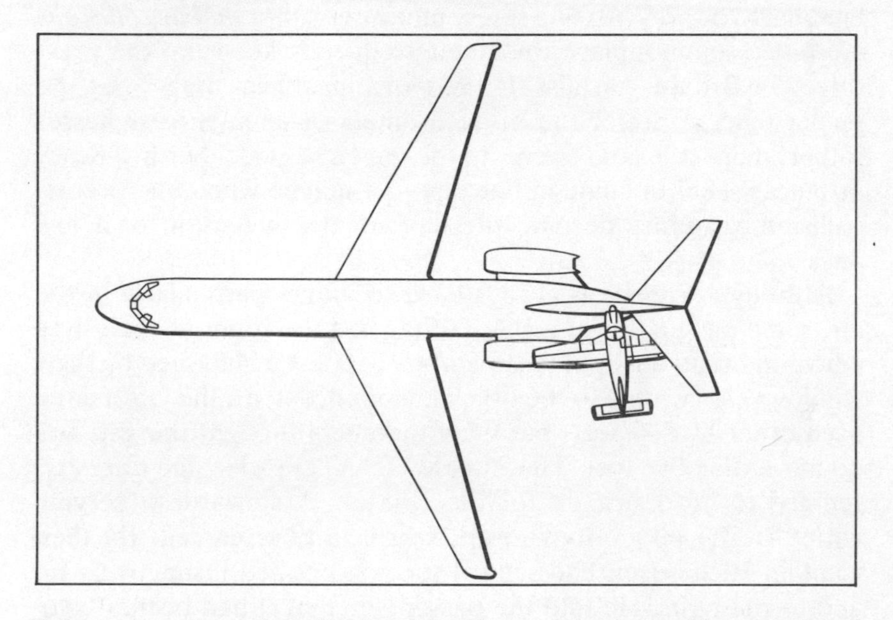

Estimated positions of the DC9 and the Piper, at the point of impact, viewed from above. The reconstruction is based on wreckage examination, particularly the damage the Piper's propeller caused to the leading edge of the DC9's vertical stabilizer. (Source: Accident Report by the NTSB.)

The ultimate condemnation of the entire system was that *it had chosen to live with the known risks:* since 1967, the NTSB had bombarded the FAA with '116 recommendations as a result of its investigations, special studies, and special investigations of midair or near midair collisons. A review of these 116 recommendations identified fifty-six that are pertinent to the accident at Cerritos.'

The sky gets crowded around airports and along air lanes like a revolving door between a spacious hotel lobby and a leafy road. It is mostly light aircraft that tangle with each other, as well as commuter, military and business jets. Teeming traffic in the busy American airspace alone produces some *eight hundred* such accidents every year – mostly in clear skies, under visual rules, often almost head on. Like fires, collisions conjure up images more horrible than death itself. According to an international poll conducted by Lufthansa, airline pilots themselves rate collisions as the most feared threat.

Cerritos involved a passenger jet, and – because public outrage is still proportionate to the numbers killed in one go – became a landmark for preventative safety measures. *Near midair collisions* are more commonplace, but their frequency keeps up the pressure. (In Britain, 'airmiss' is the word in official usage, but the media tend to prefer the more dramatic *near miss* as if misses rather than collisions were to be given a wide berth.) Scary instances seem to come in batches – or maybe when one occurs, collision headlines become the flavour of the season for a few days.

Although distance is always hard to judge, particularly in the air, some passengers of a BA TriStar had the fright of their lives when, in broad daylight, a Bulgarian Tupolev 154 flashed by their windows. According to the official announcement, the jets missed each other by 300 feet, but crew members thought the gap was no more than five feet. The Tupolev (NATO codename *Careless*) seemed to head straight for the TriStar. A stewardess, serving coffee to the pilots, looked up, saw it, and screamed. By then Captain Richardson had seen it too, and pulled instinctively up and to the right. He told the passengers that it had been a 'very close shave'. With altogether 553 people on board, both aircraft were running some thirty minutes late – one from Paris, the other from Sofia – as if to keep what might have been a deadly rendezvous some 18,000 feet above Lydd on the Kentish coast.

The developing close encounter had *not* been spotted by air traffic controllers.

Two years earlier, two aircrews, a hundred passengers, and the inhabitants of Stanmore, a north London suburb, were exposed to the threat of an aluminium shower when two BA flights from

CLOSURE SPEED			
DISTANCE BETWEEN TWO AIRCRAFT	960 mph	600 mph	360 mph
	COLLISION IN SECONDS		
10 m	37·5	60	100
6 m	22·5	36	60
5 m	18·75	30	50
4 m	15	24	40
3 m	11·25	18	30
2 m	7·5	12	20
1 m	3·75	6	10
½ m	1·8	3	5

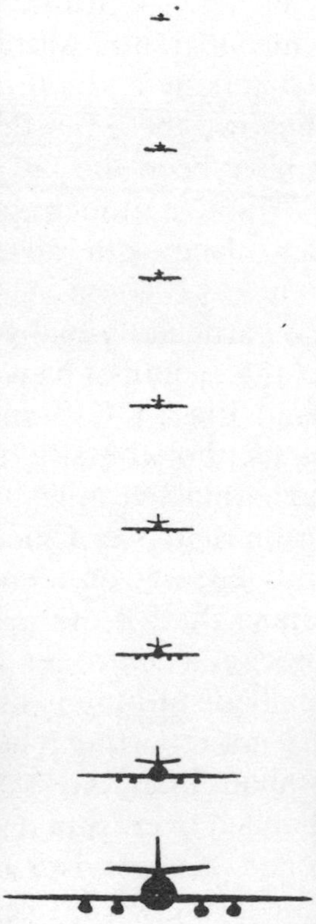

RECOGNITION AND REACTION TIMES *IN SECONDS*

SEE OBJECT	0·1
RECOGNISE AIRCRAFT	1·0
BECOME AWARE OF COLLISION COURSE	5·0
DECISION TO TURN LEFT OR RIGHT	4·0
MUSCULAR REACTION	0·4
AIRCRAFT LAGTIME	2·0
TOTAL	12·5

Collision is inevitable: the pilot has not got sufficient time to see and avoid.

Munich and Edinburgh converged, as instructed, towards an uncharted yet all too real point 7,000 feet up, from where they would turn and take the well-trodden path to Heathrow. (See chart above.) A collision grew more and more inevitable with every split second. Their combined closure speed was 600 mph. If, at that speed, a pilot can notice an oncoming aircraft two miles away (where it appears to be as small as a fly) he has twelve seconds – half a second less than he needs – to save his life. For he has to recognize that the speck of dirt he sees is an aircraft, guess that it is on a collision course, decide what to do, react, and wait for his aircraft to respond.

On this occasion, terrified air traffic controllers noticed the Munich flight, and sent out frantic warnings much too late: the two blips were already merging on their radar screens.

But, by then, somehow, the Munich pilot had spotted the danger in what must have been the last second to take evasive action. He forced his 737 into a sudden, steep climb, and virtually leapfrogged the other plane with an estimated fifty feet to spare.

1987 seemed to be a particularly bad year for U.S. carriers. It was then, for instance, that, south of New York, a PanAm Airbus A310 and a Venezuelan Viasa DC10 missed each other by 300 feet vertically and 500 feet horizontally; that two Delta 727s had only one instead of the normal five miles separation; and that only luck prevented a collision right over Chicago when two departing flights had inadvertently been given identical transponder codes for radar identification so that, if the instructions and the navigation were equally precise, a controller could have steered them into each other. 'Equally disturbing is that American called the FAA and said, "We're not reporting a near midair collision, but we want you to know about it anyway".'*

When aircraft are handed over from one to another controller, separation is often eroded. A supervisor at O'Hare once recorded in his log: the technique is 'the worst I've ever seen . . . this isn't a safe operation'.

It was in the same year that 650 people came within thirty feet of participating in a potential record death toll. A Continental Airlines Jumbo and a Delta L-1011 were flying westbound at 31,000 feet. As they followed parallel North Atlantic air lanes, there should have been ample horizontal separation between them, but the Delta strayed more than sixty miles from its assigned course, and almost rammed the Jumbo. Such gross deviations are

* Speech by Jim Burnett to Allied Pilots Association, Arlington, Texas, 1987.

unusual but frequent enough to pose a threat, said the Canadian and American investigators. Seven years earlier, the FAA had emphasized that certain procedures were essential to achieve accurate transoceanic navigation, but its circular was only advisory, not mandatory. The investigation of this near-disaster revealed that Delta crews were 'not supplied with oceanic charts to graphically display the coordinates and path of the assigned route to track', as suggested by the above circular. When the L-1011 crew misprogrammed the flight computer, they could not spot the error because they performed only one of the six recommended safety procedures, and 'did not plot their present and predicted positions' for track verification en route. Finally they tried to hush up the whole affair.

The Jumbo-pilot refused Delta's request to keep quiet about the near-collision: 'I have passengers pounding on the door and crying that they saw the whole thing out the windows.' Several other crews flying in the area then joined in on VHF to air 'many unprofessional remarks' and urge the Jumbo-pilot not to report the incident. But, unknown to anyone, all the conversations were recorded by a U.S. Air Force jet nearby.

Air traffic controllers had no idea of what had happened. The Jumbo did report the incident – but only after it landed. For the Delta flight, it would have been not only correct but also vital to confess the truth right away to Gander Oceanic Control in order to obtain guidance for finding its way back on course. The failure to contact the controller led to yet another potentially hazardous situation: the crew would have no idea whether their delayed reappearance on the assigned track would find the air lane clear of other traffic. (At the time, Delta's crew training and operational practices were under investigation because of various errors, such as the inadvertent switching off of both engines of a 767 after take-off from Los Angeles.) The Canadian investigators strongly criticized the airline, whose flight manual had to be revised, and the crew and all others who had tried to hush up the whole affair. What is a near-collison? How close must aircraft come together to qualify? And are scare-statistics due only to conscientious reporting by airline pilots and frightened but often mistaken private fliers?

The FAA defines 'a near midair collision as an incident in which a collision was avoided merely because of chance', and the case is regarded as *critical* when the separation between two aircraft is a hundred feet or less.

Britain's JAS (Joint Airmiss Section), an organization run by the RAF and the CAA to investigate all such reports, defines an airmiss as an occurrence 'when a pilot considers that his aircraft may have been endangered by the proximity of another aircraft'. The key word is *considers* because Group Captain John Maitland, head of the JAS, knows from experience that in many reports the assessment of the hazard is often subjective and exaggerated. 'Our investigation never tries to apportion blame,' he said. 'Our aim is to learn useful lessons, to clear up discrepancies produced by honest mistakes, and to establish facts that are hard to come by: even a radar trace recording won't always yield perfect evidence because when two symbols merge on the screen, we can't tell whether the two flights are fifty or 150 feet apart. Some countries on the Continent, including France, Germany and Belgium, have short-term conflict alert built into the control system. Their computers can recognize potential hazards of an aircraft climbing through the path of another, for instance. That's easier in areas where most aircraft are in transit on level flight unlike in Britain, where there's a huge amount of climbing and descending traffic at all times. I'm glad to report, however, that the system is now being introduced gradually in the U.K., and the results are promising.'

The lack of hard evidence leaves plenty of room for rumours and speculation. Both the JAS and the CAA have persistently denied all allegations of any crisis in British airspace. The CAA statistics indicate some improvements and, in the absence of a truly independent body to verify their data, it appears that they find two airmisses a week an acceptable level of risk. (Airlines experience one risk-bearing airmiss per month, down to almost a third of the late 1970s level.)

In America, authorities as well as the public find the available data – accumulated perhaps through more reliable reporting systems – rather alarming. 'Passengers should expect to see blue sky when looking out the window, not the registration number of another airplane,' said Jim Burnett. He was referring to the fact that in 1986, there were 343 near midair collisions involving passenger aircraft (forty-two per cent up on the previous year), and fifty-seven of these were critical. Yet there was an identical increase the following year, and, together with the density of traffic, the hazard continued to grow. One could perhaps be forgiven for suspecting that the chief difference betwen British and American reactions is not a matter of calm versus hysteria but a single tragedy, known as *Cerritos*. If hope against hope will not prevent a major British midair collision over populated areas, the outcry will sweep numerous occupants out of cushy offices.

Air traffic controllers, who sometimes work a sixty-hour week, cannot be fully relied upon to prevent mishaps. As the system is implicated more and more frequently in creating or failing to prevent dangerous situations, good fortune remains the chief defence against midair collisions. Equipment is becoming obsolete, unable to cope with the volume of traffic; controllers become overworked and fatigued under stress only to be bored to inattentiveness at slack times. They attract growing criticism in Spain, Italy, Yugoslavia and the rest of Southern Europe as well as third world countries. A Sardinian court jailed a controller for three years for his part in causing a DC9 crash and thirty-one deaths. In Thailand, in 1987, a 737 fell into the sea at Phuket, the holiday resort, killing eighty-three people. The Thai aircraft was closer to the runway but higher up than a Dragonair 737, also on approach; controllers probably favoured their own nationals – a frequent mistake in many countries – and told the Thai flight to land first rather than to go round and follow the Dragonair in. The proximity of the Dragonair below worried the Thai pilots, who watched it so intently that they failed to monitor their own air-speed indicator and slowed down below stalling speed. After the crash, there were disciplinary hearings, controllers were transferred, and the installation of better radar was strongly recommended.

A special risk of mutual inattentiveness in the air and on the ground occurred in October, 1985. Flight JL 441, a Japan Air Lines 747, was on its way from Tokyo to Paris via Moscow. Off the Japanese coast, the captain encountered some turbulence, and chose to go round it by switching to heading mode. After the detour, he reset his course, but forgot to re-engage the navigation mode of his INS (Inertial Navigation System). A little later, when he carried out a routine check of his position with the aid of a radio beacon, and noted some islands showing up on his radar, he made a stunning discovery: he was flying sixty miles off course, closing in on Soviet military installations of great secrecy on Sakhalin Island.

Any reader who experiences at this point a spinechill of *déjà vu* should feel perfectly justified. Flight 441 was now almost exactly at the point where Flight 007 of the Korean Air Lines had been shot down so callously by a Soviet air-to-air missile two years earlier.

The Japanese pilot changed course immediately, and tried repeatedly but in vain to contact Khabarovsk. The Soviet controllers were not listening. Cut off from everyone in Soviet airspace, the pilots must have feared the worst. They would never even see where a missile might be coming from, though Japanese defence monitors detected that two MiG fighters had already scrambled 'for some unknown reason', and were on their way towards Sakhalin. To avoid yet another botch-up, Soviet military controllers kept calling the Jumbo on 121.5 Mhz, the international emergency frequency. When their efforts also failed, probably because the pilots were desperately busy trying to make their own contact, the local commander must have sought urgent orders from his superiors. The unanswering intruder would soon escape from Soviet airspace – just like the Korean aircraft seemed to do. At least the caution taught by the 007 tragedy delayed decisive action long enough this time to save four hundred lives. When the pilots managed at last to raise Khabarovsk, Flight 441 was already close to its right course, and got permission to continue. As a result of the dangerous incident, JAL introduced some changes for safety in its operations manual, the captain apologized for his inattention but, as usual, nothing became known about Soviet reaction to their controllers' slackness.

Air traffic control errors are multiplying everywhere – though they are not as readily admitted anywhere as in the United States where, because of the fast increase in mistakes (or greater awareness of their regularity), the alarm has been sounded many times by the NTSB, ALPA and other authorities. Usually, the system is just patched up here and there to keep ATC in business. In 1988, for instance, after the serious operational errors at the busy Chicago O'Hare control tower and radar room, a special NTSB investigation and tight monitoring by the FAA (with daily problem reports to the Administrator himself) led to greater safety through ad hoc band-aid of the sort the third world and even Europe tend to apply if and when necessary. Now America is about to take two revolutionary though belated steps to reduce the collision hazard and, at the same time, prove that those densely populated skies could be made even more crowded without compromising safety.

TO SEE OR NOT TO BE

Pilots are drilled to scan the skies and their instruments alternately and almost incessantly, but boring routines dull the senses, and *looking* does not always equal *seeing*. What pilots call the *Mark 1 Eyeball* is an imperfect instrument with a retina that loses its sensitivity outside a small central area.

If you look at a bookshelf without *any* eye movement, you can read only one or two titles – depending on the distance. An aircraft that maintains a collision course in height and speed with your flight, coming in towards your path at a fairly shallow angle from the direction of the corner of the eye, may remain almost invisible to the regular forward scan until it is too late to avoid it. Raise your arm sideways to shoulder level, stare straight ahead, and fly a finger slowly towards the limit of your reach in front – you will see how late the blurred 'intruder' will enter your field of vision and take a recognizable shape. Scanning, instead of staring, particularly if you turn your head, will reduce the time it takes to spot and recognize, but your eyes will make the sweep jerkily because, in a room, various objects will frequently hold and delay them, and, in the empty sky, there will be nothing to guide them. Yet under Visual Flight Rules, the see-and-avoid principle prevails. Flight instructors and specialists offer an abundance of valuable advice* like 'remember that the aircraft that you're likely to collide with is the one that appears to be stuck in the same place on the windscreen. If it is moving, you'll probably miss it.' Whereas strict adherence to the best advice will not guarantee protection (investigators found that, at Cerritos, 'both

* 'How Not To Have A Mid-air' by Roger Green, Principal Psychologist, RAF Institute of Aviation Medicine. The article was originally published in *Air Clues*, July, 1982.

airplanes were within the pilots' fields of vision for at least one minute and thirteen seconds' before impact), the already available on-board warning devices could do just that.

The idea for TCAS, the Traffic Alert and Collision Avoidance System, has been around for some thirty-five years. Its full development was slow, it was seen as a costly extra gadget and, as usual, airlines would equip their fleets with it only if and when they were forced to do so simultaneously with their competitors. By 1986, some versions of TCAS were there for the asking but, not for the first time, it took yet another terrifying accident to provoke the outrage that gave the impetus to new safety legislation. Following the Cerritos disaster, the investigation concluded that, had there been such an instrument aboard the Mexican flight, the alert would have made the pilot actively search a certain section of the sky, and the probability of spotting the Piper in good time would have increased from thirty to ninety-five per cent. The recommendation to introduce TCAS was repeated by the NTSB yet again, and the FAA ruled that, beyond December 1991, every domestic and foreign passenger aircraft with more than thirty seats must have TCAS mounted in the cockpit if it flies in U.S. airspace.

With varying degrees of sophistication, TCAS can detect an intruder in an aircraft's protective separation envelope, warn the pilot about its proximity and potential collision course, alert him when it becomes a threat, and suggest ways to avoid it. The three manufacturers of the equipment stand to make two billion dollars because, at least initially, the cost of installation per aircraft will exceed $120,000, but if an airline with a fleet of one hundred managed to avoid just one collision at the cost of twelve million dollars, its savings would be tremendous. Some operators argue, however, that the frequency of the threat does not justify the investment – at least not as rapidly as all that. While Northwest Airlines became the first in 1989 to sign a contract for retrofitting its entire 309-strong fleet and equipping all its new aircraft on order, most airlines and IATA, their trade association, still fight for at least a year's extension of the deadline, which they are likely to gain. The resistance to TCAS is probably strongest in Britain where the CAA is anxious about accuracy. 'What may be viable in the wide open skies of the Continent or America, is not necessarily going to be useful in Britain,' they told Harvey Elliot of *The Times* on June 6, 1989, and they expressed fears that 'a false alert can lead to distrust of the system and it may be eventually ignored'.

Crowded skies need more advanced technology. As long as safe passage through the airspace relies on human judgment, collisions remain a major hazard, particularly while there is an almost critical shortage of well-trained pilots and air traffic controllers everywhere.* It is estimated that it will take two more decades to achieve world-wide, computerized '4D controls' (time being the fourth dimension), but the second part of the two-pronged American attack on the problem, the introduction of a national *traffic flow management*, may help to reduce the collision risk, cut out eighty per cent of controller errors by 1996, and prove that more can be crammed more safely into limited airspace.

It was a breathtaking experience to be the first 'non-industry' visitor to the new National Airspace System in Washington. In the bureaucratic citadel where the FAA lives, a single step led from a featureless corridor of power into the twilight of a science fiction film set of the next millennium. The Centre was in its final, experimental stage. Controllers were still learning to use their futuristic paraphernalia, the banks of multicolour computer screens that could conjure up the entire airscape of the United States. The two-dimensional pictures were full of dots, each representing an aircraft in flight. In some areas, the traffic was so dense that the dots merged into blobs to blank out whole states like Pennsylvania.

'The flow management concept was developed, in fact, as a sheer necessity during the long strike when a handful of people managed heroically to do the impossible, and kept the traffic rolling,' said John 'Jack' Ryan, Director of the FAA Air Traffic Operations. (In 1981, President Reagan declared the controllers' strike to be illegal, and sacked 11,483 of the 16,375 staff. Augmented by managers and the military, the remaining less than thirty per cent of manpower ran the system despite the growth in traffic. Their achievement also revealed that the original structure

* Whatever indignant denials one hears in various quarters, it is most revealing that massive advertisements to recruit pilots and controllers for national airlines and authorities have lately begun to appear outside the trade press, even in daily newspapers and on television.

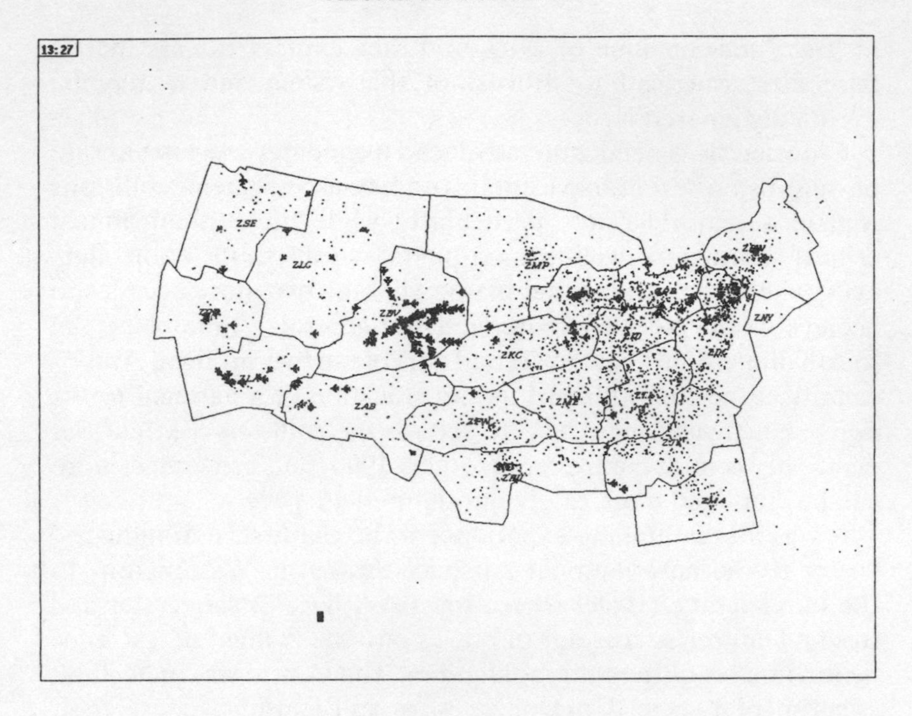

Airflow is U.S. airspace at 13.27 (top) and 13.48 (below). Not so much dense as jammed. (Source: FAA.)

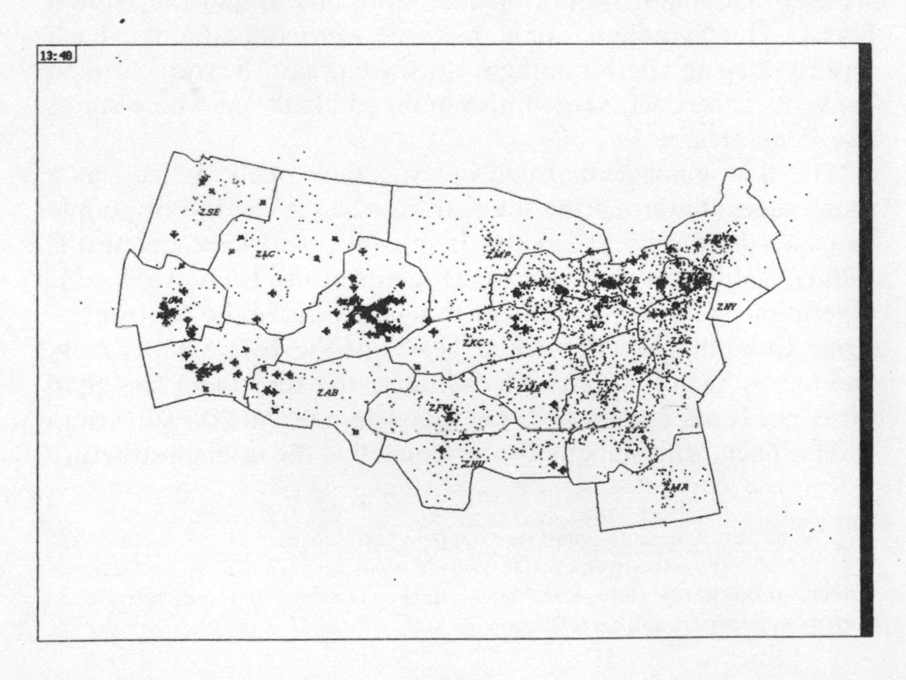

carried a great deal of fat. Since then the work force has been increased gradually, but at the end of the 1980s – with traffic up to 400 million passengers a year, a fifty per cent increase in a decade – it has still not reached the pre-strike level in numbers and training. The long crisis also proved that overcrowding at hub airports, the product of deregulation freeforall, could be coped with through voluntary agreements between airlines – re-regulation by self-control in everything but name.)

'The strike began in the week of Thanksgiving, the busiest time in the United States. At five a.m., reports began to come in that people were not turning up for work. By eight-thirty, we had to put one of our contingencies, the *fifty plan* into operation: we told all airlines to cancel fifty per cent of their flights. It had to be as drastic as that. Four days later, we were up to seventy per cent. We had to make complex estimates and create a sophisticated space allocation system. It was probably the real beginning of ASD, Aircraft Situation Display, which we'll now have at long last on a national scale.'

The Display, on one huge screen and numerous 'working' screens, feeds on automatic updates every five minutes from twenty air control centres, so that each FAA operator in Washington can see instantaneously the whole tactical situation in U.S. airspace at any time. It is then possible to zoom in on a single zone or an airport, or select up to sixteen airports and identify all aircraft en route to those airports by painting them in one of sixteen colours at the touch of a button. And that is not all. 'Look,' beams a fifty-year-old toddler who has been let loose among the Christmas offerings of this super-toyshop, 'at that red dot there!' A few dancing steps by his fingers on the keyboard, and an identification tag appears next to a miniature red aeroplane. 'It's the Flying Tiger Flight 275, a DC8, cruising at 35,000 feet, with twenty-six minutes flying time from its present position to San Francisco because its current ground speed is 453 knots. The only thing we don't know is what size shoes the captain takes.

'Now that green dot there . . . Eastern Flight 829, a 757 coming from Salt Lake City, descending through 35,300 feet, twenty-nine minutes to Denver. Let's look at Mexico . . . that blue dot, there . . . it's LI6365, a 727, climbing through 26,000 . . . Where else do you want to go? Here's the North Atlantic traffic. That's the

Irish coast . . . that's a TWA over England, going to Heathrow . . . Great pity that, so far, Europe has refused to send us flight plans, so we know only all the U.S. aircraft going to anywhere over there, but don't know in detail what's coming this way until they get into radar range from Boston or New York, and we can start tracking them. And we're only thirty seconds behind real time because the information comes in via satellites. Soon we'll have winds and weather data, too, on the screen to let us be more precise with our projections.

'The most important is our monitor alert capability. Look, we can pick out any of the 653 sectors of our airspace, any one of the airports, and put range rings on the screen. There . . . you now see two rings round Chicago . . . and all the aircraft flying there just now. Some are within the sixty-minute ring, the rest in the ninety-minute range. That tells us how heavy the traffic congestion will soon be there. If Chicago wants to restrict arrivals to fourteen in any fifteen minutes, the computer will be programmed to turn that sector red every time a fifteenth arrival is projected within that quarter of an hour that may be three hours from now. The controller may then discuss with us how to prevent a bottleneck and avoid the need of wasteful stacking. We may tell some aircraft to slow down or delay departure. We may even be able to convince airlines that a slightly longer than usual route may be cheaper by avoiding delays and burning up fuel in the stack. There . . . Orlando, Florida has an alert! It's foreseen that forty minutes from now, too many aircraft will want to land . . . '

Washington will have twenty operators, one for each regional centre which, in turn, will have the same display on the local screen, so they all will see what they discuss. This airspace management will not only lead to better exploitation of a very limited commodity, but can also help controllers to learn from their own mistakes: as everything is recorded, they can replay on video an eight-hour traffic movie in fifteen minutes, with dozens of aircraft gliding across the screen, and showing how things went wrong.

It is inevitable that, one day, a similar system will be introduced at least in Europe to replace the current steam-age technique of flow control, and overcome the problems due to low-capacity ATC and inadequate radar cover all over and south of the busy Mediterranean routes. It appears, however, that, despite growing international cooperation, the conflicts of national interests and pride may continue to constitute an even greater obstacle than the cost of hard- and soft-ware. At the time of writing, Europe's leading airlines have launched an ambitious plan to merge all the independent ATCs of their twenty-two countries, and virtually privatize the handling of ample yet underutilized airspace. Their aim is eliminate chaos, delays, collision risks, and perennial crises of crowding. The American taxpayer will fork out some eighteen billion dollars in all for the new national system. The European airlines propose to fund the Euro-network by 'user charges' so that 'governments would be relieved of the need to provide money for ATC investments'. Though the expense may look enormous, here is a thought to counter it: if Europe could eliminate most of the delays that cost some three billion pounds a year, that saving alone could soon pay for the necessary investment. (In 1987, Lufthansa, for instance, burned up fifty million Deutschmarks [some seventeen million pounds] worth of fuel by having to *hold* aircraft for 5,200 hours over just three key airports – Frankfurt, Munich and Düsseldorf.)

The CAA and other national authorities 'expressed interest' and found the plan thought-provoking. Keith Mack, formerly of the CAA's top ATC echelons, flaunted his new hat as director of Eurocontrol, and came out in favour of the idea saying that 'there's undoubtedly a need for a unified system. We already have the equipment and the expertise to provide the basis and we have been trying for many years to achieve the kind of harmony the airlines are now asking for.'* But the airlines have been asking for the same, albeit on a smaller scale, while bodies like the CAA have kept asking: 'What crisis?' for just as many years. And even if a new, truly European view evolved, how far would it be from implementation? In 1988, more actions were started against EC member states than ever before for blatant non-compliance with Eurolaws they themselves had voted in.† That, of course, would

* *The Times*, 8.9.1989.

† Italy, for instance, received 107 warnings, fifty-two notices of intent to prosecute, and was prosecuted fourteen times in that year. 1988 was Britain's best annual record ever with thirty-three warnings, fifteen threats, but no actual prosecutions against her in Eurocourts. Belgium was taken to court ten times, Greece fourteen times. Concerning the completion of the 'single market' by 1992, only seven of sixty-eight legally enforceable directives have as yet been fully implemented by all members.

not surprise anyone familiar with the frequent disregard for the law that English is the language of the air (perhaps the oldest law in international aviation). Law-breakers also deviate from standard, prescribed forms of clear communications, and the result can be painful, ridiculous or potentially disastrous.

*VAT'YA SA' MUCHACHO? – OVERSHOOTEZ! OVERSHOOTEZ!**

TCAS, satellites and ASD are expensive gadgets for the 'fat cats' of aviation – speech costs nothing, but its careless usage can cost lives.

The problem is as old as aviation itself. That it has been allowed to grow as old as that is almost criminal.

In 1971, it was still allowed in Sydney that aircraft should taxi, backtracking on a runway that was also in use for landings and take-offs. On a clear but dark night, a DC8 landed. The controller instructed the pilot to 'take Taxiway Right – call on 121.7'. This was acknowledged with 'roger' but not read back. Eventually, the four-man cockpit crew (all from another English-speaking country!) would declare unanimously that what they heard was 'Backtrack if you like – change to 121.7.' They made a 180° turn and started back down the runway. The controller thought he saw the DC8 turn on to the taxiway nearby, and cleared a 727 for take-off. There was a collision, but, somehow, nobody got hurt. It is not impossible to mishear 'take taxiway right' as 'backtrack if you like', and the accident report acknowledged that 'accent and idiom' might have contributed to the misunderstanding. But had the pilot given or the controller demanded a proper read-back of the instruction, the danger would have never arisen.

In 1974, a BOAC 747 (callsign Speedbird 029) was approaching Nairobi. Its controller was working under supervision, but the supervisor left the room for four minutes during which a clearance was issued to the aircraft: *speedbird zero two nine . . . descend seven five zero zero feet. The QNH is one zero two zero decimal five.* The pilots heard *'five zero zero zero feet'* and that was what the co-pilot read back fast and without any hesitation: *Roger speedbird zero two nine cleared five thousand feet on one zero*

* Agitated exchange between Brazilian pilots and French controllers.

two zero decimal five. The mistake was not spotted, and the acknowledgment was not challenged by either the flight engineer on board or the radar controller because, as it often happens, both might have heard what they *expected* to hear. It is probable that the word *'seven'* was indistinct, the *'five'* was given greater emphasis, and the pilots were concentrating on counting the zeros. Setting their target height, the pilots now had a chance to pick out the error: 5,000 feet is 500 feet below ground level at Nairobi! There were, however, slight problems in the cockpit. The crew, on duty for nine hours, were at the 'lowest point of their normal circadian rhythmic cycle' because at home they would now be asleep. The co-pilot had a bowel infection for which he had treatment from his own doctor without informing the airline, and the medication might have reduced his efficiency just when he might have been keen to impress a very senior captain of great repute with whom he was flying for the first time. Perhaps he acted briskly, without pause for thought which might be interpreted as indecision. The captain tended to do too much by himself, GPWS was not yet employed, the company monitoring system, which would soon have to be improved, was not followed properly, and some instrumental indications of the problems were overlooked or ignored. When the aircraft broke through the clouds – and the co-pilot shouted: 'Give full power . . . give full power! . . . Check height!' the ground was coming up at them at 217 knots. By the time they managed to climb away, they were only seventy feet from impact.

If one *tried* to stage the worst tragedy with the biggest toll on record, one would need the computer power behind the Voyager space venture, and the outcome would still be dependent on a timely puff of wind – or someone deciding to ask a pilot 'what did you say?'

On March 27, 1977, two Jumbos were heading for Las Palmas in the Canary Islands. Both were on time. The PanAm came from Los Angeles, the KLM from Amsterdam. But at the airport a bomb explosion was followed by the threat of another, so all incoming flights were diverted to Los Rodeos airport at nearby Tenerife. There the runway centre lights happened to be out of order yet again, the runway intersections were not lit or even marked properly, and the weather was playing one of its treacherous tricks, a local peculiarity, with drifting fog that would produce dramatic changes of visibility from one moment to the next. The airport was soon overcrowded with diverted flights. As the PanAm had pulled up just before the KLM, it got hemmed in at the end of a parking area.

Waiting, waiting, everybody itched to go and make the short hop to Las Palmas. The controllers waited to see the back of them all. The KLM was meant to return to Amsterdam the same evening. The pressure was on the pilots: their duty-time would be exceeded, and everybody would be stranded overnight if they could not leave Tenerife pretty soon. When at long last the 'all clear' came from Las Palmas, the weather deteriorated, and the KLM captain decided to refuel in case he could not get into Las Palmas after all. That held up the American pilots who knew that their 380 tired passengers were keen to start their twelve-day luxury cruise. They walked impatiently around the Dutch plane to measure if there was enough room on the apron to squeeze through and go first, but the gap was too narrow. They just had to wait.

The two Jumbos would have priority to leave, but the taxiways were still chock-a-block with parked, diverted aircraft. As both captains were ready to depart, the tower gave them both permission to taxi almost in tandem, backtracking on the main runway. The KLM would go to the far end, turn back, and take-off. The PanAm would follow it but only as far as the third intersection where it would turn off, and hold.

There was some confusion in both cockpits. The Americans were unsure which intersection they were supposed to find in the fog. That was sorted out, but they had only a small map of the unfamiliar airport, and they missed the unmarked turn-off. Still, they could take the next exit.

The KLM crew rushed through their pre-take-off checklists as they taxied down the runway. They wanted to finish the job in hand, so they delayed asking for their route clearance until they reached their take-off position. They were already lined up and keen to roll when the co-pilot warned the captain 'to wait a minute' because they had not yet received their route clearance. (That was already a little more interference than the co-pilot was keen to exercise: his captain was not only very senior to him, not only the face in company advertisements, but also the chief KLM flying instructor who had type-rated him, and who, therefore, *must* be right at all times.) The captain told him to get the clearance. The scene was now set for some transmission congestion.

The Tower reeled off the route to follow *after* and subject to take-off clearance. While the co-pilot was reading it back, the brakes were released, and the captain increased power. The crew overheard the communications with the PanAm flight giving rise to some doubts concerning the whereabouts of the other Jumbo. The flight engineer asked repeatedly: 'Is hij er niet af die Pan American?' (Isn't he off, the PanAm?) The answer he received from the captain was an 'emphatic' but most ambiguous 'Jawe' (Yes). The co-pilot ended his read-back with the non-standard, meaningless words: *'We're now at take-off.'* The controller did not ask for any clarification because he interpreted that to mean that the KLM was now at the point from where it would be ready to leave as soon as the second, the vital, *take-off clearance* was granted.* So he just said 'Okay,' which in itself was incorrect. But two seconds later, perhaps a little bothered by that peculiar sentence, he added: 'Stand by for take-off. I will call you.' He did not know that the actual take-off procedure had *already* been initiated some six seconds earlier.

The PanAm crew did not like the sound of what they heard. All that talk about take-off . . . Lost and insecure in the *soup*, they chose to warn the other pilot directly. The simultaneous radio transmissions coming into the KLM cockpit now caused a temporary blocking of the frequency as they fused into a four-second shrilling squeal. The word *'take-off'* had been mentioned. Raring to go, and under the influence of anticipation, the Dutch captain's longing to hear that clearance must have turned wishful thinking into reality – and the 747 began to accelerate.

The Americans could not see the aircraft bearing down on them through the fog, but sensed the menace. 'Let's get the hell out of here,' said the captain.

Co-pilot: 'Yeh, he's anxious, isn't he?'

Captain: 'There he is . . . look at him . . . goddam . . . that son of a bitch is coming!'

* Spanish, American and Dutch investigators who listened, eventually, to the voice recording agreed that the co-pilot's words would not imply anything like taking off without further ado.

The PanAm tried to turn and get out of the way.

When the KLM captain saw the obstacle in his path, he had two seconds to save himself. He yanked at the wheel, and tried desperately to become airborne. The excessive rotation managed to lift the nose, but the tail just dragged along scraping the runway. The collision could not be averted. Almost 100,000 litres of fuel poured out of the ruptured tanks. The whole airfield seemed to be afire.

All 248 souls aboard the KLM perished. Seventy of the 396 people on the PanAm flight emerged from the inferno alive, but all survivors were badly injured, and nine of them would soon swell the death toll to 583.

The investigators clearly blamed the captain who had taken off without clearance and disregarded both doubts and instructions. (A Dutch inquiry insisted that circumstances and coincidences should exonerate the crew.) The old culprits, 'inadequate language' and the ready acceptance of non-standard communications, were named prominently among the factors which had contributed to the tragedy.

Forerunners, repetitive errors, that painful ring of familiarity? There is just no end to them, it seems. Fatigue, stress and noise play havoc with eager expectations. That is how sprinters on the track or competitors in a swimming pool jump the gun and make a false start. That is how it is easy to fall prey to selective hearing, like a military pilot who absorbed only *'cleared . . . for take-off'* out of *'cleared into position and hold, stand by for take-off';* or the captain who misheard the non-standard phrase clearing him to *'runway 09 holding position',* and read back *'runway 09, hold in position',* which ought to have been picked out by the controller – that is the point of the exercise – but the man in the tower, anticipating no more than an echo of his own words, failed to spot the slight change.* The problem is that even trained crews find it difficult positively to expect the worst, i.e. mistakes all the time, and therefore a chasm can often be found between *readback* and *hearback*. Misunderstandings can occur anywhere in a similarly innocent fashion. A legendary example from the First World War concerned a telephone message from the front: *'Send reinforce-*

* Examples used by Seth Golbey in article in ISASI *forum*, April, 1988.

ments, we're going to advance' was heard as *'Send three and fourpence, we're going to a dance,'* and dismissed with a laugh. A more recent example occurred during a Panamanian coup in 1989. An American officer reported to the Pentagon and the CIA in two separate phone calls that *'the rebels won't turn General Noriega over'*. On both occasions, the message was understood as *'the rebels want to turn General Noriega over'*. Reading is usually more infallible than listening, and yet in the course of turning peacefully the pages of a book, undisturbed by noise, adverse circumstances or the monotony of monitoring, one can glance at the first words of a well-known proverb and overlook the misprint . . .

is no reason to search for errors or expect to see any.*

Airborne misunderstandings work the same way. As reported to CHIRP, a British aircraft in U.S. airspacc received the instruction *'to continue climbing to six zero'*. The word *to* was somewhat elongated by 'the usual heavily accented growl', and so the co-pilot confirmed *'continue climbing two six zero'*. The differencc was a mere 20,000 feet, and, on its mistaken way up, the British flight barely avoided a very close encounter with another aircraft. Even the standard, ICAO approved phrase *'Line up and wait'* has produced confusions when pilots heard it as *'Line up one eight'*. It is a daunting fact that seventy per cent of the fifty thousand safety reports in the NASA database involve some oral communication problem.

At Tenerife, an apparently innocuous exchange between Tower and PanAm pilots could have created a fatal trap for the KLM captain. The squeal in the cockpit blanked out a PanAm transmission but, one second after the noise had ended, the controller

* Illustration from Roger Green's lecture at the Norfolk and Norwich Institute for Medical Education.

said: *'Papa Alpha one seven three six report runway clear.'* It was audible in the KLM cockpit, but was it a statement in pidgin English or a request? If it was taken as a statement, the Dutch crew might have stopped the conscious effort to listen any more to chit-chat that seemed to be none of their business. The PanAm answer was *'Okay, will report when we're clear'*, but would the busy KLM crew fully absorb any but the anticipated last words *'we're clear'*?

Very frequently, the root of the troubles is a distinct *lack of discipline*. The callsign of an aircraft consists of the airline RT (Radio Telephony) name and the trip number, e.g. SR123. When a controller is talking to two flights almost in one breath, each pilot may think that an instruction applies to him, if the controller omits the *boring* repetition of the callsign, and the acknowledgment compounds the mistake by saying only *'Roger'*. Just imagine the potential hazards caused by such radio shorthand when two Flight 123s – SR123 and SAS123 are flying side by side. In 1988, flying at level 230 (23,000 feet), a captain was told to *'maintain two eight oh'*. Preoccupied with some turbulence, he read back *'cleared to maintain two eight oh'*, and began to climb. Reaching 24,000 feet (two four oh), the controller called again, rather calmly: *'Return to 230 – that 280 was airspeed'*. And Roger Green told me about the pilot who in Spanish airspace, approaching Tenerife, could not make out fully what the controller had said: 'Instead of calling back and telling the chap to offer a procedure he could understand or else they go home, he just discussed it with his co-pilot, and proceeded with what they *thought* they had been asked to do – which is not very reassuring when flying at five hundred miles per hour.'

BABEL IN THE AIR

Non-standard phraseology and the undisciplined use of *language* are perhaps the most common causes of creating totally unnecessary risks. While English remains the only accepted means of communication between any two nationalities in the air, ICAO – keen to please all its masters – declared Russian, French and Spanish, too, as official languages which can be used within a company or within national airspace. So if a pilot flies only domestic routes, he needs no English. That in itself creates confusion when domestic and international traffic are mixed. It also seems to be a licence, particularly in French- and Spanish-speaking airspace including much of Africa, to use the local language whenever possible. 'Japanese pilots suffer a great deal from the impenetrability of Russian and Chinese used without a care for others over vast expanses,' Captain Hiroo Moroboshi, a JAL vice president told me. 'In the Caribbean, you tend to get some rather funny clearances and your best bet is take advantage of the good weather, take a good look around – and do your own thing by talking to other pilots in the area,' said Captain McBride of Canadian ALPA.

Flying south in Europe, pilots hear incessant chatter in Spanish, and suspect, not without good reason, that controllers connive this way to arrange queue jumping for local flights. It is also a dangerous practice because non-Spanish speakers in busy terminal areas are unable to monitor traffic and stay aware of what goes on around them. Sometimes they have to beg the controller to repeat what has been said. IFALPA concluded, from numerous complaints, that if non-Spanish pilots try just to eavesdrop, it is easy to mistake the instruction *'espera'* (wait) for the equally common *'despegar'* (take-off). Greece is no better in this respect, and there are frequent complaints by pilots about Germany, too, though German authorities encourage even private and business pilots to use English at all times. The French are among the worst offenders. In Quebec, speaking local French seems to be a matter of national pride. Elsewhere, French tends to get mixed freely with English, and, to top it all, they may slip into *Franglaise*. In 1988, due to delays caused by an ATC computer break-down at Gatwick, planes crowded the French airspace to wait for the lifting of British night restrictions; eight tired pilots were in a stack over Abbeville, and the confusing instructions given by a panicky controller in two languages almost engineered a multiple pile-up.

A beautiful example of *'le mix'* with a happy landing was used in a training film by Captain Manfred Reist, deputy head of Swissair's flight safety.

SR236, a Swissair DC9, was waiting to take-off from Geneva. An Air Afrique flight, callsign RK31 was on its final approach.

TOWER: SR236 cleared for rolling take-off, wind calm, traffic passing outer marker.

RK31: OG en final . . . trenteun.

TWR: Oui continuez, un DC9 décolle, le vent est calme, vous êtes numéro 1. (Continue, a DC9 is taking off, wind is calm, you're Number 1.)

RK 31: Oui.

TWR: 236 confirm for rolling TO.

SR 236: Cleared for rolling TO SR 236.

RK 31: (*alarmed*) Il est toujours sur la piste, votre gars, hein! (He's still on the runway, you guys, hey!)

TWR: 236 stop departure – stop your departure 236, brake . . . 31 overshootez, montez vers Passeiry sept mille pieds. (Overshoot, climb towards Passeiry, 7,000 feet).

The controller's phraseology is not exactly out of the text book, but his resourceful instructions avert a nasty situation.

A Cambridge University team, including French and Australian specialists, found in 1988 that the mixing of languages, RT abbreviations, departures from standard forms and straying into idiomatic, inexact everyday English have been key factors in many accidents. They contributed heavily to the misunderstandings, for instance, that precipitated the death of 180 people when a Yugoslav DC9 crashed into a mountainside in 1981. Old ICAO hands gesture in despair: 'We can only publish the agreed, accepted standards for good practice – we haven't got the right or the means to shoot down offenders. The recently introduced RT phraseology may bring some improvements, but again, it's a matter of discipline we cannot police.'

Unfortunately, natural English speakers do not make life much easier for the rest of the world either.* They assume that everybody else must be in full command of their language when, in fact, many controllers and pilots have difficulties with anything beyond standard expressions of air traffic and, in panic or when baffled, may revert inadvertently to their own native tongue.

* The FAA recorded a delicious specimen of U.S. *air-talk*. Fighter pilot to Tower: 'This is chrome-plated stovepipe, triple nickel eight ball, angels eight, five in the slot, boots on and laced, I wanna bounce and blow.' Unlike the old-timer in the Tower, few Americans, let alone foreigners, would have understood it to mean that the jet's callsign was 5558, flying at 8000 feet, five miles from the airport, with gears down and locked, requesting to do a touchdown and go exercise. The Tower answered in style: 'Roger. You got the nod to hit the sod.'

'In many parts of the world, pilots and controllers are taught only very basic English, just enough to pass their exams,' said Mike Perry, an ATC consultant with many years of experience at Heathrow. 'Until their standards can be improved considerably, strict RT phraseology is our only safeguard. A dog trained to respond to the command *"sit!"* will be at a loss if somebody invites him to *"take a pew"*. It is often difficult to differentiate between a statement and a question. That's why one really ought to say "interrogative" before a question. Pilots are taught to pronounce "Mayday, Mayday" clearly, but sometimes that's not enough. The controller always needs to know the nature of any emergency. We had once a Pakistani pilot who wanted to land in a semi-emergency just after getting airborne. He knew what was wrong – the aircraft yaw damper was unserviceable – and he tried to explain at length, but it was quite impossible to understand him because of his heavy accent. It was useless to ask him a lot of questions, because he wouldn't understand colloquial English. So we restricted communications to basics, and eventually it transpired what the problem was. Because he couldn't damp out the yaw, we knew he'd need a long, something like a fifteen-mile straight-in approach, and for that we'd have to clear a lot of airspace for him. Though it may hurt a Londoner, it sometimes helps to learn a little pidgin English.'

Captain K. Blevins recalled the tale of 'an old, pompous, superior type gentleman pilot' of an English airline initiating the following exchanges:

'Brindisi control, this is *(callsign)*, good evening to you, sir'.

'Buona sera, sir, you cleared to *fly* level 370 reporting to . . . '

'Thank you very much, Brindisi, understand we're cleared to 370 . . . er, as we're going southbound, all the way to sunny Africa where we shall endeavour to indulge in some local style Christmas jollifications, we'd like to extend our compliments of the season to all your staff at Brindisi control, and take this opportunity to thank you for your help and cooperation all the year round'.

After a long, pregnant pause: 'Say again, please'.

The pilot obliged. In full.

'Oh . . . what problem, sir?' A cascade of Italian words, and then: 'You declare emergency? Yes?'

'Never mind, Brindisi. 370 Roger. *(callsign)* Out'.

Interpreters may be available, but, in a genuine emergency, the use of such help could waste invaluable seconds. Yet interpreters might be welcome when pilots talk to Irish, Scottish, Welsh or even Cockney controllers. In American airspace, not only foreigners but also the natives can be puzzled by great variations of accent, incessant irrelevant chatter on the air, CB language, non-standard phraseology with a heavy sprinkling of slang and verbal shorthand in *NewYorkSpeak*, or crucial instructions rattled off at machinegun speed in local jargon around Los Angeles and Chicago.

It is no good, however, blaming undisciplined individuals alone for misunderstandings. Controllers file regular complaints about callsigns. It is not surprising that in 1987, for instance, the wrong aircraft kept answering calls when a British Midland DC9 and a DanAir 1-11 flew on parallel radar headings barely five miles apart: their callsigns were BD-082 and DA-082. Sometimes four dots appear on the radar screen answering very similar callsigns such as BA5662 and BA5642. It may not help even to spell out the full names because the words may sound similar, such as Air India and Air Indiana (both flying out of Chicago) or Swissair and US–Air. Changes of names are out of the question because of fierce safeguarding of national pride and consumer goodwill attached to established names.

The struggle against callsign confusion has been long, brave, determined, idealistic and, yes, quixotic. Throughout two full decades, the battle flag has been carried by Captain Denis Leonard, a totally caring man, whose reward is to be left with an enormous amount of computer data, tons of well-argued submissions, and bitter disappointment. Backed by numerous colleagues, BALPA and IFALPA, he and his allies devised a new, complex yet apparently fail-safe '*alpha-numeric system* of callsigns which, although not receiving universal acceptance, does provide a system to eliminate callsign confusion'.* The alpha-numeric names would also help ATC to recognize right away that, for example, it is a lighter aircraft in the wake of a heavy one, and alert the controller to give it sufficient separation on approach

* A Crisis of Capacity: A Pilot's Point of View, by Capt. A. Liddle of BALPA, at the Guild of ATC Officers' *Convex*, 1987.

and arrival, thus protecting the small plane from the vortex behind the jet.

Leonard's dedication to the now probably lost cause stems from life-long airline flying and several moments of personal exposure to callsign confusion. Various tests of his system were most promising; but he feels that some other, half-baked trials were unfair to the potential of his concept; and when his proposal was just about winning vital support, there was a change at the top of ICAO. 'At best one may conclude that the new Secretary failed to understand the project and by declining to consult the rest of the study group, he produced a substantially inaccurate report,' he wrote to me in 1989. 'Others might say that there was a deliberate attempt to smother this project. Perhaps the truth lies somewhere in between. The discussion appears to be closed but the problem still remains and mistakes are still being made every day.'

An ICAO working group acknowledged the great effort, but concluded that the alpha-numeric system was 'inappropriate', and that resources would buy greater value if they were directed towards the development of a better, 'more widely acceptable and less disruptive solution'. Therefore, when the proposal was finally submitted to the 159 members states, only three reacted favourably. They then passed their own amendment to *Annex 10*, which becomes operational in 1990. It aims to improve callsign standards and will be compulsory – well, as compulsory as these things usually are. All a dissenting state needs to do is to file its intent to differ, though it is hoped that after years and years of wrangling, this will not happen, at least not very frequently. The three-letter designations will certainly be better than the two-letter ones though nobody believes that this change alone will achieve perfection. To help reduce confusion between flight level and flight number, ICAO now recommends that no flight number should end with five or zero because one of these is often the last digit of the assigned flight level.

Most probably, the final answer is to be found in automation. The *direct data link* which is under development will cut out much of the voice communication (and the mistakes that go with it) but at the moment nobody cares to forecast what technical mishaps and human failures will create new-style electronic misunderstandings between ATC and the 'dumb blip' on the screen – misunderstandings which may no longer be cleared up by a pilot's quick and simple: *'Say that again, muchacho.'*

– 13 –

MURDER IS AN EASY OPTION

Future dictionaries of synonyms will probably have an entry something like *LOCKERBIE – a type of disaster, death by design from the air, horrific event out of the blue, mindless slaughter of the innocent, etc.*, offering examples of worldwide usage such as: *We must never have another Lockerbie.*

Strictly speaking, *sabotage* is a police rather than air safety problem, but it defeats the *raison d'être* of the industry to provide fast and safe air transport, and it could produce the biggest 'aluminium shower' if that is the saboteur's choice. Its *modus operandi* must be uncovered in order to plug security loopholes on which sabotage thrives. Aircrash detectives must never accept sabotage as an easy answer, as a substitute for 'we are not sure about the cause of this accident'. Genuine safety risks, such as structural weakness, must never be lazily mistaken for sabotage.

'I sometimes wonder,' said NTSB Board member Joseph Nall, 'how many old turboprop accidents and midair break-ups were rubberstamped "sabotage" only because positive proof of anything else had eluded the investigators.'

Bombing and *sabotage* are highly emotive words. If within a few hours, let alone a few days, investigators *still* fail to solve a mystery – and which crash is not a mystery for a while? – SABOTAGE makes handy headlines and good copy for politicians who want to be first to demonstrate their care for the travelling public. In 1989, for instance, a Cuban, a Norwegian and a Korean accident attracted instantaneous pronouncements which will be long forgotten by the time the investigation has taken its course. Such guesswork is so unpredictable, however, that sometimes it may even turn out to be correct. (An Italian DC9 seemed to have exploded en route from Bologna to Palermo, killing eighty-one people. It sprinkled twenty-two miles of sea with wreckage and bodies before disappearing under 9,900 feet of water. Sabotage and a straying Sidewinder missile were quickly blamed. The Air Force denied not only firing missiles but also the suggestion that the flight had been seen on radar. But the story would not die. British experts who were asked to examine a few bits of wreckage concluded there must have been an explosion. Then the flight recorder was recovered. In 1989, nine years after the crash, a military air traffic controller confessed that he had seen the plane disappear from his radar screen, and an Air Force officer admitted that he had destroyed the radar tape of traffic that night. The guesses were proved right.)

The bombing of aircraft used to be the chosen technique of the fraudster who wanted to cash in a favourite aunt's life policy; to seek a profitable divorce by taking out a massive life insurance just before the spouse's fateful flight (such cases led to the ban on airport slot machines dispensing instant cover on anybody's life in favour of anonymous investors); or to sacrifice himself to save his family from bankruptcy, knowing full well that his inflated accident cover would be invalidated by the standard suicide exclusion clause if he jumped under a train. Of course, he falsely imagines that all traces of his plot will vanish in the wreckage, preferably at the bottom of the sea. In 1988, a South African flight and seventeen lives were blown out of the sky only because someone believed in the 'perfect crime', and expected to fool everybody despite his new big life policy. And insurance was not a factor in 1989, when a Ukranian wanted only to get rid of his unwanted wife and child by planting a time-bomb in their flight bag. The bomb was found and the courts granted him 'separation' for thirteen years – in a Siberian labour camp.

The urge to perish in a spectacular fashion has also figured significantly. In 1987, when forty-three people died in a ten-foot crater dug by the nosedive of a BAe146 commuter jet, there was no mystery about the crash. Not that there was meant to be any: in November, David Burke had been fired by Pacific Southwest

Airlines for stealing sixty-nine dollars in in-flight cocktail receipts; on December 2, he made a new will; on December 7, he left a message of love on his former girl friend's answering machine; he boarded the PSA flight, and wrote a note on an airsickness bag (found in the wreckage) to a fellow passenger – the man who had fired him: 'Hi, Ray, I think it's sort of ironical that we end up like this. I asked for some leniency for my family, remember. Well I got none and you'll get none.' Burke shot his former boss, then made his way to the cockpit where the CVR recorded the rest of the drama. 'We've got a problem here,' a cabin attendant is heard saying, and a male voice, presumably Burke, answers: 'I'm the problem.' He killed the pilots with the .44 Magnum he had borrowed from a friend.

The gun was found among the wreckage, but how did it get aboard? Further investigation discovered an almost ridiculously obvious loophole, a lesson that cost forty-three lives: although the show of ID cards had long been compulsory, airport security personnel were in the habit of just waving familiar faces through the checkpoints. So, ten days after the crash, U.S. airports were ordered to check all airline staff and their luggage through the detection equipment used for passengers.

The findings came six full months *after* some shocks and eye openers contained in an FAA civil aviation security survey. And yet even this investigation was not taken as a serious *final call*, because it took a Lockerbie and another two years to tighten procedures and introduce security screening for all staff entering restricted areas at British airports.

The light, seemingly flimsy aluminium tube that flies has often misled attackers into believing that as a target it is a 'soft touch'. They are often disappointed, particularly since the size of aircraft and the damage tolerance design of their structure have grown a great deal. When a bomb exploded under the passenger seat of a Japanese boy in August, 1982, it holed the floor, killed the youth, injured slightly fifteen passengers, and convinced 285 people on board that their day of judgment had come, but the PanAm Jumbo flew on and landed twenty-three minutes later at Honolulu without further trouble. Another small bomb, of about four ounces of TNT, was also placed under a passenger seat in 1981. Blowing a nasty hole in the fuselage *(see picture in photograph section)*, it damaged fuel pipes and flying controls – but fortunately, that 727 of Aeronica was empty and still on the ground because of a fifty-minute delay at Mexico City. Four people on the ground were injured. Had that aircraft taken off on schedule, there could have been a disaster – a fine example of miscalculated or indiscriminate but always difficult timing of sabotage devices.

Those who put their faith into statistics like to believe that they ought to carry their *own* bomb, because, statistically speaking, the chances are infinitesimal that there should be *two* bombs on the same aircraft. But they would be wrong to trust that no plane would be bombed twice: the AAIB wreckage museum contains some relics of two explosive attacks within three years – by coincidence, on both occasions, a particular Philippine Airlines BAC1-11 was the target, both flights had the same captain in charge, and both explosions, in the rear toilet, claimed a single victim each . . . the two offenders.

Even a large aircraft could, however, be brought down by a small bomb in the hands of an aeronautical engineer with suitable tools and almost unlimited access to key areas where the duplicated and triplicated systems must come together to work essential controls. Unfortunately, manufacturers cannot 'design out' this potential threat nor can they build tanks that will fly as well as be strong enough to withstand impact from runaway trucks or midair collisions. (Many weak spots for fire and, and to some extent, bombing have been eliminated by techniques like separation. The 747 is among the few that have flying control runs in the roof. But to this day, a rather popular aircraft – which should remain nameless in this context – has its electrical wires running so close to its titanium hot air system that, if an accident, such as chafing of the wires, or a small bomb caused some arcing, the titanium fire could be tackled only by buckets of sand rather than the usual in-flight extinguishing agents.)

Most obstacles in the path of amateurs have been sieved. Terrorists now must have great expertise as well as all the resources of a professional organization with access to funds and plenty of high-performance military explosives. (Two kilos of any such substance, like the notorious Czech Semtex, would blow an aircraft to bits wherever it was placed.)

Why do people sabotage aircraft? Eric Newton, one of the British AAIB's legendary investigators, an expert on air sabotage – who still tends to get dragged out of his idyllic south coast retirement for consultations – calculated that, between 1949 and 1989, the motives were: suicide – seven per cent, insurance fraud – sixteen per cent, unknown – twenty-eight per cent, and political – forty-nine per cent. In the 1980s, politically induced sabotage accounted for an even higher percentage.

According to Newton, there have been at least seventy civil aircraft damaged or destroyed by explosive sabotage, murdering 1,898 people, in the last forty years. Since the introduction of some new security measures, the successful bombing average decreased to two per annum (not a reassuring achievement from the passenger's point of view), but as aircraft become larger the toll they take multiplies, and there were years like 1985, when there were nine airborne sabotage cases and eight attacks on airports in addition to 28 hijackings. Terrorists seem to be influenced by fads and fashion. Following each devastation, there is a spate of copycat threats, including dangerous hoaxes which waste time and weaken the alertness of the defence.

While fraudsters want to hide their crimes, political terrorists do not. If anything, they are proud of them. When they time a crash to occur over unpopulated areas, they show only 'goodwill' to the innocent on the ground – if they think about them at all. When they time a crash to happen over deep seas, they are trying to hide not the cowardly act itself, but the technical 'signature' of the bomber. That makes it even more imperative that every attack should be fully investigated, almost regardless of costs which will be enormous if they involve deep-sea search and recovery.

Since most sabotage cases today are political, rumour is generated even faster when the crashed aircraft had at least one politician on board – a not very unusual occurrence. 'Witnesses hear a bang, see bits falling from the air, note smoke, presume fire, and cry murder,' explained Eric Newton. 'They wouldn't know, of course, that a main wing spar failure would produce something like a cannon shot followed by disintegration, or that the *smoke* could be grey kerosene clouds escaping from the tanks.' He said that the murder of 171 people by the destruction of the Paris-bound French UTA DC10 in September, 1989, was a typical chance for self-appointed experts to jump to the sometimes correct conclusion. 'After a stopover in Ndjamena, the capital of Chad, the plane was first reported to be missing. Its broken sections were found scattered all over the Sahara, indicating some sort of midair mishap. The press got hold of the passenger list – it included a Chadian Minister and an American ambassador's wife. Add to that the immediate revelation by the airline that there had been a threat allegedly from the Islamic Jihad, and people did not hesitate to declare that it was sabotage – long before any traces of explosives would be found in the wreckage or anybody would have a chance to follow the wreckage trail, see what part of the aircraft was first to hit the ground miles back from the crash site, and eliminate the possibility that, for instance, a large door might have failed in flight like at Ermenonville.' (Sabotage was, in fact, proven when traces of Pentrite were found among the debris of the UTA DC10. The same powerful plastic explosive had also been used during an indiscriminate Arab bombing campaign in Paris.)

On the other hand, the presence of a politician among the victims could become a time-consuming red herring for the investigation. In 1980, a Cessna C421 crashed soon after take-off from Lisbon, killing Prime Minister Carneiro and six others. Various faults and technical causes were discovered, and the wreckage was detained. Despite this, accusations of a cover-up would not die away. In 1982, two years after the official investigation, a TV station flew out Newton who declared: 'I have thirty years of experience. I am proceeding from the scientific facts alone and am not interested in political disputes.' He found absolutely no evidence of sabotage, yet some people still chose to ignore all the expert findings, and returned to battle by innuendo.

U.N. Secretary General Dag Hammarskjold died violently in an air tragedy. In 1961, at the height of bitter fighting with mercenary involvement in the Congolese province of Katanga, he flew to have private peace talks with Moise Tshombe. His DC6 crashed in bad visibility near Ndola, in what was then Northern Rhodesia. Dozens of interested parties had good reason to dislike both Hammarskjold and his plan. The likelihood of sabotage was obvious to everyone. Circumstantial evidence galore fuelled the guesses, and nothing more so than rifle bullets found among the wreckage and in some bodies. Now what could be better material evidence of a horrendous crime? It was well worth a full examination – but that produced contrary conclusions. An RAF pathologist and British investigators established that the bullets lay superficially in tissues near the skin, the wounds were not associated with any discernible bleeding, there was no sign of shots being fired and, most important of all, the bullets were 'accompanied by percussion caps and debris of disintegrated cartridge cases'.* This proved that bullets had not been fired by any firearm. Both the men whose corpses contained bullets were soldiers who carried magazines of ammunition which must have exploded in the post-crash fire so that bullets *and* cartridge cases penetrated their flesh.

The case was closed but the rumours survived it. In 1982, a massive Swedish television investigation assembled an impressive array of witnesses with allegedly new or hitherto suppressed statements. Secret agents claimed to have information about plots, the theft of the DC6 pilots' maps, false ATC transmissions to the flight disguising the presence of a mountain en route. Here we have the ingredients of a gritty thriller – but nothing beyond circumstantial evidence. Who knows, if the bullet theory failed under close scrutiny, allegations might rewrite the history books. Meanwhile, people like Group Captain Tony Balfour, Wing Commander Ian Hill (both of them pathologists at the RAF Institute of Pathology and Tropical Medicine), and Eric Newton find it infinitely preferable to stick to known facts until the unlikely emergence of any scrap of genuinely new and tangible proof.

It is interesting to note that, while the wilful destruction of aircraft can be accomplished in many other ways, most saboteurs resort to bombs, the most obvious, relatively easiest, technically least demanding, but most spectacular means. Perhaps they also find a touch of heroic grandeur in the fact that explosives are involved. But actual attempts on politicians' lives are more likely to rely on technical expertise because, it may wrongly be hoped, the crash would disguise the clues of unlawful interference. In 1981, an inspection discovered that four vital control cables had been cut deliberately on an Air India 707 which would take Prime Minister Mrs Gandhi on a foreign tour. But was that grave discovery itself politically tainted? The revelation was made *sixteen* days before the trip. Despite their technical knowledge and free access, the saboteurs, if that is what they were, seemed to have overlooked the fact that the aircraft would always be withdrawn from service for a specially thorough final inspection *three* days before a VIP flight.

* Lecture at the British Association of Forensic Medicine, July, 1967.

Pakistani dictator Zia died in an aircrash, and the suspicion of sabotage arose automatically and understandably. He and his entourage were all killed on August 17, 1988, soon after take-off in a massive C-130 Hercules military transport plane, known as the Presidential Pak One. Suspicious circumstances began to pile up right away: he had already been the target of several assassination attempts including plans for missile attacks on his aircraft; many of his enemies might have had easy access to Pak One just before departure from an army base; some influential officials had excused themselves from flying with him on the fatal trip. The KGB, its Afghan allies, and even the Indian authorities had ample reasons to kill him. Of the 500 people interrogated at the airport, a hundred were arrested. Benazir Bhutto, the daughter of the man Zia had deposed and hanged, was his chief opponent and eventual successor. She would claim in her autobiography that the dictator's death was 'an act of God'. But her supporters had never ceased to conspire against him, and some of his enemies went as far as claiming his death as their victory – at least until it transpired that the American ambassador was also among the victims.

Accident investigators are not great believers in divine intervention – not if they are given a full opportunity to exercise their skills. In the Zia case, however, limitations were imposed on their work. Only military investigators were given access. Even their American helpers had to be military personnel. While, with Lockheed's assistance, they could eliminate the potential mechanical causes, engine malfunction, missile attacks, pilot error and fire on board from a bomb or any other source, Islamic laws requiring immediate burials were invoked to prevent the performance of autopsies. (Normally, pilots are excepted from Islamic custom but, this time, even the plastic bags containing the bodies were buried before pathologists could examine them. Post mortem was carried out only on the U.S. ambassador, and the absence of soot in his trachea proved that he had stopped breathing before the fire broke out following the impact.)

Three weeks after the crash, officials in Islamabad announced that six Pakistani and American investigating teams had 'reached near consensus' that a bomb was responsible for the thirty-one deaths, and predicted that full details would remain a permanent state secret. Four weeks later, on the eve of the elections which would bring Benazir Bhutto to power, a committee of inquiry published part of its report that phosphorus and antimony were found in the wreckage, that the boxes of mangoes loaded for the flight showed high concentrations of potassium and chlorine, and that the conclusion was that odourless poisonous gases, concealed in drinks *or* fruit boxes, had incapacitated the passengers as well as the crew *within minutes*. The holes in the investigation as well as in the report were wide enough to let through cartloads of theories about both the saboteurs' identity and technique. Rumours are still rampant about traces of explosives. As poison gases do not kill with the first breath, how come the pilots stopped answering calls long before the aircraft went out of control? If there was a low intensity explosion in the cockpit, how come nobody claims to have found any sound trace of it on the air traffic control tape? Or was that tape mislaid? Who is covering up what and for what reason? Dictators love secrets, and dislike revealing circumstances of violent deaths – especially their predecessors' way of dying. But a democracy cannot afford to be seen leaving a crash investigation incomplete and the result undisclosed, for, if nothing else, international air security may learn something useful from it.

'These days there are fewer oportunities to practise the investigation of midflight failures, and to keep alive vital skills, like the interpretation of wreckage distribution patterns on the ground, because aircraft are now more reliable,' said Newton. 'But sabotage is one of the easiest to investigate – provided that all the wreckage is found. This is why it's absolutely essential to recover and preserve every bit of it.' When a 707 broke up above the Ionian sea in 1974, sheets of paper were sighted floating furthest back along the flight path. Had any of those been recovered, they might have told the investigators which part of the aircraft broke up first. When wreckage is scattered over forty or fifty miles of land, souvenir hunters with morbid interests pick bits up. Even the horror of Lockerbie failed to deter the looters, and it was only because of the massive security precautions that at least some of them were caught.

Having eliminated all other possible causes of an in-flight break-up, the sabotage investigator must work in close cooperation with pathologists and those who read the flight data and voice recorders, but his ultimate proof is usually forensic and metallurgical evidence. Whenever some of the wreckage cannot be recovered, chemical identification of explosives may not be possible, but there can be dozens of telltale signs to reveal that there was an explosion ranging from corpses to small, fragmented traces of metal. The signs, of course, must be noticed and read properly.

'The fact that there was an *explosion* does not mean that there was sabotage,' Eric Newton argues. 'A lightning strike can produce an explosion, but it's very different from a bomb. The disintegration of a high-speed turbine will produce a big bang, and shrapnel type pieces of metal will be found, but closer examination will reveal what sort of blast we're dealing with. But even fracture analysis can be misleading, and will work only in conjunction with other evidence because, like heat and blast effects, it may not be classified as proof positive. Perhaps the most conclusive evidence is found on metal surface specimens that reveal what we call cratering. Magnified maybe 1,500 times by the electron microscope, cratering creates a typical pockmarked moonscape, (*see: picture in photograph section*) though it must be remembered that the word "crater" may mislead the layman into believing that it is the product of an eruption from inside. The sort we see on a bombed aircraft could more precisely be described as "pitting", with rolled edges, with burns at its bottom, containing specks of molten metal from the explosive or its casing. It can be caused only by modern, high performance, military-type explosives that create and spray everything with minute, very hot fragments travelling at up to 28,000 feet per second. The smaller the fragment, the higher performance the explosive must have been. It's a kind of signature of the stuff. Gunpowder won't do it, household mixtures won't do it, and even bits of a broken turbine won't travel that fast with that sort of effect. But the evidence won't be found without well-trained and meticulous examination.

'The presence of a pathologist is no guarantee of real assistance unless he's experienced in this kind of work. On an African case, when we didn't get enough of the wreckage, I knew that corpses could provide the answers. A very keen pathologist told me sadly that he had found no trace of what I was looking for. But what was I looking for? He thought I was searching for shrapnel, the sort you come across at battlefields. When I explained that a miniaturized version of rabbit skin peppered with shot, small clusters of red dots with almost invisible skin punctures were the target, he lit up: oh, yes, he found plenty of those! When the tiny fragments embedded in the skin were dug out, we had the proof of sabotage.'

The modern sabotage investigator is aided by a whole range of forensic techniques. 'If, for instance, we find that the metal stamp of a serial or part number has seemingly been fully obliterated by scoring or filing, we can recover it by acid etch and heat treatment, like we did with a helicopter part,' Ronald Hayman, director of the superbly equipped CASB 'lab', told me in Ottawa. 'Paper examination techniques can also be hugely helpful in cases of sabotage or, indeed, any accident investigation. We can analyse burned, charred, shredded, sun-bleached and deliberately altered documents. Where there seems to be nothing on the paper, ink trace analysis can produce a reading. We had here documents that had been in water for fifteen years. Luckily, they were recovered with great care, but we couldn't handle them because of their porridge-like consistency. So we froze and then freeze-dried them to remove all moisture, and restore the paper structure. Sometimes we have to resort to infrared reading of logs, loading sheets, pilots' computations. Once we managed to read a pilot's apparently illegible log in which he objected to malpractices by his employers, a small airline, and revealed that he had taken drugs for a medical condition. It gave us an extra angle in the investigation.'

Such methods are widely available, but many specialists rightly complain that 'a little learning' can be extremely dangerous in the wrong hands. Having read brief descriptions or examples of techniques, somebody who, in the absence of genuine experts, has been put in charge of an investigation, may misread the signs, jump to the wrong conclusion or convince himself that what he sees is *proof positive* rather than a red herring or merely circumstantial evidence. That is why many people on both sides of the Atlantic hold some very critical views about the investigation of one of the greatest catastrophes ever attributed to sabotage.

WHERE WAS THE 'SMOKING GUN'?

At 07.14 GMT on June 23, 1985, Air India Flight 182, a 747, disappeared from the ATC radar screens without any warning. It was soon established that it had crashed into the ocean some 110 miles west of the Irish coast. There were 329 people on board. There were no survivors.

Naturally, sabotage was high on the list of suspects, particularly because, the same morning, within an hour of this disaster, a suitcase just unloaded from another Air India flight exploded on a baggage cart in the transit area of Tokyo's Narita airport, killing two and injuring four people. Both planes carried mostly Indian passengers (coyly described by Canadians as 'members of one of our visible minorities') from Vancouver. Both had some two dozen people with the popular Sikh name of Singh on board. A special alert, in force because of rebellious Sikhs' political threats to Air India, would have applied to both. If it was a bomb that destroyed the Jumbo, it might have been timed to detonate after the landing at Heathrow, and so to coincide with the Tokyo explosion – but Flight 182 was running late. A similar miscalculation had once turned an intended airport attack into slaughter in the air in an African case.

The suspicions and theorizing were therefore well justified – at least as far as policemen and politicians were concerned. But evidence was hard to come by. Most of the Jumbo was several thousand feet under the ocean, though searchers were amazed 'how much of a 747 would actually float'. India seemed to conclude at once that it had been sabotage, and so Ottawa received a letter from the Indian Prime Minister complaining about Canadian airport security. Not that Canada disagreed: 'A team from Toronto descended upon London with a ready-made bomb scenario,' Geoffrey Wilkinson, then in charge of the AIB, told me. 'I couldn't blame them. The bomb theory was the front runner, a soft option that suited everybody including Boeing, the politicians, the airline, and our visitors who were only detectives and spooks, after all. But we had no facts. Nothing to satisfy an aircrash investigator. All we knew was that, at some point, there had been a sudden interruption of power to the flight recorders, and that, subsequently, the aircraft broke up. If it was a bomb, it had to be very large or something planted in a crucial position.'

Britain, Canada and America joined forces to assist the Indian government which appointed a layman, a high court judge, to be in overall charge of the investigation. The underwater mapping and recovery of pieces of wreckage was a mammoth task, even with the help of video cameras aboard SCARAB, a French submersible. Canada spent millions on the job. But, despite the international effort, only some five per cent of the wreckage was ever recovered.

Some people believe to this day that, irrespective of the cost and political pressures, the work should have continued because no 'smoking gun' was ever located.

Two months after the accident came a blood-soaked red herring – the JAL 123 crash in Japan. Those who loved to jump to conclusions were ready to pronounce that the two cases must have been due to an identical cause, the rear bulkhead failure. While this was gradually discounted (albeit without material proof from under the ocean) fine Canadian police work amassed circumstantial evidence which was irresistibly tempting to all but professional crash detectives. For who could be anything but suspicious seeing the revelation that suitcases had been accepted for flights without being accompanied by their owners?

Some of the details were hair-raising. Not only had luggage been transported without the owners but it also transpired that some of the airport security staff had never been through the inspection training programme, crews had not been subjected to any security checks, warnings had been ignored, the baggage X-ray machine at Vancouver had worked only intermittently 'for a period of time', and the baggage of passengers at Toronto transferring to the Air India flight was not X-rayed 'as it was presumed to have been screened' at the original overseas departure point.

CANADIAN REPORT SUBMITTED TO INDIAN INVESTIGATING JUDGE

Flight CPA 003:
Vancouver–Tokyo, connects at Narita, Tokyo, to Air India 301 to Bangkok

Flight CPA 060:
Vancouver–Toronto, connects to Air India 181 to Montreal where it becomes AI 182 to Heathrow and Delhi

June 20

A man makes two reservations by telephone: a single ticket for Jaswand Singh to Toronto by CPA 060, accepting to be wait-listed for AI 181/182 to Delhi; and a return for

Mohinderbel Singh Vancouver–Tokyo–Bangkok by CPA 003 and AI 301.

June 20 (several hours later)

Indian man pays $3,005 in cash for the above tickets at CPA Vancouver ticket office, but makes some changes in the reservations: Jaswand Singh's wait-listed seat will be taken by M. Singh; and instead of Mohinderbel Singh, L. Singh will fly to Tokyo–Bangkok – but only one-way.

June 22

Mr M. Singh phones to confirm reservation, and asks for his luggage to be checked right through from Vancouver to Delhi. He is told this cannot be done for he is only wait-listed from Toronto onwards.

Three hours later

Mr M. Singh checks in for CPA 060, holds up long queue by arguing for luggage to be checked through; tumult at counter, and breaking the regulation, passenger agent relents: Singh's luggage is tagged through all the way to Delhi.

Twenty eight minutes later

CPA 060 departs for Toronto: M. Singh is not in his assigned seat.

Two hours later

L. Singh checks in: suitcase is tagged CPA 003 & AI 301 to Tokyo–Bangkok.

20.22, June 22 (All times GMT)

CPA 060 lands at Toronto, passengers and some luggage transferred to AI 181.

20.37 June 22

CPA 003 departs for Tokyo: L. Singh is not in his assigned seat.

00.15 June 23

AI 181: late departure for Montreal.

02.18 June 23

At Montreal, AI 181 becomes 182; it departs for London ninety-eight minutes late.

06.19 June 23

Thirty-eight mins. after the arrival of CPA 003 at Tokyo, baggage cart explodes in transit area. (Bag is not identified.)

07.14 June 23

Sixty-one mins. before scheduled arrival at Heathrow, AI 182 vanishes from radar.

It sounds convincing, but it is not *material evidence*. Yet, under the guidance of a judge, in utter disregard of aircrash investigating principles, it was accepted as such. And it seems no coincidence that the eventual report – leaked four weeks in advance – begins like a fairy tale:

'. . . The aircraft in question – Kanishka, was named after the most powerful and famous king of the Kushanas who perhaps ruled in India from AD 78 to AD 103. Besides being a great conqueror, he was an ardent supporter and follower of Buddhism – a religion which preaches non-violence. Emperor Kanishka, however, met a violent end. After twenty-five years of reign he was killed by some of his own subjects. His life was thus brought to an abrupt end. It is indeed ironical that the Jumbo Jet which bore the name "Kanishka" also met with a violent and a sudden end on that fateful morning . . .'

Like the king, whose non-violence could not have gone beyond *preaching* it in the process of becoming a great conqueror, the court had to make certain compromises to promote selected views, and present them as facts to prop up its verdict based on circumstantial evidence.

Geoffrey Wilkinson: 'The few pieces of wreckage that had come up showed no bomb clues, no high-speed particles, no indication of an explosion at all or that one had occurred in the alleged location. The Court was unwilling to hear any evidence that would not support its theories.'

The Canadian submission recognized the considerable circumstantial evidence, and declared, correctly, that the evidence is not conclusive regarding sabotage though it does not support any other conclusion either. The Canadians' work was not made any easier by the court because, as Ronald Hayman told me, 'many key parts of the aircraft were recovered but India would not release them to us for analysis'.

Eric Newton, whose help is acknowledged in the Indian report: 'Too little, no more than three or four per cent of the wreckage was recovered. I examined it all, and I'm satisfied that the wreckage I scrutinized contained no physical evidence of a bomb. No cratering, nothing. Scientists at Fort Halstead found no positive trace of any explosives. An Indian metallurgist thought that several holes, going from inside out of the lower fuselage skin, were a proof of bombing. But those were too large, one-inch holes, due to low-velocity impact penetration made by floor support struts (parts of which were still in some of the holes) when the lower fuselage was crushed flat as it hit the sea. Nothing like a bomb. The worst mishandling of evidence appears in the "Analysis and Conclusions" of the report. The metallurgist who contributed it had read my paper* on explosive sabotage investigation which described an important type of clue called "twinnings". At normal rates of strain, metals would distort by movement. When there's an explosion, thermally activated very high rates of strain occur, giving no time for normal distortion, so in some metals such as copper, iron and steel, the crystals of the metal get deformed by "twinning", that is to say by parallel lines or cracks cutting across the crystal. That phenomenon occurs only in extreme circumstances induced by explosion so, under careful examin-

* *The International Journal of Aviation Safety,* March, 1985.

ation, it can be the proof we look for. To support the basic theory about the cause, the Indian metallurgical report simply and wrongly applied the principle to aluminium, in which there can be no twinning. And in fact, no supporting evidence of explosion was submitted by the Indian explosives experts.'

The 131 bodies that had been retrieved from the ocean were subjected to meticulous autopsies. Wing Commander Ian Hill observed all that work, and analysed the findings: 'We found no blast injuries, no medical evidence of sabotage. Most victims didn't suffer what I'd call massive injuries, some had virtually no injuries and died from drowning. The only firm evidence I found was that the accident must have been very sudden, with no intimation of impending disaster, because many bodies revealed that no seatbelt was worn, i.e. the pilot gave the passengers no warning. If it was sabotage, it must have been a very effective bomb in a particularly dangerous position. After all, the Korean airliner the Russians shot down was hit by *two* missiles, and it was still flying. A suitcase bomb that can bring down a 747 right away would leave plenty of blatantly obvious evidence. The case illustrated the problem that can occur when you have only small samples to work with. You'll have a biased view if you fail to look at *all* the evidence. If you cannot find the proof of the link between cause and effect, and if you accept shortcuts, you may end up redesigning the camel – with its humps underneath the belly.

'The unequivocal tone of the Indian report leaves a big question mark and a great deal of unease. Was the real cause discovered? What if some engineers are right suspecting to this day that the Karishka failed like the Japanese Jumbo? Was the investigation brought to an end too abruptly? They simply decided what the answer was, and the court didn't listen to or ignored those who might contradict that. The report states that a bomb was near the front of the aeroplane. That is totally at variance with the evidence. Once it was even suggested to me that I should interpret my pathological findings to fit in with their theories! Some people were surprised that I refused the idea.'

The treatment of the cockpit voice recording was another example of the court's handling of evidence. The tape was analysed endlessly by numerous specialists of international renown. Bernard Caiger, now a consultant, told me in Ottawa: 'The last sound the cockpit area microphone picked up was a *loud noise* that must have been connected with some event that quickly incapacitated the entire voice recording system. It is unlikely that any conceivable structural failure, such as that of the rear bulkhead, would cause such a total and immediate disruption. A bomb? Could be. But analysing the tape signal is a most complex affair, and it was impossible to identify positively that "loud noise" which lasted no longer than two hundred milliseconds.'

The Canadian National Research Council and the British AIB analysed that loud noise in detail, and concluded it was not the sort that 'would be expected from the sound created by the detonation of a high explosive device'.

Hans Napfel, Engineering Director of Fairchild, the leading manufacturer of aviation recorders: 'Unfortunately, India would not release the original recording. We had to work on a poor copy. A special problem is that voice recorders are not designed to pick up the sound of a bomb: usually there's up to a second's delay before the data is sensed and safely stored on tape, by which time there may be no power available.'

Ray Davis, the AIB specialist, also had to work with a poor quality copy. He testified to the court that he had not found any proof of a bomb being detonated. He could not rule out fully a detonation 'in a location *remote* from the flight deck', but that would be considered 'most unusual, if not unique, in that we have never failed to detect sounds of structural failure, decompression, explosives etc., on any accident CVR, even though the event occurred at the rear of the aircraft'. His conclusion was that, in the absence of proof from CVR, wreckage and autopsies, something other than a bomb would have to be established as a cause. The court then pressed him, only to state yet again that the lack of proof did not rule out the possibility of an explosion.

A submission by the Indian Bhabha Atomic Research Centre went against the grain, with S. N. Seshadri claiming to have proved by his own research that the essential clues (low frequencies the others had looked for in vain) would not always and not necessarily be present in the recorded sound of an explosion (meaning 'a bomb, a very fast device'), and that, in his view, an explosion did occur at forty to fifty feet from the cockpit.

The final report pronounces that the Kanishka was sabotaged. It enumerates the circumstantial as well as the preferred material evidence in support of that. The ultimate item in that list, the

'CVR tape analysis', refers only to Seshadri's view which corroborates the 'evidence that there was a bomb in the forward cargo hold of the aircraft'.

The Indian judge concluded that it was a bomb. This was *probably* correct, wherever it might have been placed, but as Ian Hill said, a Scottish-type *not proven* verdict would have been more appropriate and more in keeping with vital, traditional principles of aircrash investigation.

On the positive side, the investigation made everybody wake up yet again to the threat of sabotage. Canada, to her credit, made no attempt to cover up the blunders, and tightened up security. The tragedy sent such tremendous shock waves round the globe that it had to become obvious to every caring person that a repetition would never be allowed to happen, not by luggage travelling without its owner, not by a duped innocent serving as a *mule* to carry the bomb, not by any other means. Famous last words? Sadly, yes.

Before the reverberations generated by the Indian report died down, there were other cases of sabotage, and, in the wake of warnings by the NTSB, a congressional investigation discovered that, on average, the American airport weapons screening system for carry-on baggage was only an unacceptable eighty per cent effective. At Anchorage, Alaska, only one per cent of the test objects were missed, but at Phoenix, Arizona, the tests showed a sixty-six per cent failure rate. A startled FAA followed up with its own survey, and collected $228,777 for security violation penalties in the first half of 1987 – a sizeable increase over 1986, but a mere pittance in airline and airport budgetary terms. The survey criticized both airlines and airports, in America and abroad, and promised to watch the development of detection equipment. In Britain, the House of Commons Transport Committee found numerous loopholes in airport security – but the government rejected all its recommendations.

Just when the FAA survey was published, in November, 1987, a high-ranking, Libyan-trained but disenchanted Japanese Red Army terrorist made a confession – and Tokyo warned the world to beware of attacks on airliners. It seemed only reasonable to expect the tightening of security everywhere, particularly since terrorists had already threatened to disrupt the Olympic Games to be held in Seoul the following year. Four days later, a South Korean 707, en route from Iraq to Seoul with 115 people on board, went missing over Burma. Even before it was discovered that it had actually crashed, a completely independent drama was developing in Bahrain. Airport officials, acting on a tip-off, arrested a seventy-year-old Japanese and his companion, a young Korean woman, who used false passports and intended to fly to Rome. Under guard, the man bit into a cyanide capsule concealed in a cigarette, and died within seconds. The woman tried to do the same, but an alert policewoman snatched away the capsule virtually from between her teeth. Even the small amount of cyanide knocked the Korean girl unconscious, but she survived. It was then established that the pair had left the Korean 707 during a stopover in Abu Dhabi, from where they had flown to Bahrain, intending to lie low for a couple of days.

The body and the woman were flown to Seoul, where she broke down and confessed to being Kim Hyon Hui, a North Korean agent whose mission was to blow up that airliner. As the real-life spy story developed (child actress to secret agent, martial arts expert, fragile beauty with the arms of a stevedore), political jockeying in the glare of publicity diverted most of the attention away from Kim's description of the mere technicalities, and the ease with which bombs could be smuggled and left on board. The infernal machine (timer and plastic explosives concealed in a transistor radio, and a 'duty-free' whisky bottle containing liquid explosive) were collected from local agents in the Metropolitan Hotel in Belgrade. From the Yugoslav capital, one of the key terrorist bases in Europe, the pair took a flight to Baghdad. Security failed to detect the bomb in their hand luggage. The cabin crew insisted that, as a security precaution, the radio batteries should be removed and handed over for safekeeping. A stewardess returned the batteries to the two passengers in Baghdad.

Twenty minutes before boarding the Korean 707 for Seoul, the timer was fixed to activate the bomb nine hours later – well after the scheduled stopover in Abu Dhabi. As the pair went through the Baghdad checkpoint, the bomb remained unnoticed, but the question of the batteries was raised once again. The experienced agent simply turned on the radio to demonstrate both his annoyance and the innocence of the set. Kim's confession sounded ambiguous when she said the Iraqi airport officials 'acted as if they were sorry' about the inconvenience they caused. Whether duped or part of the plot, they let them board with the agent's vinyl bag, containing the 'whisky' and the Panasonic radio complete with batteries. The bag was stored in the overhead rack, and was left behind when the pair, carrying only Kim's cabin bags, disembarked at Abu Dhabi.

What happened? A series of blunders or acts of connivance? The basic facts were disseminated worldwide without delay to make radios and items like tape recorders even more suspect than ever before.*

Undeterred by that alert, other terrorists were busy by then, as we now know, preparing another atrocity. The Lockerbie bomb was in a radio-recorder. The terrorists must have realized that the new safety measures were hardly better than lip service.

All that Lockerbie teaches us should have been learned long before, without waiting for yet another final call.

Acting on an Israeli tip-off in February, 1988, West German police and intelligence began surveillance on a group of known Palestinian terrorists. Their telephone calls, including those to Syria and Cyprus, were bugged; they were seen buying radio cassette recorders, switches, digital clocks, batteries; and even the identity of the gang's bomb-maker was identified. Police thought that an Israeli handball team or U.S. bases in Germany might be their targets.

On October 26, twelve apartments were raided, sixteen arrests were made, an arsenal of automatic rifles, rifle grenades, mortars, pistols, blocks of TNT and six kilos of Semtex were found. One of the gang was arrested in his car: the search of the boot yielded a radio-recorder that was already a complete, sophisticated bomb with a timer that would be activated by a barometric trigger, i.e.

* Kim has been sentenced to death by hanging. She appealed, and her chances to be pardoned were said to be good because 'she is repentant'. In 1990, South Korea chose to forge political capital out of leniency: a special amnesty spared Kim's life, and declared that the North Korean President and his son were the real criminals. Thus Kim could become a propaganda asset to the government against the communist north. Which raises a difficult question: will a few more revealing statements, ignored by the north, be a better deterrent to terrorism than uncompromising pursuit of retribution for multiple murders?

it would go off at a predetermined height. Though from additional evidence it could then be assumed that an Iberia flight was the target, there was no danger: the whole gang was in jail. But not for long. Through some scandalous bungling (or, possibly, some high level diplomatic pressure) all but four of the suspects, and even the bomb-maker, were released within twenty-four hours, and two more were freed within two weeks.

The fact remained that the police saw evidence that five bombs had been made, but only one had been found. One of the released men left Germany carrying a radio-recorder. He was tailed but lost in Belgrade.

November: Germans showed the captured radio-recorder bomb at an international detective conference; black-and-white photos were sent to relevant authorities, including the FAA and the British Department of Transport; a mock-up of the bomb was built at Heathrow.

December 5: the U.S. Embassy in Helsinki received a warning – a woman would unwittingly carry a bomb on a PanAm flight from Frankfurt to New York via London within the next fortnight. Admittedly, if every hoax were taken seriously, air traffic would grind to a halt. That warning was not issued to the public, but it was considered to be serious enough to inform U.S. airlines and diplomats without urging special action.

December 9: the Department of Transport prepared a bulletin on radio bombs with recognition clues for airport security, but it would not be issued for a month, awaiting colour photographs.

December 15: a German press conference revealed with pride that a major terrorist network had been broken.

December 21: when passengers checked in for a PanAm 727 to London, less than fifteen per cent of the luggage had been hand searched at Frankfurt. All pieces went through an X-ray machine which was known to be ineffectual as far as the detection of certain types of explosives, particularly Semtex, a Czech product to which terrorists had easy access through the good offices of their Libyan patrons. In London, many of the passengers from Frankfurt joined others to take *Maid of the Seas*, a PanAm 747 to New York. Luggage from connecting flights such as the PanAm 727 was not X-rayed, the security procedure was not designed to identify passengers who might leave the flight even though their luggage had been checked through to New York, and bags to be carried into the cabin were hand searched only if the passenger *looked* suspicious – a precaution hardly ever applied to those travelling on American passports.

That evening – or possibly even earlier – a suitcase housing a radio-cassette bomb was placed in a sensitive area in the hold of the Jumbo. At 19.03, it exploded at 31,000 feet above Scotland. The aluminium shower covered some 840 square miles of country-side and the village of Lockerbie, killing everybody on board plus eleven people in their homes and cars, taking a toll of 270 in all.

At the time of writing, the police inquiries are still incomplete, though astonishing details have been uncovered about an Iranian-backed terrorist network and its meticulous planning – with Belgrade figuring once again as the quartermaster's base where false papers, explosives, cash and escape routes were made available to the bombers. Legal wrangles have only just begun, with wild allegations to shift the blame on to anyone and everyone except each lawyer's client. The survivors' relatives, who allege criminal negligence, should have a good case against a whole range of authorities if not the entire industry. But here we wish to concentrate on future air safety. Like any foreseeable but unforeseen, man-made tragedy, Lockerbie should not qualify as an 'accident'. Any security specialist who wishes to plead ignorance of the risk ought to plead deafness and illiteracy, too, for it was impossible not to know about the much-publicized loopholes highlighted by all the previous examples of sabotage. As Thomas Hinton, the CASB director of investigations, put it, 'with the Air India case, Canada lost her virginity regarding aviation security: we ought to have done more about it before, we're obliged to do more now.'

Israeli security has, after all, developed the art of security over three decades, and demonstrated its efficiency – albeit at a cost, paying the penalty in 'wasted' hours of long check-in time for safety. Isaac Yeffet, former security director of El Al, argues that virtually all American carriers hire security firms solely on the basis of the lowest bid, and that those guards are insufficiently trained. Two years before Lockerbie he carried out a survey for PanAm, and told the airline that it was highly vulnerable to terrorist attacks, particularly to explosives concealed in cargo, and that it was merely providential that no such major disaster had occurred to that date. After Lockerbie, he told an excellent

BBC investigation* that PanAm did not implement his recommendations against just that type of attack because it felt that its operation was too big, that El-Al-style searches would cause too much inconvenience, and that delays would drive passengers away to its competitors.

Although, in 1989, the FAA proposed to impose a fine exceeding half a million dollars on PanAm for lax security at the time of Lockerbie, it would be grossly unfair to suggest that PanAm alone was guilty. The industry worried about the *negative effects* of security. Gunter Eser, Director General of IATA, suggested in 1986 that the way to combat terrorists was to deny them safe havens anywhere. He found that the elaborate security was 'adding to the cost of air travel while making it less attractive to fly', and said that the measures that turned airports into fortresses were 'frightening passengers' and contributing to the then expected decline in traffic.†

Emotionally charged or not, the word *'sabotage'* seems to cause cramps in people's guts rather than in their pockets. Though SAS would eventually blame Lockerbie and a cluster of bomb threats for a huge drop in its profits for the first half of 1989 (copycat hoaxes always burgeon after every tragic success or well-publicized bomb alert), airlines like BA and TWA reported no noticeable change in the volume of post-Lockerbie ticket sales and reservations and, just a week after the slaughter, people were still flocking to fly from Frankfurt and London to New York with PanAm because it offered the cheapest fare. Though security requirements were tightened everywhere, at least on paper the FAA went on record saying that, despite its newly enlarged inspectorate, it could not be sure that the latest procedures were fully implemented by everyone, and Raymond Salazar, its director of security, seemed to throw his hands up in despair saying: 'Unfortunately, we build a twelve-foot fence and the criminal finds a thirteen-foot ladder.'‡

Lawyers for airlines, airports and regulatory authorities have been arguing endlessly whose responsibility the passengers' safety is,

* Gavin Hewett, *Panorama* 25.9.1989.

† Quoted by *FLIGHT INTERNATIONAL*, 5.4.1986.

‡ *FLIGHT INTERNATIONAL*, 7.10.1989.

but, in the final analysis, it is the industry as a whole which has failed again and again – often for reasons of cost alone. Even without actual disasters, the symptoms of laxity, the manifestations of ignored warnings are there for all to see. It is not only at third world airports that one notices that bearers of apparent badges of innocence, such as food trolleys, cleaners' buckets, and loaders' thermos flasks, are allowed to approach aircraft on the tarmac without anyone taking a closer look. How frequently are the contents of fire extinguishers aboard fuel trucks checked?

At Cairo, I walked twice back and forth between the terminal and the 'secure' zone, through the waist-level swingdoor next to the metal detector gate, without being challenged. Rome, Athens, Vienna, Belgrade, Bangkok, Karachi, numerous African and Middle East airports have frequently been implicated in cases when bombs or hijackers' weapons had been smuggled on board with ease. After Lockerbie, some American and European airlines began to list airports of 'dodgy' security for special attention. But do they ever go as far as El Al, which actually suspended its flights to Vienna in February, 1989, because it found the new check-in procedures there unsafe? How many airlines will stop flying to Colombia only because in November, 1989, somebody (possibly the drug barons) could place a bomb aboard an Avianca 727, apparently with the greatest of ease, to kill 107 people just after take-off from Cali? Yet grave warnings may work. When after a Greek blunder, President Reagan advised all U.S. passport holders to avoid Athens, the impact was tremendous.

Singapore runs a system of security rating. 'We classify airports from green to red,' said Captain de Silva. 'We review their technical facilities and discipline – do checkers really keep their eyes on the X-ray screens? – and we watch out for political upheavals. That's how the volatile situation puts Colombo into our *red* category that calls for extra measures including the rule to open and physically check every bag. We had special measures for Seoul during the Olympics. Athens has all the equipment, but there're doubts about the adequacy of personnel using it, so oncc again, we insist on our own thorough inspections and actual rummaging through luggage. Now there seems to be some improvement, so our station manager has required a downgrading from *red*. We'll see. Karachi is another *red*. So are Indian ports due to social and religious unrest, Sikh rebellion, etc. When

France had a spate of bombings, we immediately made it *amber* for a while.'

'All current boarding gate search equipment was developed in the 1970s to deal with the hijack threat' – not the modern terrorist, wrote Professor Paul Wilkinson in *Jane's Airport Review* in 1989. So the bombers' skill and the equipment available to them can keep well ahead of the defence. Against barometric triggering devices, some airports, such as Frankfurt, pass luggage through low-pressure test chambers which mimic the conditions of aircraft flying at the usual altitudes, and could thus activate a bomb hidden in a suitcase. But professional terrorists circumvent this by using the barometric sensor to trigger merely a timer which explodes the bomb a few hours later in the air rather than in a secure test chamber.

Since modern plastic explosives are so malleable these days that they can easily be finger-sculpted into any shape including toys and electronic components, 'bombs can come in any shape or size,' said Newton. 'We've seen bombs in the shape of a modified commercial catalogue, a carrier bag, and a domestic soapflake box. Even the timer and the detonator can be disguised in a similar manner. Concealment may be by wrapping the bomb in paper, dirty rags, objects of everyday use like tins, boxes, oil cans, washbags, containers of cosmetics. Aboard aircraft, the waste towel bin in the toilet has long been a favourite.' In 1989, a warning was circulated that tubs of Halawi, a marzipan paste from the Middle East, were being used to hide bombs or smuggle Semtex (the only difference between the two odourless substances of similar colour and texture is that the explosive is not sweet).

Short of the arduous El-Al-style questioning, hand search and visual inspection, security procedures are dependent largely on X-ray and electronic 'sniffer' machines which are still inadequate for the magnitude of the task. It is perfectly possible to conceal a bomb in the false bottom of a suitcase so that it can remain undetectable by both standard X-ray or casual search. Well-trained dogs may still out-perform electronic sniffers in the detection of explosives. The latest, revolutionary *thermal neutron analysis* device which costs a million dollars, does more for the authorities' nerves than for security. It was installed and tested at airports like New York and Gatwick, but in September, 1989, the

American *Science* magazine revealed that the standard sensitivity of the machine would not have detected the Semtex bomb that destroyed the Jumbo over Lockerbie, for Semtex is virtually undetectable by the equipment in current use. The new Czech government may get embarrassed at last by its communist predecessor's success and generosity with the product, may yield to pressure to build a detectable *fingerprint* into the explosive, and may stop its free flow to the terrorists' Syrian, Iranian and Libyan armourers.

The world-wide rethink, and the new tightening of aviation security, have also initiated fresh research into safer systems such as a computerized bar coding, now under development, which may keep track of luggage even within the hold, and help to pinpoint the passenger who gets separated from his luggage on disembarkation. There is much resistance to 'causing passenger inconvenience' by more meticulous checking; unaccompanied luggage is known to be one of the gravest bombing hazards; yet airlines tend to dismiss with not too sincere apologies the fact that more and more suitcases are misrouted – an 'inconvenience' the wily saboteur might well engineer for his own purposes.

Most airlines accept that a radio or electrical equipment should only be carried as hand-luggage, and the passenger must be able to demonstrate that it is in good working order, 'but have those caring airlines just conveniently forgotten what fat lot of good such a demo did in Baghdad?!' an exasperated security director exclaimed, and not without good reason. For whatever technical marvels will become available, security ultimately depends on people who can be bribed, blackmailed or duped, people who may be airport employees or even security guards of low standing, employed without much ado, travellers who can claim all the rights of the passenger even if being foolish enough to carry sealed and gift-wrapped packages for strangers, and also people who run the system with a shrug of the shoulders.

In January, 1990, passengers to Luxor were fully scrutinized at Gatwick. Apart from electronic checks (even a small bunch of keys triggered the gate alarm), there was a body search, and thorough rummaging in hand-luggage. The return flight on the twenty-third was a different story. The security gate was inoperative. There was no search. The X-ray for hand-luggage was out of order. Bags were opened but their contents were not examined at all.

Noreene Koan, who chairs the U.S. Association of Flight Attendants' safety committee, attributed ineffective security screening to cost considerations: 'Why not increase the salaries and improve the training of these screeners, who now receive minimum wages? Because of our inattention to these crucial people, some airports have an annual one hundred per cent turnover in personnel'.*

'Search by TV is perhaps the worst invention,' said Newton. 'To watch bags going through for hours on end is not only soul-destroying, but also an almost impossible task. That's why you see at so many airports uniformed guards chatting away merrily – with their backs turned to the screen.'

It is a vicious circle, indeed. Offering better conditions would enable employers to be more choosy in staff selection; better paid personnel would be more receptive to better training and would stay on longer – and would become even more bored and complacent. Yet certain human qualities would seem to be essential. Apart from thorough training, security staff need some understanding of psychology, an appreciation of certain passenger behaviour patterns and, above all, a fair amount of alert intelligence – a somewhat tall order of requirements for such a menial job. But how else did an Israeli guard realize, for instance, that when a passenger had completely emptied her hand luggage, as requested, it still felt too heavy for its size? That is precisely how an innocent and romantically duped *mule* was found to be carrying a bag with a false bottom, given to her by her Palestinian fiancé. The cleverly constructed, invisible compartment was full of explosives which failed to show up on the screen.

Increased spending on security, now demanded by passengers as well as regulatory authorities, could perhaps provide finance to plug the loopholes in countries which plead poverty as the reason for doing so little. France introduced a security tax in 1987, and passengers hardly noticed the extra few francs added to the price of their tickets. The International Foundation of Airline Passengers' Association urged the industry in 1989 to set up a massive security fund by imposing just a one-dollar tax on each

* National conference on 10.6.1988.

airline ticket, and a similar levy was proposed in the same year by a British parliamentary all-party committee.

But it is only a global approach which may really improve security, and ensure that Lockerbie will not be forgotten as readily as the lessons of other bombings.

ICAO has now formed a security branch to advise member states on policy as well as the selection and training of staff. The British government has issued new directives, and tighter regulations. Operations could be suspended if lapses are discovered, and infringements by individuals could become a legal offence. In America, stricter controls have also been introduced. But in January, 1990, Billie Vincent, a former head of FAA Security, warned the congressmen investigating Lockerbie that 'the security procedures and processes that have been implemented since (the tragedy) are not sufficient to prevent' a repetition.

Better mechanical, human and organizational measures will not be enough. The laws against air terrorism must be tightened everywhere to protect the innocent. Each country must impose its maximum legal sentence for such crimes, with no pardon even for alluring propaganda advantages. The 'no hiding place for criminals' principle would have to be enforced with automatic extradition of the culprits. The same principle would also have to be applied without any exception to bombers and skyjackers who are sometimes viewed sympathetically as political refugees, and may now get only light prison sentences as a prelude to asylum.* Wherever one's sympathies may lie, tough laws, which cost nothing, must be courageously enforced. Gunter Eser was right. Piracy on the high seas used to thrive on safe havens, and died out only when nobody would accept a share in their ill-gotten gains – or forge political capital out of their marauding.

* In April, 1990, the Tokyo High Court refused to grant political asylum to a Chinese man who, fearing persecution after demonstrations in Peking, hijacked an aircraft with 233 people on board. Extradition could equal a death sentence, but a long jail sentence could show that the world will not tolerate such crimes that could endanger hundreds of lives.

– 14 –

HAS IT GONE WITH THE WIND?

In the early days of flying, life was simple for the investigator. Any World War One pilot who survived a crash would be quickly dispatched on another mission to restore his nerves. The problems arose with the discovery that it might be possible to save lives by learning from other people's misfortune, provided that the real causes of the mishaps were unearthed.

Recognizing the limitations of what the *jigsaw wallahs* (or their American colleagues, the *tin kickers*) could decipher from the clues in the tangled, burnt wreckage, investigators realized the importance of evidence which has gone with the wind but which might have lived on in people's memories alone. Human recall is, however, an unreliable instrument. Witnesses who volunteer may be too keen to be helpful by trying to say what they think the investigator wants to hear. People honestly believe that they saw fireballs in the air rather than on the ground, after a crash; this is an easy way to sow confusion in moments of excitement.

Interviewing a witness is a highly developed and finely tuned technique to verify statements without overt suspicion, without ever making people of goodwill feel like criminals in the dock. Yet certain questions must be asked. Did he see it? Could he have seen it? Did he see what he says he saw? Did he see what he thinks he saw? What makes him think he saw what he says he saw? Could he have seen things he does not recall readily?

Some like to be *experts* ('I heard the pilot gunning it'), while others tend to recall emotive details ('it was obvious he'd crash –

I just felt it in my bones'). The simplest, least sophisticated people are usually the most reliable witnesses. Children are often the best, though they can be rather impressionable. There have been cases when children changed their perfectly accurate accounts only because their parents remembered some events differently. A boy, who witnessed the crash of a light aircraft only a few feet away, told his mother that he had seen a spiralling dive, but offered the investigators a different tale a day later: his account was heard by his schoolmates who embellished it and discussed it *ad infinitum* until the witness himself began to remember not his own story but the consensus of his class.

A British accident produced an all-time classic. Just minutes after the crash, a woman reported that she had seen the aircraft wavering up and down quite violently. That implied some control problem, so it was a potential lead. The investigator in charge rushed to the woman's house to check what she had reported. To hear her statement, he himself sat in thc armchair where she had sat knitting at the time, and, yes, the air lane the crashed pilot had used was now in view. While she was describing her observation, he heard the approach of an aircraft. He looked up, just as the woman must have done, and to his horror, he saw another plane 'wavering up and down quite violently'. It was, of course, a flaw in the undulating glass, through which the flight appeared to be out of control.

In Washington, after the Air Florida icing accident, two people reported that they had stood together, watching the take-off and witnessing the crash. They both knew exactly the spot where the aircraft had lifted off the runway. The investigator took them back to their vantage position, and 'asked them to point on the count of three to the spot where (the jet) had broken ground. One, two, three . . . both pointed simultaneously – in different directions'.*

After a helicopter crash in Borneo, it was known that the rotor blades had come off in the air but it was vital to establish whether the separation was the cause or the effect of some control problem. A Dayak man claimed to have seen it all. Interviewed by Don Cooper of the AIB, he used a model to show level flight, the main rotor coming off, then the aircraft falling to earth. Through an

* *Air & Space*, August/September 1987.

interpreter he said: 'It came out with its stalk'. Yes, the *stalk* was the shaft, but did he really see it all from three miles away? Cooper asked him casually what happened to the rotor blades after they had come off. The man spread his arms, and raised them until he could clap his hands above his head. Now that was hard evidence because that is precisely how the blades would behave – it is called *coning* – and the Dayak could not possibly have known that. Detailed examination of the wreckage then confirmed it all: there had been no control problem before the main gearbox failed and allowed the whole rotor to separate from the helicopter.

The investigator likes to have some facts before he turns to witnesses, so that he can evaluate what they say. His early questioning is also restricted by hospital rules and other considerations. Delays, however, can be a serious risk: media reports and various unauthorized leaks often colour the memory of the witness. Some experts would love to use hypnosis as an aid. As far as I know, only once did they resort to a 'truth serum' – but the effect was then quite stunning. In 1962, a Lockheed Constellation crashed on Canton island, killing the crew and one of the passengers. An FAA doctor who sat behind the captain in the cockpit to conduct observations, witnessed the full development of the emergency. He survived the crash with serious injuries that blotted out his relevant memories. When the investigation ran into blind alleys in all directions, the doctor consented to narcosynthesis. Under the influence of sodium amytal (the occasional tool of medical and intelligence interrogators, as well as the bread and butter of spy fiction), he recalled amazing details of the drama, performing the role of the yet to be installed flight and voice recorders. Later when he listened to his 'confession' on tape, he confirmed every detail. Investigators then matched his renewed memories with the known findings, and pieces of the jigsaw fell into place.

The modern *electronic witness* is known as the *black box*. The name indicates a box of tricks with the conjurer's powers of darkness, though it is painted bright orange or yellow to be more noticeable, and marked with reflector tape to help recovery. With their handles, the boxes look as if they belonged to a filing cabinet. Weighing between twenty and thirty pounds, they are some five inches wide and seven inches high, and the flight recorder, at some twenty inches in length, is seven inches longer than the CVR. Both enjoy tremendous crash and fire protection, and must contain a battery-operated 'pinger' which is activated by immersion in water, and will transmit a signal for up to forty-five days, again to help recovery at sea.*

Initially, there was a great deal of confusion because there were a dozen different types of foil, wire and magnetic recorders, many of which were easily scratched, torn and damaged in a crash, rendering them almost illegible. Although the antiquated machines are on their way out to be replaced by digital types (with *solid state* memory) which can record a hundred or more aspects of engine and flight performance, some countries, including Germany, continue to follow the slow-moving American example* and allow even Jumbos to fly with black boxes which use foil and record only five parameters, one of which is time. For two decades, the chief argument has been cost: it is true that rewiring an aircraft for complex recorders is expensive, but the cost diminishes when it is part of the manufacturing process. Most modern aircraft have some thirty parameters recorded. But, under the influence of the powerful aviation lobby's false economic considerations, new American aircraft may still roll off the production line with ancient recorders until the early 1990s – even though the same model is delivered to some other countries with dozens of recording channels at no extra cost.

In Britain even business jets will soon have to install recorders, but in the United States, where foil recorders can remain in use until 1994, the NTSB is still battling to make it compulsory for commuters to carry black boxes. Bernard Caiger said in Ottawa: 'Canada was a late starter, but began with a requirement for about twenty parameters on commercial aircraft, even business jets, over 12,500 pounds in 1969. This prevented any use of foil recorders. Air Canada had installed much more comprehensive recording well before that initial legislation. Following that good start, however, the requirements remained static for many years – and in fact, they were even relaxed to permit the use of foil-type recorders on turboprop aircraft. It became an embarrassing topic

* It has long been suggested that an explosive device be added; this would be activated similarly by immersion in water to eject the recorder, make it surface and let it float – but the risk is that the explosive may cause accidents.

† As described in Chapter 5.

for discusssion at international meetings.' All this would appear to be less surprising in view of the fact that even the ICAO proposals for the last decade of the century remain somewhat vague and ambiguous.

Like the watch, which stopped because of impact and thus revealed the time of the crash, damaged flight instruments and needle positions, dial settings and filament bulbs in warning lights used to retain telltale clues. With the spread of the 'glass cockpit' concept, the data displayed on a TV screen will just vanish when the cathode ray tube goes blank or is smashed, so recorders will become even more essential tools for investigators. Fortunately, the companion concept of 'fly by wire' makes it easier to record additional parameters, and now a new idea is gaining ground: if video recorders were installed in the cockpit, the camera could watch and 'remember' all the elusive electronic data on the screen.

Flight and voice recorders play much the same role as the truth-drugged doctor: their revelations hardly ever provide full answers, and need to be proven consistent with evidence from independent sources. Specialists mock their work rather than their 'toys' when they say that a flight recorder is like a bikini: it reveals all the important parts, but conceals the vital ones. Such remarks are no irreverence towards the dead – merely a way to safeguard the crash detective's sanity when he is wading knee-deep through gore and horror only to resurrect nightmares. Perhaps even more than the sight of mangled corpses, it is a hugely depressing task to listen repeatedly to the last minutes of a CVR, to the final moments of banter and routine exchanges, to the sudden realization of impending death. Investigators are often warned: 'Joke if you must on site, but don't ever be caught by the camera, for a laughing face among the wreckage in next day's *Daily Mirror* is an insult to the relatives of the victims.'

Ex-AIB Ray Davis, an international consultant whose fascination with black boxes has never waned through hundreds of cases, prefers to recall the amusing stories like the one about the captain who was heard on CVR lecturing his co-pilot: 'He was having a real go at the man. "Pay more attention . . . no-no, that won't do . . . I'll take over, just to show you how it's done", he said and with that he proceeded to crash-land short of the runway. Or there was the flight to an African airport. In poor visibility and pouring rain, the co-pilot is heard protesting all the way down:

"We're not going to make it, no way, we're much too high." The captain keeps telling him: "Don't worry, it'll be okay, you just watch it". He then lands two-thirds of the way down the runway, goes off at the end into fields, and sounds surprised when he says on the CVR: "Now why should we let this happen to us?"

'The beauty of the equipment is that the DFDR tends to tell us first what didn't happen – and saves a lot of time. Then it indicates *what* happened, while from the CVR we may learn *why* it happened.' The CVR also provides investigators with additional evidence, and even yesterday's wind cannot just disappear and be gone forever: how strong was it? did its direction slow down or speed up the aircraft? 'We often wish to know, for instance, the pre-crash prevailing wind conditions experienced by the pilot at altitude. In one accident, the CVR offered us some reference points in the pilots' words about passing a DME (*Distance Measuring Equipment*), talking to ATC, and calculating distances. That helped us to time and reconstruct precisely when the flight reached two definite points above the ground, giving us the groundspeed, the actual rate of progress. The True Airspeed was already derived from the recorded impact. Now the difference between airspeed and groundspeed revealed the headwind component of the actual flight.'

The relaxed conversation of the pilots might reveal that they had no indication of a problem. Apart from the spoken words, the CVR can come up with subtle clues for clever interpretation which would delight Sherlock Holmes. The captain's coughing throughout a flight was proof that he was not very healthy. In another case, Ray Davis called in medical specialists who confirmed that the pilot's breath rate had increased seriously, a symptom of undue and inexplicable stress, long before any other sign of an emergency, and so helped to pinpoint the very beginning of some control problems.

The analysis and filtering of background noise may tell the investigator about mechanical malfunctions the crew themselves might not have been aware of at the time. It also reveals the exact moment when the pilot does something. If, for instance, a flap is deployed, a spike will show up in the graph or data print-out as long as that motor is running. After a helicopter accident, the CVR was recovered long before the wreckage from the seabed. Some noises revealed which gear had failed at what precise moment – all to be confirmed later by jigsaw wallahs. On the CVR from a Danish helicopter that had crashed into the North Sea, the even sound of the tail rotor was heard until it gave a problem signature: suddenly the noise trace went up, then dropped, and disappeared – indicating that the drive shaft must have broken. When that clue was checked, the story found in the wreckage was consistent with the CVR.

After the Continental DC9 crash at Denver in 1987, Jim Cash, a CVR specialist of the NTSB, managed to filter and single out of the *noise cocktail* the soft, repeated thumps contributed by the turning of the nosewheel. From that alone he could calculate the actual take-off speed of the DC9. In 1985, two much sharper thumps, followed by violent vibration on take-off, persuaded the captain of a Lockheed Electra to reduce power and return to land. He stalled, and crashed. The flight recorder was out of order. All the material and operational evidence could show was that the turboprop engines as well as all the systems and controls had worked faultlessly. So was it pilot error? Two thumps on the CVR yielded an essential clue: the noises were traced to an improperly sealed access door on a wing, and were the cause of the vibrations the pilot tried to cure by an excessive cut in power. On the CVR, the sound level of the engines drops, but the co-pilot's warnings rise to a crescendo of: 'Pull up! Pull up! Max power!' by which time it is too late to save the aircraft, and the recording ends abruptly.

In the CVR of the Air Canada crash at Cincinnati, the sound of fire bells was additional evidence that there had been fire in the air. But replay after replay detected another peculiar noise in the background. Having eliminated all the familiar audio clues, the investigators identified that odd sound: some circuit breakers were popping right behind the cockpit. Logic and an examination of the electrical layout did the rest – those breakers belonged to the restroom circuit; therefore the restroom area had to be suspect; that was, in fact, where the fire had started.

Few investigators will ever be as lucky as those who were borrowed from Britain to help Malaysians dealing with a crash. The CVR revealed that the Malaysian Airways 737 had been hijacked and diverted. Flying over Jahore, a man was heard shooting the pilot. But for the professional investigator, that was no proof of what actually caused the crash. The wreckage was in a swamp. A hundred men were employed to keep digging for it in atrocious conditions for two weeks. An engineer suggested that the swamp could be drained, but that was a hopeless proposition.

Meanwhile, witnesses were interviewed, including people from a Malay wedding party. Allegedly they saw the crash. A Boeing man said they might have been drunk, and not much worth listening to. True, what they said sounded doubtful, and couldn't be treated as evidence. Yet their testimony was noted: they claimed that the aircraft had dived out of the night sky, pitched up, recovered twice, and then plunged into the swamp in a nose down attitude. Their account was elevated to the status of hard evidence only when the recorder read-outs were completed, and the air traffic controller, who had watched the 737 on radar, confirmed the evolving flight profile.

'The area mike in the cockpit picked up all the noises, and so reduced the clarity of speech, but the hijacker could be heard on CVR entering the cockpit,' said Ray Davis who worked on the case. 'The pilot did exactly as he was taught. He was polite, played it cool, right down the middle, and reassured the gunman by explaining what he was doing whenever he touched a switch, changed some setting or increased power. The hijacker didn't sound like a fanatic or a madman at that stage. There was no indication of sudden fright or confrontation and, in fact, they had just asked for and received coffee, when suddenly he went berserk. He might have suspected poisoning or seen some imaginary threat, but he started shooting. Bang, bang, bang. Three shots. We could not tell if he shot himself, too. But was that *the* cause of the crash?

'When we synchronized the voice and flight recorder read-outs, we began to see what actually happened, and that the Malay wedding guests were good witnesses. The flight data showed that, right after the shooting, the path of the aircraft became kind of wavy. Slightly out of trim, the nose goes down, gathers speed, begins to lift causing a decay in speed, goes down again. On the CVR you can hear the changes. Some noises tell us that people from the cabin are breaking through into the cockpit. We have no idea what they do there, but the nose comes up again. With that the airspeed is reduced. When it slows to a dangerous level, all the background noise disappears, and you can almost hear a pin drop. It's a stall. Then the noise increases as the aircraft dives, accelerates to five hundred knots, and one could almost calibrate the build-up to the speed with which it crashes into the swamp.

'It all happened in thirty minutes, just the time span when the looped tape would return to the beginning and start erase/recording another half an hour.'

The read-out of the flight data is an extremely painstaking, time-consuming exercise, full of pitfalls for the inexperienced, unwary investigator, tempting him to misread red herrings for genuine clues. He* might even pooh-pooh frequent warnings that the transcription of a CVR tape is more an art than a science, because it is all too easy to mishear words or misinterpret sounds. Sometimes it is useful to invite a friend or colleague of a dead pilot to listen, try to catch a mumbled word, help to decipher the correlation between half-words and intentions; but, on the other hand, his presence may compromise confidentiality.

WHOSE WORDS ARE THEY ANYWAY?

Is the CVR a 'spy in the sky'?

As Ray Davis knew only too well, 'even some British airlines tried to use flight recorders for such spying. Some pilots were fired or disciplined but had to be reinstated when the recorded data was proved to be faulty. But such defence is always an uphill struggle. And let's not forget it, what other profession would tolerate constant monitoring? It shows the pilots' dedication to safety that they put up with it. But they must have enormous trust in the system, and they must never be let down.'

The two extremes of attitudes to confidentiality were probably represented by Australia and the U.S.S.R. In the former, no use of the CVR was once permitted if the crew survived a crash, while in the latter, black boxes are often used for monitoring pilot performance. (Russian inspectors pull out and run recordings right after ordinary landings to conduct spot checks and initiate disciplinary procedures – just the sort of application of the device that pilots' unions in the western world have rightly fought tooth and nail from day one.)

* For the invariable use of the male gender, my apologies to the first woman ever to be employed as a British aircrash investigator. She won a Civil Service competition for the position to specialize in flight and voice recorders.

Spanish and Italian authorities are known to have leaked deliberately some recordings that would incriminate foreign pilots only to whitewash their own ATC shortcomings. In some countries, like Germany, the judiciary investigator in charge can *demand* that the recording should be handed over. 'In Canada, CVR is claimed to be confidential but isn't fully protected in reality,' said Captain McBride of CALPA. 'Fortunately, there has been no serious breach of confidentiality . . . so far. But if, for instance, a newspaper can convince the CASB or a judge that it is in the public interest, the transcript can be revealed or published in full, even though a few *asides* or totally irrelevant banter can cause a lot of unintentional damage to reputations, and discredit crews, dead or alive.' Gary Wagner, an Air Canada pilot, added that 'it is quite disgusting when the Americans leak the CVR within hours of a crash. It opens the way to speculations like those that filled news bulletins after the Detroit Northwest accident.'

Ex-NTSB Chuck Miller said: 'My pilot friends will hate me for this, but I believe that *all* recordings are relevant, factual information that must be made public – but not without informed analysis, and not without protecting the feelings of a dead man's family. For example, I see no need to play screams of about-to-die pilots on the evening news. Otherwise, every detail helps to investigate and understand the case. You run into problems only when people, especially the media, try to prejudge what is and what isn't relevant. While I was with the Board, we once made a study of a hundred recordings to see in each case if the CVR helped or screwed the pilot. It was a subjective assessment as these things tend to be, but we found that CVR cleared the pilot twice as often as it indicted him.'

In Britain, the use of CVR is very strictly controlled. If the crew survive, they are invited to listen to the tape, but management representatives may join them only with the pilots' permission. If all the cockpit crew are dead, the management can help with the transcription, but the CAA, the regulatory authority, must not. Crash reports may quote certain relevant excerpts, but non-pertinent conversation, idle chit-chat, private remarks, gossip about a stewardess or something nasty about superiors are always kept confidential.

The protection of privacy notwithstanding, lawyers might have a bonanza one day if it became necessary to define fully the copyright in CVR. The AAIB keeps the tapes from crashed aircraft in its library, and would not release them – but its right to do so has not yet been tested in the courts. If the transcript was stored in computers, the Data Protection Act would apply – but it is not. Theoretically, the tape is airline property though, once it is removed from the aircraft, it may never be put back to use. The AAIB could perhaps buy the tape itself from the operator, but that would still leave the juicy question of copyright: to whom does it belong? the crew? the airline? Should it be treated like a letter, with copyright vested in the writer, and possession in the hands of the person who received and kept it? And what happens when the transcript – held by foreign investigators, and seen but never released by the AAIB – is bought by a newspaper or a TV company to be re-enacted by actors? (Unfair to the crew or not, it did happen after the Zagreb collision as well as after a 727 accident at Tenerife, and it could happen again if it involved countries where fair play and confidentiality have a different meaning.)

Aviation lawyer Tim Scorer sees it as a very tricky problem: 'In a liability case, requests to the AAIB for a full text have been met with resistance. It is said that disclosing transcripts of CVRs would destroy the trust between the AAIB and the pilots who, if they felt that the trust was threatened, might resist the installation of CVRs or otherwise inhibit their use. There's also the question of copyright because the pilots, being the authors of the words, might claim the rights in them although arguably, since the words were spoken in the course of the pilot's employment, the transcript could be said to belong to the aircraft operator. This is yet another area where the interests of safety on the one hand and legal liabilities on the other can vie for priority, and the outcome can restrict the advance of aviation safety.'

Undoubtedly, some seemingly light exchanges can be crucial evidence.* BALPA has fully binding agreements drawn up with some British airlines governing the permissible uses of FDR and CVR, and people like Sydney Lane firmly believe that 'to some extent, the international aviation community is to be blamed for the irresponsible exploitation of the recordings in some countries: through ICAO they could and ought to strengthen *Annex 13*, the

* E.g. CVR excerpts from the Air Florida icing case, see Chapter 8.

investigators' bible, to protect pilots against practices that could be seen as self-incrimination.'

In the United States, the prevailing concepts about air safety and the Freedom of Information Act steer a different course. As Jim Cash summed it up: 'At the beginning, we have the transcript to ourselves, when in principle it can be used only to help the investigation. After sixty days – when indelicate, non-pertinent conversation as well as expletives have been deleted – this edited, sort of sanitized version must go into the public docket. The problem is that all interested parties who may participate in the investigation have access to the CVR right from the start, and it's during those crucial sixty days, when we're just gathering the facts, that the damaging, out-of-context leaks occur only because it's in the interest of one party or another to be quickly exonerated of all possible blame.' Steve Corrie, an ex-Marine Corps, helicopter and fixed wing pilot, who was an investigator with the NTSB for eighteen years, supports the policy of openness, and would like to see 'much stronger protection by law against premature leaks', yet recognizes that such laws 'may reflect only wishful thinking. It is difficult to keep the participants of an investigation in line. Tightened NTSB procedures have probably gone as far as they can. Much depends on the integrity of the investigation itself. Cooperation with the various parties can be damaged by premature release of "raw" information.' Though similarly, most American investigators advocate some temporary confidentiality, and a restrained process of delayed disclosures, they are said to be 'sensible enough' not to ask their foreign colleagues for a full transcript which would have to be revealed back home to anyone who wished to see or publish it.

An inadvertent, and therefore unforeseeable, demonstration was set into motion at Detroit Metropolitan Wayne County Airport, when the engines of a Northwest Airlines DC9-82 were started up forty seconds after 20.34 on August 16, 1987. Jack Drake, in the middle of his second decade as a crash detective, remembers that day just as clearly as most people can recall what they were doing in the moment President Kennedy was murdered in 1963.

'It was a Sunday with plenty of sun and fun,' he told me in his cramped little NTSB office which was so packed with stacks of documents that it would have been no surprise if the walls began to show an advanced state of pregnancy. 'After a swim and dinner with my parents north of Baltimore where they keep their boat, we kept laughing with our three children as I was driving the family home that evening. It would be a relaxed, forty-minute journey to Columbia, Maryland.' (As if to emphasize the mood, the last word sounded like *Merryland*.)

Having wasted a few minutes by missing the assigned turn-off at taxiway three and needing redirections, Flight 255 was cleared for take-off at 20.44. Less than a minute later it was airborne with 149 passengers and a crew of six on board. Witnesses saw it climb steeply, roll to the left and right, level out, and bank to the left until the wing hit a light pole in a rental car parking lot. It disappeared from some people's view, but others saw it strike a second pole in another lot, then the sidewall of a roof and, still rolling, impact on a road and cars beyond the airport boundary, slide along, disintegrating all the way, until the crash into a railway embankment demolished it. As the aircraft had spent only twenty-seven seconds in the air, there was some confusion among witnesses whether they saw 'orange flames' in the air or only on the ground.

Just past nine o'clock, some twenty minutes after the crash, Drake's pager buzzed. It killed the jokes in the car, for it could be anything but good news. 'I stopped at a store that had a phone outside, and dialled the most likely callers,' he said. 'Alerts would normally come from the field operations duty officer, but this time it was Terry Armentrout, Director of the Bureau of Accident Investigations. I told him it would take me twenty minutes to get home and pick up my bags, then another hour to National Airport to catch a flight to Detroit, via an FAA airplane.

'For years I used to work solo on some four hundred smaller accidents. When you're on the Go team in Washington, cases become much more complex. I've never kicked the habit of having a hefty go-kit. Apart from clothes, and boots, and the usual things like basic tools, camera, ident badges, tags for wreckage, tape measure and basic guidelines for setting up specialist groups, I carry all sorts of forms, reminders, grease pencils, magic markers, everything others might forget to bring. Working solo taught me little tricks of the trade. In my wallet I always have a business card with a measure printed on the flip-side: it's useful when, for instance, I put it alongside something I photograph because it gives an immediate idea of the size of some possibly unrecognizable piece of wreckage. I also carry a card with L on one side and R on the other to identify wreckage from the Left or Right side.'

Droves of people descended upon Detroit. There were sixteen NTSB investigators, Board member Lauber, three public affairs officials, and dozens of interested parties, more than a hundred people in all, to be organized into thirteen specialist groups. Additionally, apart from the media and the FBI (interested only in sabotage), two dozen pathologists and dental forensic experts arrived to identify the bodies: of those aboard the DC9 only a seriously injured four-year-old girl survived, and seven people were killed or injured on the ground. The investigation would turn into a massive, fascinating textbook case.

Rumours began to feed the media right away with *revelations* about sabotage, windshear and, above all, engine failure. But some of the speculative reports smacked of deliberate plants by interested parties who were keen to exonerate quickly their own people or product. When the bomb-on-board theory abated on the eithteenth, the NTSB was allegedly looking into 'airline labour troubles', and violent accusations against the engines flooded the headlines with demands to ground all aircraft with that model. If the engines were at fault, then operators, airframe manufacturers and all other parties would be cleared at Pratt–Whitney's expense.

On that Tuesday, the second day of the investigation, when the NTSB labs in Washington had just received the *black boxes*, the engines were, indeed, the *cause du jour*. The artistic accident reconstruction for television showed the DC9 going down with its left engine ablaze. The banner headlines kept changing throughout the week, and, by Saturday, the investigation was supposed to have 'stalled' due to conflicting evidence. The truth was, of course, entirely different. The spotlight of suspicion fell on one potential cause after another, and, as more data were gathered, pet theories were eliminated in turn.

The examination of the wrecked cockpit offered a most unlikely clue: the flap handle was found in the UP/RET position. It might have moved during the crash sequence after the crash. Had the pilot tried to take-off with flaps up and slats retracted? Nobody would do that. On both take-off and landing, the extension of flaps enlarges the wing surface and, aided by slats, provides added lift and stability at low speeds. 'Flaps' is a standard, key part of the checklist, and, even if somehow the pilots omitted it, an automatic warning would sound in the cockpit to give them an urgent reminder on take-off.

'One of my specialist groups which was checking out all *systems*, documented the flaps and slats positions,' said Drake. 'They looked at various parts of the badly damaged operating mechanism – pulleys, tracks, rollers, etcetera – that would show scratches and impact marks. With the enormous initial workload, one thing we overlooked in the first two days was the condition of the cables that drive flaps and slats up or down. But we noted some conflicting evidence: on an outboard wing section, evidence indicated that the slats may have been in the extended position, but elsewhere they were not. That would prove nothing because, on impact, when the operating cables were severed, the suddenly unrestrained slats could begin to move freely up or down.'

Meanwhile, the first replays of the CVR turned up more evidence: by the sound of it, the pilots neither called for nor accomplished the *taxi* checklists in accordance with the airline's *Handbook*, and simply failed to extend the flaps and slats. That would explain why the handle was in the UP/RET position, and would have certainly imposed severe limitations on the climb performance of the airplane. But if that was the case, and if the DC9 was thus improperly configured for take-off, how come there was no recording of the automatic alert system screaming at them? Did they extend the flaps silently, without the required oral demand and verification? Or was the audio warning faulty?

Unlike Boeing (some of whose executives used to reject complex recorders with the arrogance of 'we build good airplanes, we don't need Mickey Mouse devices'), McDonnell Douglas have long favoured sophisticated FDRs, and Flight 255 had fifty-six parameters recorded. 'While this was extremely useful to us, it magnified our task enormously,' said Dennis Grossi who was in charge of the read-out that would translate endless columns of data into intelligible guidance for the men on site. 'We got the flight recorder on Tuesday morning, and worked with it non-stop. At first I suspected some stabilizer problem, but, as we looked at the read-out, it seemed immediately obvious that the airplane was not properly configured because the flaps and leading edge devices were not deployed. It was late at night by the time we had checked and double-checked everything, and we delayed relaying the information to Detroit only to let those equally overworked guys over there have a full night's sleep.'

The news reached the investigators on Wednesday morning – and it was leaked to the media by the evening. 'We never knew where the leak came from,' said Ron Schleede, 'but it was most advantageous to the airframe-, and particularly, to the engine-manufacturers to show that the pilots rather than the hardware were responsible for the crash. It was extremely annoying, but instead of trying to change our principles about openness, we learned from it: by the time we'd have our next big accident at Denver, we'd know how to play a little bureaucratic shuffle and prevent similarly premature leaks.'

On August 21, the first correct headline appeared, but even that failed to stem the flow of speculations because the media rightly suspected that the leak had been just one of many that served conflicting interests. Drake, however, took Grossi's news more seriously: 'I went to see those outboard slats on the left wing. They're worked by the up and down cables that form a continuous loop. First I saw only one of the cables. It was hanging loose, and it was severed. It must have been cut by some impact. [The portside wing had hit a light pole at the beginning of the crash sequence.] So I used one of the old tricks from my solo days, and looked for the other cable which I found hidden inside that wing section. When we pulled it out – we knew we were staring at some of the first hard evidence.

'When the slats are operated, the two cables in the loop move in opposite directions. In the SLATS RETRACT (UP) position, one cable goes inboard, and the other goes outboard. If the loop was cut, as it had to be, by a single impact, the severed points would be misaligned by some fifteen and a half inches. But now, when we pulled the cables taut, we could align them only if the slats were fully RETRACTED. The cables were separated at the point where that wing section was broken off. But did that prove that we had found the probable cause of that accident? Sadly, no. My photographs of the matched cables only proved beyond doubt that *at the point of impact*, the slats were not extended for whatever reason. They told us nothing about the configuration *before the moment of impact*, and could not prove that Flight 255 had actually attempted to take-off without flaps or slats. We usually have material evidence as well, but on this occasion the proof had to come from the recorders without which some people would still be debating to this day who or what caused the deaths of all those people at Detroit.'

Like the CVR, the DFDR worked on a continuous loop so, having reached the end, it would continue automatically by starting to erase/record at the beginning. Flight 255 thus left behind a full history of the last twenty-five hours in the life of that DC9-82. Grossi could read out its configuration during previous take-offs, cruises and landings. Flap-and-slat positions could be deduced from different sensors four times every second. It was seen that the flaps and slats had last been retracted during taxiing after landing at Detroit. From then on, their positions never changed until the crash, i.e. the end of the recording. That verified Drake's evidence about the cables of the left wing, and proved that the position had been identical on the right wing about which there was no conclusive material proof. But that was not all. The DFDR provided detailed information about the performance, pitch and bank attitude, of the DC9-82. The way the aircraft had struggled into the air during its last, tragic, twenty-seven-second flight was fully consistent with the performance that could be expected only if the flaps and slats had not been properly deployed. That calculation showed to what height the aircraft could climb in such configuration after take-off – the exact height of the first impact on the tip of the light pole in the parking lot. With at least the slats extended, the aircraft would have cleared that light pole by five hundred feet.

'The oldfashioned recorders, which as of May, 1989 no longer meet U.S. recorder requirements, would have given us no more than a vague indication of a *possibly* improper configuration during take-off,' said Grossi. 'But reading a modern DFDR in conjunction with the CVR gave us the full story about Flight 255.' The labs traced the crew's work patterns during the flight into Detroit as well as during their departure, and found many examples of non-standard performance: the weather radar was left on well past the completion of the *after landing* checklist; they missed a gate upon arrival; they made mistakes with radio frequency changes; they failed to locate the assigned taxiway to the take-off runway; and they experienced difficulties with engaging the autothrottles for the last take-off only because they had omitted yet another item on the *taxi* checklist. All this added up to a lack of disciplined and rigorous adherence to prescribed checklist procedures which are designed to protect the crew from making mistakes.

But why was it that the automatic warning system failed to come to the crew's assistance? For some undetermined reason, the system was inoperative on that occasion. That was the final coincidence.

Like the China Airways *aerobatics* with a Jumbo,* Flight 255 was re-created by *computer animation* using CVR, DFDR and recorded radar data with actual air traffic control recordings dubbed on. 'Instead of indigestible paper mountains including a one-and-a-half-inch tome of DFDR figures of airspeed, altitude, heading and the rest, we could present our case clearly in this form, and in real time,' said Cash who produced the videotape that put all the data in context, and conveyed the predicament of the pilot who had barely three or four seconds to react.

There is banter and laughter, just the sort that could be exploited and misinterpreted if leaked. The aircraft misses the taxiway. The pilots struggle with the autothrottle which won't stay on. # – an expletive. A laugh. 'Vee one . . . rotate . . . ' In full knowledge of what is to come, one feels like shouting at the dead pilots: '*Flaps!*' They seem to have all the time in the world, but it is too late . . . the sound of the stickshaker . . . stall warnings . . . 'ah #' – you know the light pole must be looming up – and then the sounds of the first, second, third, fourth, fifth, sixth and seventh impact is in seven seconds.

It is sickening to watch it, but the video presentation is not an exercise in cheap titillation. It helps everybody to understand that the fate of all aboard is sealed long before the nosewheel leaves the ground. It is a tool to train people because the lesson in the raw accident data may be too elusive even to-pilots. The more complex computer reconstruction the Canadians are experimenting with will go even further when showing, for instance, instrument 'close-ups' and allowing the operator to *fly* the model: extra power, a change of configuration – and the model on the screen will respond like a real airplane. If done in that way, Flight 255 could be steered out of trouble. Though that would not

* See Chapter 1.

bring back the dead, at least it might save other lives. The full exploration and video presentation of the Detroit crash could help to achieve what the NTSB report recommended, that the warning system should be made more fail-safe and that, above all, pilot training should be improved in some specific ways to ensure full adherence to prescribed procedures and achieve more effective cockpit resource management.

Ultimately, the tragedy of Flight 255 was a painful reminder of the grave dangers in premature leaks. And if Detroit was not bad enough, worse was to follow right away. In 1988, an American court ordered that the CVR tape from a Delta crash at Dallas must be released to the press without any delay. The FAA and the pilots' union deplored the decision, and ALPA threatened to retaliate with what would hurt air safety most: its 40,000 members could 'pull the plug' and disable all voice recorders. Pilots have a very well-developed sense of responsibility, so the threat was not carried out – but it might still happen someday.

In 1989, the NTSB, so firmly dedicated to 'sunshine hearings and investigation', felt compelled to join the protest and vote for new legislation to keep the CVR tapes secret at least from the unfair and unnecessary exposure of human behaviour. The NTSB reaction was taking not only a moral but also a firm self-defensive stand: the loss of CVRs or any undue restriction on their use would deprive investigators of a vital weapon in their fight for the fuller understanding of the frailty of human performance – a danger about which they have frequently issued final calls after final calls.

– 15 –

YOUR FLIGHT IN THEIR HANDS

This is just a random sample of news items from the arbitrarily chosen months of September and October, 1989.

Pilot error, concluded the Greek investigators who looked for the cause of the accident that killed thirty-four people on August 3, 1989, when the Olympic Airways Shorts 330 crashed into high ground on the island of Samos. The visibility was bad, yet the radar was off as the pilots flew into instrument rules conditions on a visual rules flight plan.

Pilot error, said the Cuban State Commission investigating the crash that killed 150 and destroyed dozens of houses near Havana airport. It exonerated all airport facilities and the Il-62 with the finding that the extremely experienced pilot had underestimated the risk of strong winds and misjudged the bad-weather performance of the old Russian aircraft.

Pilot error was blamed for the crash-landing of a Varig 737 in the Amazon jungle. The pilot (first hailed as a great hero for bringing down the aircraft in one piece, more or less, and saving all but thirteen lives on board) was a long way off his route, and was not the victim of some mysterious total failure of navigation instruments – he just mis-set his heading while listening to the Brazil–Chile football match.

Pilot error, was the off-the-cuff verdict after the new European A320 Airbus had crashed at a French airshow near Mulhouse, killing three passengers. The senior captain was blamed publicly, and suspended at once without anybody waiting for any investigation, despite his complaints about his computerized controls which have since been criticized as troublesome by other pilots.

Pilot error started a public outcry when a U.S. Air 737 aborted take-off at La Guardia, New York, and slid into the East River impaling itself on a concrete pier at the end of the runway. All but two people aboard escaped with their lives.

Pilot errors? Yes, probably, but should one not remember some circumstantial details? At Samos, the co-pilot had only three months' experience, and the captain might have been sick. (There were complaints about his 'bizarre flying', and a stewardess, who had broken off her relationship with him, was advised by a psychiatrist to report his threats to crash into her home.) The Cuban pilot might not have been trained properly for that Ilyushin. And why did the Brazilian crew fail to warn the captain? Were they also listening to the match? Why did the air traffic controller (who gave the pilot the frequency for the match) fail to note that the flight was way off course – was he perhaps watching another screen? French investigators exonerated the aircraft and blamed the crew for the Airbus crash, but had the pilot with no *display flying* experience received a sufficient preflight briefing for the show? And what caused the error at La Guardia? The inadvertent operation of the rudder trim system? (Since the accident, at least a dozen such incidents have come to light, and there may well be a design problem that sets a trap for the crew.) And what other contributory factors exacerbated the hazard? That the runway was wet? That it was short and built right into Flushing Bay? That the two-pilots did not seem to work together? That the captain had only had 138 hours in charge of the type? Or was there some dark secret that urged the pilots to walk away from the plane right after the crash, and remain unavailable for toxicological tests for a long period? (Should a car driver do that, it could almost amount to an admission of being under the influence of drugs or drink.) Problems concerning the human ingredient are much too complex to be dismissed lightly with vague pronouncements of *aircrew error* – the cause of about seven out of every ten fatal aircrashes.

We trust the pilot – we have no choice. But the last thing we want from him is heroics on the flight deck. We must live (hope to live) with the assertion that the pilot is a highly-trained professional, who can do his largely monotonous job, cope with his own very ordinary human failings, and work in constant readiness to deal with the unexpected. Captain Don McClure, chairman of the ALPA Accident Investigation Board, said: 'When I reach the threshold of that runway and start to roll, do I have a sense of burdens, being responsible for the three hundred lives that follow behind me? Sure . . . but first and foremost I myself want to live. That's why they trust me, because I must trust myself and those who work with me.'

Captain Reist, deputy head of Swissair flight safety, told me: 'We can change switches and instruments, but not human nature. We're all just normal neurotics who must be taught to know and live with our problems and weaknesses. Heroes? Beware of heroes in the cockpit!' Which sounds like the reincarnation of a fine definition: 'Aviation character is the triumph of humility and common sense over arrogance and overconfidence.'*

Although a McDonnell Douglas review of 196 widebody aircraft incidents and accidents attributed only twenty-three per cent to so-called pilot errors (had there been no maintenance or turbulence problem 'in the first place, no accident would have occurred'),† a Lufthansa study held crew members (their lack of caution, vigilance and discipline, their failure to be the last line of defence) responsible for some three-quarters of all accidents, and the NTSB flight and voice recorder specialists attributed ninety per cent of the problems they encountered to the human element. Allan McArtor of the FAA, offered these American figures in 1988: 'sixty-eight per cent of aircarrier, seventy-nine per cent of commuter, and eighty-eight per cent of general aviation fatal accidents are caused by pilot error'.‡

An already mentioned Boeing project, working with an international sample of twenty American, European and Asian-Pacific airlines (all Boeing users with better than average safety standards), noted that, ever since the introduction of jets, seventy per cent of the major accidents were caused by the flightcrew, and that half of their sample cases were due to the the most disturbing failures: the pilot deviated from basic operational procedures, or the crosscheck carried out by the co-pilot was inadequate.

Undoubtedly, the performance and behaviour of the flightcrew represent the greatest hazards in the air. But *pilot error* used to

* J. W. Ray: Aviation Character, An Essential Ingredient, *Mac Flyer*, June, 1977.

† F. Wiegers and E. Rosman: A Safety Profile of Widebody Commercial Aircraft, paper given to the 39th annual Seminar of the Flight Safety Foundation.

‡ Speech to National Aviation Club, Crystal City, January 21, 1988.

be the easy answer when investigators were baffled by an accident. Though it revealed ignorance and prejudice, it remained the last resort of blinkered investigations until it began to transpire that there must be more to an error than meets even the jaundiced eye. So a new concept was coined: the *human factor*. It not only sounded much more scientific as well as compassionate, it could also help to explore and interpret even the most complex events. It was therefore probably the most profound innovation in the history of accident investigation.

But in the honest process that followed, has the expression the 'human factor' become a bit of a euphemism for unadulterated human mistakes for which pilots should take the blame? Reading through dozens and dozens of accident reports, one senses the emergence of a potentially grave risk. Will the human factor be allowed to be barely an excuse just like the society/upbringing/education complex can serve as a tripartite let-out for the juvenile mugger, granny-basher, crack-pusher? And can the 'human factor' deteriorate into one of the numerous *causal factors* ('just one of those things') that may be used to cloud the vital determination of *The Cause?* Could it sometimes disguise somebody's fault which turned a containable emergency into an inevitable disaster? Mistakes tend now to be called bad habits, lack of discipline, lack of proper training, lack of supervision. Should it not be admitted that an error is an error is an error? Wouldn't *Reasons of The Cause* better describe the road to tragedy?

Anything connected with a flight may be a causal factor. When an engineer's mistake slips through the net of an inadequate and badly supervised maintenance programme, when the resultant malfunction in flight is exacerbated by the weather conditions, and when the pressure not to miss an allocated landing slot combines forces with a mild flu as well as home problems on a tired and relatively inexperienced pilot's mind, there comes a point where a disaster could or could not be averted, depending on the captain's abilities. If it becomes an accident, even though another pilot could be expected to overcome those particular difficulties, it will be a pilot error – that might have been made easy by several adverse coincidences. Even the most sympathetic and considerate investigator must be firm in differentiating between naming the cause and making recommendations, i.e. the recognition of a mistake (even blameworthiness) and the suggestion of preventive measures against any repetition. Fudging the issue will not lead to a cure.

Nobody would denigrate the efforts to treat the cause rather than the symptom. The old attitude, defined sometimes as *prescribing rubber gloves for leaky fountain pens** is clearly untenable. Yet the pilot who fails to wear the available rubber gloves, however ridiculously makeshift they may be, will still be responsible for not keeping his hands clean. For ultimately, research and noble intentions to eradicate the root of the problems must not overlook the risk of human fallibility. Sadly, this fact seems to be in need of frequent rediscovery as if it has never been heard of.

'It's true,' said Harold Marthinsen who heads the American ALPA's investigation department, 'it's difficult to maintain the right balance. Not very long ago, everything was a pilot error because it helped the NTSB to make its case. Now the human factor is sometimes overplayed. If, for instance, a pilot makes a stupid mistake because he's badly trained and not up to the job, we sure have to review the training, but it's still a mistake.' Captain Caesar, the Lufthansa safety chief agreed: 'Pilots are trained to deal with emergencies. Say a twin-engine aircraft makes a critical take-off, fully loaded, on a hot day – and an engine fails. The crew is not there just to fly from A to B, they're taught to cope with the unexpected, and a situation like that is not even unforeseen. So why that engine failed, if the captain can't deal with it, though the aircraft is fliable, the accident will be his fault. We may perhaps sympathize with his predicament in difficult circumstances, but that cannot fully exonerate him.'

Take the case of the Air Illinois HS-748 which suffered a double generator failure.† The report blamed the pilots for various errors and, above all, for not turning back, but a leading British investigator would have given them the benefit of doubt. In the Detroit case‡ the crew simply did not bother to read and follow the checklist. A mistake? Undoubtedly. Yet I have heard some investigators blaming the poor man's training alone. Identical mistakes (manifestations of indiscipline, if preferred) have been committed many times before and after Detroit. Captain Frank Hawkins

* Safety Versus Economics by A. W. 'Tony' Brunetti, *Flight Safety Digest*, October, 1986.

† See Chapter 9.

‡ Reviewed in the previous chapter.

summed up the three main reasons: 'saving time', driven by corporate considerations of false economy; excessive 'trust in memory, as a matter of professional pride, a dangerous philosophy'; and 'reducing workload', yet another innovation from cloud cuckooland to defeat the point of the exercise – it is the adherence to the checklists that *reduces* the workload.*

The intense study of human factors in the last three decades has led to crucial improvements. Smaller, poorer airlines, for instance, expose their pilots to errors by the simple expedient of buying several second-hand aircraft with different cockpit layouts. A third world pilot who flies only 727s may find that something is different in the cockpit every time he takes off. In an emergency, his reflexes may be tailored for the wrong aircraft. Better understanding of human factors and the force of habits has called attention to the importance of standardization. It has also spotlighted instruments that were far from pilot-friendly.

Misreadable altimeters used to lead regularly, and for many years inexplicably, to what the trade called *controlled flight into the ground*'.† Despite the vast amount of accumulated experience, manufacturers can create traps – and airlines fail to spot them. A pilot's letter to CHIRP was a good illustration: 'Following the start of the second engine, the "memory" checklist calls for the APU to be switched off. As I reached up for the switch, there was some distraction. When I turned the switch I looked up again, and I had in fact turned the Standby Power Selector to "off". No real problem, but the error brought home to me the poor ergonomic layout of the 757 overhead panel, where the multitude of identical switches serves different systems.'‡

'Unfortunately, airlines don't very much like confidential publications like *Feedback* and NASA's *Callback* because they claim they don't get enough detail out of the system,' said Roger Green who runs CHIRP. 'If you read between the lines, what they really say is they don't get anything about the person who writes to us, and whom they would dearly like to pull in for questioning.'

One of the worst mistakes a pilot can make is to shut down the wrong engine. Hair-raising as it may sound, it has been a not

* Checklists and the Human Factor, *Flight Safety Focus*, October 1987.

† *Aircrash Detective* listed numerous examples in the chapter on so-called pilot errors.

‡ *Feedback*, December, 1987.

infrequent event. 'In the old days, the first pilot to recognize a malfunction or fire would stop the engine right away,' said DC10 simulator instructor Jack Shirwell. 'Because it was too easy to shut down the healthy one on a multi-engine jet, British and many other airlines introduced a safety procedure. If a pilot wants to shut down an engine, he'll say: "Number One to close", touch the throttle, and say "Number One throttle", but do nothing until someone says: "Number One confirmed". Same with fuel levers and fire handles. Name it, get it confirmed, announce "off" – and move it only then. Unfortunately, that doesn't prevent unintentional shut-downs.'

In August, 1983, both engines of a United Airlines 767 suddenly died over the Rocky Mountains. The plane began to glide down, losing 8,000 feet of altitude in four minutes. The captain called for crash-landing preparations. His 197 passengers put their heads in their laps, and prayed. At 14,000 feet, the height of some of the peaks all around, he managed to restart the engines to make an uneventful landing. There was nothing wrong with the engines. The scare remained a bit of a mystery. It might or might not have been the forerunner of an occurrence in June, 1987. A Delta Airlines 767 was climbing out of Los Angeles. At 1,700 feet, a fuel control warning light came on. The captain reached down to deactivate the switch – and accidentally caught the switches that shut down both engines. His immediate 'restart engines' drill succeeded but not before he had lost some 600 feet. The mistake came within a whole series of errors by Delta pilots (landings on the wrong runways, straying badly off course and causing near-collisions, etc., all within a fortnight), so the pilot was suspended, and the investigation concentrated on Delta training as well as the potential booby-traps in the cockpit. Within a week, the FAA ordered some emergency modifications on all 767s. BA went even further when it fitted protective perspex covers over the fuel control switches on its fleet of two dozen 757s.

Once a helicopter pilot, feeling too hot, chose to switch off the heater – and only when all went quiet around him did he realize that he had turned off the fuel supply. A first officer, about to land an aircraft, called for lowering the flaps – and the captain switched off the fuel. 'How can pilots so clearly intend to do one thing and then find that they have done another? In exactly the same way, presumably, that you may have thrown the sweet in the bin and put the wrapper in your mouth,' said Roger Green in an article on the human factor.*

Accidental shutting of fuel cocks also happened repeatedly on BAC1-11s. When CHIRP received the first reports of such occurrences, British Airways grumbled that it would be more helpful if they could talk to the pilot rather than read an anonymous account. 'We needed essential details,' said Roy Lomas, head of the Air Safety Branch. 'When did it happen? Day or night? How did it happen? Was the pilot exceptionally busy with RT or some problems? Was he using the modified switches that were supposed to prevent such mistakes? Like a little learning, limited information can be misleading and dangerous. So initially, all we could do was to alert all pilots to the hazard, and to seek more precise information.' But the pilots must have felt uneasy: their honesty about mistakes could dim their promotion prospects. One anonymously published report was quick to prompt several others. Swallowing their dislike of the indirect route of anonymous notification, BA investigators were equally quick to respond. Though the danger was very serious, the solution was simple: a bright yellow rubber sleeve was fitted on each fuel cock switch for easy identification, to ensure that no engine would be starved inadvertently. But Lomas is uneasy to this day: 'Our managers must strive to have their people's full confidence. For whatever form of carelessness causes an incident, we must find ways to make it ever harder for pilots to make mistakes.'

'SHUT UP, GRINGO!'

In 1983, an Avianca 727 was preparing to land when, flying over Mount Oiz, the CVR recorded that the ground proximity warning began repeating monotonously: 'Pull up . . . pull up . . . pull up.' Allegedly, the pilot got fed up, shouted at the busybody gringo invention to shut up, and proceeded to crash, killing 148 people.

'Perhaps even more than mistrust of the equipment, over-reliance on modern instruments is a minefield in the man-machine conflict,' said Captain Pritchard who chairs the Air Safety Group. 'A pilot must monitor the systems and know where he is at any

* Faith In Flying – How To Control The Human Factor, *The Listener*, 16.5.1985.

given moment, for if he's complacent, and something goes wrong, there's no time to go back to basics.'

'Modern aircraft are a dream,' said ex-AIB John Owen. 'The use of the Inertia Navigation System, what I call *flying on numbers*, takes the drudgery out of the job. You punch in where you are and where you want to go, and let the boxes of tricks get on with it. That's fine. That's probably how that Korean pilot strayed into and got shot down in Soviet airspace. Notwithstanding electronic wizardry, I'm a firm believer in having an "idiot pilot" who double-checks everything . . . just in case. Regarding *wizardry* I'll never forget the story of a TWA Tristar. As it was thundering down the airway to Frankfurt, it slowly made a 180° turn back for reasons best known to itself. When the pilots realized what was happening, the young co-pilot, brought up on electronic goodies, immediately started to press buttons to sort out what went wrong, but the captain, who had learned flying by the seat of his pants, stopped the laddie: "First we point the nose the right way, then we'll see who has screwed up – me or George."*

'Whether the equipment malfunctioned or it was wrongly programmed, it would have been a pilot error to carry on in the wrong direction a moment longer. Wisdom in retrospect is a useful thing for the future, but in a moment of emergency, it's just one of the four most useless things for pilots – the other three being, of course, concrete, i.e. runway behind you, altitude above you, and fuel in the depot.'

Apart from the obvious lack of discipline, errors in the cockpit can produce an infinite variety. They range from the tragic to the extremely embarrassing like landing on the wrong runway or even at the wrong airport, not to mention the case of the two-man crew who locked themselves out of their own cockpit during the 'uneventful' cruise phase of a flight.† Bad habits can be disastrous, as in the case of an RAF pilot who had acquired the habit of removing everything uncomfortable during flights, and jumped out in an emergency without wearing his parachute. Once a first officer, flying the sector, suspected some engine problem when the aircraft was climbing sluggishly after take-off. At 500 feet, he called for the flaps to be retracted – but nothing happened. Both

* The old nickname for the auto-pilot.

† *Flight Safety Focus*, May, 1987.

pilots had forgotten to select flaps for take-off. As he 'confessed' to *Feedback*, the first cause of the mistake was some distraction at a time when the checklist should have been completed; he could have double-checked everything 'instead of looking out of the window' during the five minutes he had to waste waiting at the holding point; 'however, (in my defence) I must add that in the previous weeks I had flown a lot with another captain who always insisted on selecting flap himself despite it being the co-pilot's job – perhaps I had got out of the habit of doing it myself.'

Distraction is dangerous. That is why the FAA 'Sterile Cockpit Rule' prohibits all non-essential activities – such as eating, drinking, reading, chatting up stewardesses or passengers, using the intercom to promote the company or point out sights of interest – during critical phases such as taxiing, take-off, landing, and all flight operations except cruise below 10,000 feet. The reasons for this were frequently demonstrated in *Callback* to show that the atmosphere in the cockpit can become too relaxed. (After a brief visit to the cabin, a captain once returned to the cockpit only to find that the aircraft was cruising seventy miles off course: his crew, including a stewardess, were listening avidly to the co-pilot's war stories.)

It may be hard to imagine, but distractions can be so bad that pilots fail to pay attention even to their lifeblood – fuel. There were red faces when, in 1983, that happened to Republic Airlines pilots twice within two months. Once the crew of a DC9 were so preoccupied with some minor problems that they forgot to switch fuel supplies until the tank in use was exhausted: the engines flamed out, and had to be relit. On the second occasion, after an overnight stay in Fresno, the airline agent forgot to refuel the aircraft before departure – and the crew did not notice it until they were in the air. They made it to the nearest airport, with just five gallons to spare. (As David Learmont promptly pointed out in *FLIGHT International*, neither of these foul-ups was without precedents.)

'You may reach the right decision,' said Roger Green who made a special study of distractions, 'but some sort of hidden trigger mechanism may activate the wrong motor. You go to the bathroom to wash your hands, and find yourself washing your teeth. A fighter pilot I know landed safely, decided to raise the canopy to cool off – and raised the undercarriage. Subconsciously, he might have been rethinking the mission he had just completed. Professor Reason of Manchester University found the following gem. On a Saturday morning, a woman bought a dress. By mid-afternoon, she was astonished to find herself in bed. Then it slowly came back to her: she returned home, got through various chores, then went to the bedroom to try on the new dress, so she undressed – but that simple act fired the wrong motor because in her mind, *un*dressing was firmly associated with going to bed for the night. The mechanism is the same in the air, just somewhat more dangerous. Particularly at times of stress, you revert to the best drilled motor programme.

'That's why skill acquisition, the development of correct automatic reactions through training, is so essential for pilots. Having made the right decision in an emergency, instinct and training take over to do the job automatically. It saves valuable time. But if the wrong motor is activated, and if, because of some distraction, the pilot cannot monitor his own actions, he may remember his correct decision, say, to lower the landing gear, but not that his hands have attained a certain independence and retracted the flaps instead.'

Many such pilot errors have strictly human origins that have nothing to do with design, layout or misreadable instruments. Are they due merely to training problems? Why is it that a perfectly good aircraft is suddenly mismanaged? And how come, in public air transport accidents where no-pilot could have worked on his own, there was nobody in the cockpit to shout: '*Hey, watch it, this is no way to talk to a ground proximity warning system*' or just: '*Are you sure you're doing the right thing?*'

'The investigation of the Potomac icing accident was probably the great watershed in this respect,' said Ken Smart of the AAIB. 'It uncovered inexperience, unquestioned procedures, too rapid expansion, accelerated upgrading of pilots, and managerial blindness to risks, and the NTSB was quick to respond with the recruitment of several human behaviour specialists.'

Board member John Lauber, who has a PhD in experimental psychology, has spent his entire professional career exploring pilot errors and their prevention. Before his move to the NTSB, he worked with NASA in charge of research into aviation safety with reference to human performance. 'The range was vast,' he said in one of the spacious, magnificent offices reserved for NTSB top brass. 'We looked at training and procedures, design of equipment, use of dangerous substances like narcotics, alcohol and medication, psychological factors, stress, fatigue, jet-lag, economic and marital problems, and one of our great achievements was in the field of making simulators even more realistic by building known difficulties into the programme. When we added heavy communication workload, non-satisfactory ATC service, weather, and even distractions like a purser who'd uncannily pick the worst possible moment to call in, airlines began to object. "We train people, they learn to handle all this in time," they said but after a few trips in our simulators, their eyes were opened. They saw that things like decision-making, crew coordination, and flightdeck communications could be learned.

'We also learned as we went along, particularly about *flightdeck management*, a newly recognized hazard, and I confess I'm pleased to think I may have invented the term *cockpit resource management*. You're climbing out of New York. Weather and ATC information is pouring in. You have a minor oil pressure problem with an engine, you go through the routine diagnostic procedure, and finally decide to shut down an engine. Several alternatives must be considered. If you decide to return to New York, you must dump a lot of fuel, an operation that must be planned with great care. During all this the purser is nagging: a number of passengers desperately need re-ticketing, would a pilot please call the company with the names? Normal situation, arises all the time. We found that some crews tried to do everything at once, accommodate everybody simultaneously, while making various errors, and landing – luckily only the simulator – with a hundred thousand pounds more fuel than they thought they weighed. Some captains couldn't delegate, and kept pestering the engineer every thirty seconds: "How is it going, Fred?" Others allowed minor problems to occupy everybody's attention. A famous illustration of such behaviour occurred when Eastern had an accident at Miami in 1972: as the aircraft was holding and circling, the entire crew devoted its attention to changing a sixty-nine-cent light bulb . . . and nobody was minding the store, nobody realized that, inadvertently, the "altitude hold" function of the auto-pilot had been knocked off.

'There were plenty of real examples we could resurrect for training and developing better ways to manage the available cockpit resources. There was the case of a 737 making a non-precision approach to Midway. A warning light overhead told the crew that the flight recorder was inoperative. It was of no operational consequence, and could have been sorted out on the ground. Yet there was endless discussion about it until they got behind the progress of the airplane. Suddenly they were too low down, and had to do a hell of a lot all at once. They decided to go round but, in their hurry, they left the spoilers extended, and with spoilers out the airplane doesn't fly very well. They crashed. Pilot error? Sure. But the root of it was bad cockpit resource management. The captain should have assigned that warning light problem to just one of the crew if any, while he concentrated on the real task in hand.'

Though similar problems have been encountered, and similar mistakes have been made many times ever since, let us go back to 1978, when a DC8 accident at Portland, Oregon, produced a perfect example.

United Airlines Flight 173 had 189 passengers and crew on board when it left the gate at Denver. The fuel requirement for the flight to Portland was 31,900 pounds, but almost fifty per cent more, 46,700 pounds was loaded to comply with FAA requirements (fuel for an extra forty-five minutes flying time in case a diversion was necessary) and the company contingency margins that cater for another twenty minutes in the air. At 17.06, the pilots had Runway 28 of Portland in sight. Flaps and the landing gear were lowered – and the captain noticed a most peculiar 'thump, thump in sound and feel'. People in the cabin heard a loud noise and felt a severe jolt. The aircraft yawed to the right. The gear position indicator lights gave confusing information. Was the landing gear down or still up? On the radio, the final approach and landing instructions were offered but Flight 173 wanted to stay up at 5,000 feet to sort out the probable gear problem. There was no hurry, with the compulsory extra fuel on board they would have at least another sixty-five minutes beyond the original estimated time of arrival.

The crew went through a diagnostic check routine, as trained. The evidence was inconclusive. It seemed possible that only the indicator lights rather than the landing gear had failed. At 17.38, while circling south of the airfield, never more than some twenty miles from the runway, they called the UAL Maintenance Control Centre at San Francisco which assured them that they had done everything right. There was no more they could do to verify the integrity of the gear, so there would be nothing to prevent their landing right away, but the captain said he would need another quarter of an hour to let the flight attendants prepare the passengers for a possible emergency evacuation. That sounded fine: he would still have enough fuel for a half an hour from 17.44 when San Francisco confirmed: 'You estimate that you'll make a landing about five minutes past the hour. Is that okay?'

'Ya, that's good ballpark. I'm not gonna hurry the girls. We got about 165 people on board and we . . . want to . . . take our time and get everybody ready and then we'll go. It's clear as a bell and no problem.'

But, even more clearly, the gear problem still worried them. The landing could be a mess. One of the crew remarks on the CVR: 'Less than three weeks, three weeks to retirement – you better get me outta here.' 'Thing to remember is don't worry.') Though at 17.48, Frostie, the engineer was asked about the remaining fuel, no calculations were made to relate fuel and maximum flying time.

Also at 17.48, a 'suggestion' was made: 'You should put your coats on . . . Both for your protection and so you'll be noticed so they'll know who you are.'

'Oh that's okay.'

'But if it gets, if it gets hot it sure is nice to not have bare arms.'

At 17.50, the captain wanted weight calculations for landing in 'about another fifteen minutes'. The engineer did not like the idea: 'Fifteen minutes is gonna – really run us low on fuel here.' (Flying in the holding pattern with gear down and flaps at fifteen degrees, the DC8 would then be burning 13,209 pounds an hour – just over 220 pounds a minute – leaving perhaps half a minute to spare on landing, far too little if they would have to overshoot and go round once more for some reason. Apparently, no thought was given to that.) The approach descent check was completed. The status of the landing gear continued to figure in the cockpit exchanges. The first officer made 'several subtle comments questioning or discussing the aircraft's fuel state', but no direct view was expressed, no positive warnings were sounded. At 17.57, when only some 4,000 pounds remained, less than eighteen minutes in flying time, the captain sent the engineer to 'kinda see how things are going' in the cabin. The 'girls' back there were never told to hurry: they would need another two or three minutes, the engineer reported at 18.02. Thirty seconds later he coolly announced: 'We got about three [3,000 pounds] on the fuel and that's that.'

At 18.03 when Portland Approach asked for a status report, the first officer summarized the abnormal gear indications and evacuation preparations. A minute later the captain added that he would land in 'I'd guess about another three, four, five minutes'. The aircraft was then about eight miles from the airport – flying away from it. Talk about the landing gear continued. At 18.06, some seventeen miles from the airport, a flight attendant reported that the cabin was ready.

'Okay,' said the captain, 'we're going to go in now. We should be landing in about five minutes.' And almost simultaneously, the first officer said: 'I think you just lost No. 4 . . . ' Some conflicting and half-heard conversation began about fuel. No. 3 engine was also about to flame out because of fuel starvation. Now they were nineteen miles from the airport. There was some confusion about crossfeeding the remaining two engines, and there was yet another attempt to check the gear indicator lights. Portland was advised that Flight 173 was now coming straight in from a long way out. 'Boy, that fuel sure went to hell all of a sudden,' somebody remarked. Emergency thinking set in. 'There's ah, kind of an interstate high-way type thing along that bank on the river in case we're short . . . ' Could they go to a small airfield at Troutdale along their path if they could not make Portland? 'Let's take the shortest route.'

All the pumps were working, but no fuel was coming through, and the last two engines were dying at 18.13.

'We can't make Troutdale,' said the captain.

'We can't make anything,' replied the co-pilot.

'Okay, declare a Mayday.'

At 18.15, sixty-nine minutes after the first sighting of the runway, the DC8 crashed into a wooded section of a populated area.

The wreckage path was about a mile long. The aircraft was destroyed. There was no fire. Mercifully, only eight passengers and two crew members were killed, twenty-three people were badly hurt, the rest of those on board got away with minor injuries, and there were no casualties in the Portland suburb.

The investigation discovered that the landing gear would have given no trouble on the runway. The disturbing symptoms were due to the excessive corrosion of a part that made one gear deploy by freefall (hence the thump), which in turn cut out the indicator light, and caused momentary yawing. The cause of wasting all that fuel and time, and so the cause of the crash, was to be found in the cockpit. The first thirty minutes' activities were well justified. But, from then on, there was nothing to do but to prepare for a possible emergency evacuation – and hope for the best. The NTSB determined that the landing gear problem, a genuine cause for concern, had a confusing and disorganizing effect on the flightcrew's performance. The captain, who allowed himself to be totally preoccupied with troubleshooting and evacuation planning, failed to monitor the state of fuel reserves, and to foresee the perfectly predictable exhaustion of the tanks. A contributory factor of the accident was that the other two flightcrew members either failed 'fully to comprehend the criticality of the fuel state or successfully to communicate their concern to the captain' until the crash was inevitable.

The Board emphasized 'that this accident exemplifies a recurring problem – a breakdown in cockpit management and teamwork during a situation involving malfunctions of aircraft systems in flight. To combat this problem, responsibilities must be divided among members of the flightcrew while a malfunction is being resolved. In this case, apparently no one was specifically delegated the responsibility of monitoring fuel state . . . The first officer's and the flight engineer's inputs on the flight deck are important because they provide redundancy* . . . In training of all airline cockpit and cabin crewmembers, assertiveness training should be a part of the standard curricula, including the need for individual initiative and effective expression of concern.'

Those last four words were a measured understatement of the need to speak up, yell or even scream: *'Hey, dammit, we've got to get down!'* That report was issued in 1979. Warnings were repeated many times. In 1988, for instance, after the Northwest

* Like the duplication and triplication of various systems add redundancy to aircraft controls.

crash at Detroit, the NTSB found that 'the absence of dynamic leadership and coordination demonstrated by the accident crew suggests there is strong evidence to support that the cockpit resource management training they did receive was deficient, and that future programs must go beyond the scope of a limited and traditional classroom forum.' Not surprisingly, UAL was among the first airlines to take the recommendations to heart, and devise special programmes to give such training to their crews. Some others, like KLM and DanAir, followed suit, yet most airlines were, and still are, waiting perhaps for regulations or at least warnings in even stronger language. 'Despite recurrent tragedies, very few airlines give due emphasis to cockpit resource management training,' said Barrie Strauch of the NTSB. 'Its value is not seen easily, perhaps because the program is preventive. Yet we know and keep saying that in most accidents, there is poor decision-making somewhere down the line, and decision techniques ought to be improved.'

Without proper attention and response, what good can thorough investigation and thoughtful recommendations do, and when will a final call be seen as a truly final call for safety? Do investigators need to shout their warnings from the rooftop of hangars?* The probable answer is *yes!* At the time of writing, the NTSB is investigating the crash of an old 707 flown by the Colombian Avianca. It had to 'hold' in the air because of fog and congestion at Kennedy. Using non-standard RT phraseology, the crew told the controller that they were running low on fuel, but did not declare an emergency which would give them landing priority. Eventually, when the flight was cleared to land, the crew missed the first approach, tried to go round again, reported losing

* The clarity and directness of safety messages would certainly be helpful at every level. The ICAO Accident Prevention Manual offered this tragicomic example: a private pilot once asked the authorities if he could save costs by mixing kerosene in his aircraft fuel. Answer by return post: 'Utilization of kerosene involves major uncertainties/probabilities respecting shaft output and metal longevity where application pertains to aeronautical internal combustion power plants.' Pilot's second letter: 'Thanks for the information. Will start using kerosene next week.' Answer by cable: 'Regrettably decision involves uncertainties. Kerosene utilization consequences questionable, with respect to metalloferrous components and power production.' Cable from pilot: 'Thanks again. It will sure cut my fuel bill.' Response by telex within the hour: 'DON'T USE KEROSENE. IT COULD KILL THE ENGINE — AND YOU TOO!'

two engines, and ran out of fuel. The 707 literally fell from the sky killing sixty-seven prople on board, including the pilot. Fortunately, it missed the densely populated areas like Manhattan. How come that during the seventy-seven minutes hold, none of the crew noticed that there would be *no* fuel left for landing?

ODD COUPLES

When human factors figure among probable causes of accidents, even the most obvious lessons seem to be too hard to learn and easy to forget. Despite the massive growth of the aviation industry and its insatiable demand for crews, it took even the FAA a few years and several slaughters, including the DC9 accident at Denver* to *ask* all U.S. airlines to avoid pairing pilots for the same flight if both have less than a hundred hours' experience on that particular type of aircraft. Airlines are training more pilots than ever before but, as we have seen, aircraft are being delivered faster than crews can accumulate experience even if the wider use of simulators has greatly accelerated at least the pilots' *conversion* to new types.

When major airlines select new entrants for their training programme, psychological aspects play an important part. 'We're very much aware of the hazards in personality profiles,' said Dr Ruedi Knüsel of Swissair. 'We look for many characteristics like inner stability, flexibility, stress resistance. Apart from tests, we watch behaviour throughout the different stages of selection, as later the teachers do all through the months of training – does he panic? does he seek excuses for mistakes? has he got leadership qualities? is he afraid of the loneliness of a leader? can he work in teams? does he feel he must succeed at all times? can he cope with making errors or does he see any failure as inadmissible? It's a long list, and still growing. Some fifteen years ago, the ideal pilot was seen as someone with a strong macho image. Not any more. The macho self-image creates tunnel vision, an urge to be right at all times. It leads to the loss of ability to see and assess

* Continental Airlines, November, 1987, though highly experienced in general, both pilots were virtual newcomers on DC9; aircraft failed to climb in heavy snowstorm, and broke up on runway.

reality. We want people who can make decisions, and even bend the rules if need be, but they must be secure enough to face up to errors and listen to critical advice.'

The constant use of the masculine gender was no accident. There are very few women who enter the profession. It is not because of lack of ability, and less and less because of sex discrimination, but because of economic considerations. Lufthansa, for instance, admits readily that, as it costs some 200,000 Deutschmarks (about £70,000) to train a cadet, and twice as much more spent on follow-up retraining, they want staff who will offer long years of service, ideally until retirement at the age of fifty-five. Without disputing women's potential flying skills, the airline says what counts against women is 'the factor of insecurity, conditioned by marriage or pregnancy, in the continuity of employment'. Other airlines say they would like to have more woman pilots, but worry how they would fit into a predominantly male environment, how they would cope with stress when family problems prey on their minds, and how premenstrual tension might reduce their alertness. Women are also seen as an extra risk to the chemistry of a flightdeck crew. For not only excessive conflict but also too harmonious compatibility can lead to human factor disasters.

There are numerous sociological, psychological, traditional and training factors to consider. The Boeing study of crew-induced accidents revealed that no matter who was flying the aircraft (captain or first officer), 'the non-flying pilot (NFP) had the opportunity to effect corrective action but failed to do so' in the majority of cases in the sample. (In fact, captains seemed to be the better NFPs.)

There is a tragic abundance of reasons why pilots fail to monitor critically the actions of a colleague, particularly one who is senior to them. Custom, religion and traditional 'status' play a strong part. It is forgotten with stunning regularity that cabin staff may find it difficult to use their own initiative – not to mention contradict a captain. Muslim captains are never criticized or corrected by the crews. According to some sources, Soviet crews are 'positively discouraged to argue with the senior man'. Challenging authority is basically against Japanese nature. (When a JAL DC8 cargo flight crashed in Alaska in 1979, and when another flight dived into Tokyo Bay in 1982, the flightcrews simply must have known that in the first case the American pilot was drunk, and that at Tokyo Bay, the Japanese captain was mentally unbalanced, yet 'they felt compelled to observe and respect rank, age and tradition rather than logic', said Eiichiro Sekigawa, a leading aviation writer in Yokohama.) At the other extreme, Australian and Irish pilots have the reputation of speaking up no matter how junior they might be. Allegedly, a young Ansett pilot will not proceed with the checklist, if at some point the captain says 'set' when he is supposed to say 'off'.

'An overdose of respect has best been illustrated by the case of a Twin Otter of an American commuter line,' said Roger Green. 'The captain was not only a very experienced, grumpy old pilot, but also the company vice president. The first officer was a very young one, on six months' probation, who knew his future would depend heavily on the impression he made. They had been flying together for some eleven hours, completing thirteen sectors, so by now he saw that the warnings he had received from colleagues were correct: the captain did not bother to give proper responses to the challenge-check routine of running through the checklist. Now they were on approach to their home base. The first officer began to make his call-outs according to the book, but got no response. Not a word. Not a grunt. He saw they were descending a bit too steeply, but he was not going to question his boss's flying abilities or to leave a final impression of being a fusspot. So rather than say anything, he kept his head down, and let the old man get on with it – all the way into the ground. The co-pilot survived, the captain did not. The autopsy discovered that the captain had been killed not by his injuries but by a massive heart attack: he had been dead, in fact, for quite some time before the crash. No wonder he didn't respond to calls.'

John Silver, a BA public relations manager, recalls the tale of a luckier young man: 'The South African Airways Viscount flight was to be from East London to Kimberley, then on to Jo'burg. The captain was a rather dominant old hand, the co-pilot a brash, cocksure chap. They had a young trainee pilot, flying as a supernumerary on the flight deck, watching the procedures eagerly. He noticed something odd being done, but asked no questions because of his junior status. The first leg to Kimberley would take about ninety minutes. Flying at night in clouds, they made a routine high-frequency navigation call when they were at about halfway, and duly got a response signal from Bloemfontein.

Some twenty minutes later, they called Kimberley on the VHF, but there was no response. Repeated attempts failed to make contact though they would have to be no more than fifteen minutes away from their destination. The Automatic Direction Finder (ADF) needle hunted on Kimberley station without locking on. The pilots were puzzled, they had no idea what was wrong or even where they might be.

'The youngster had a dilemma. He thought he knew what was happening, but he was too scared to comment. Finally, he plucked up courage, and instead of making some critical remark, he quietly asked: "Excuse me, sir, would you please tune in for Durban on the ADF, sir?" Just imagine the looks he got. Durban would be a long way off their path, north-east of East London when they were heading north-west. But were they? At a loss, they tuned in Durban, and the ADF signal came in loud and clear. What happened was due to a simple mistake: leaving East London, they set the heading 085° – the correct heading for the *second* leg, from Kimberley to Jo'burg. The Bloemfontein signal was received because their actual and imagined positions were equidistant from the station except that it was on their left, not right.

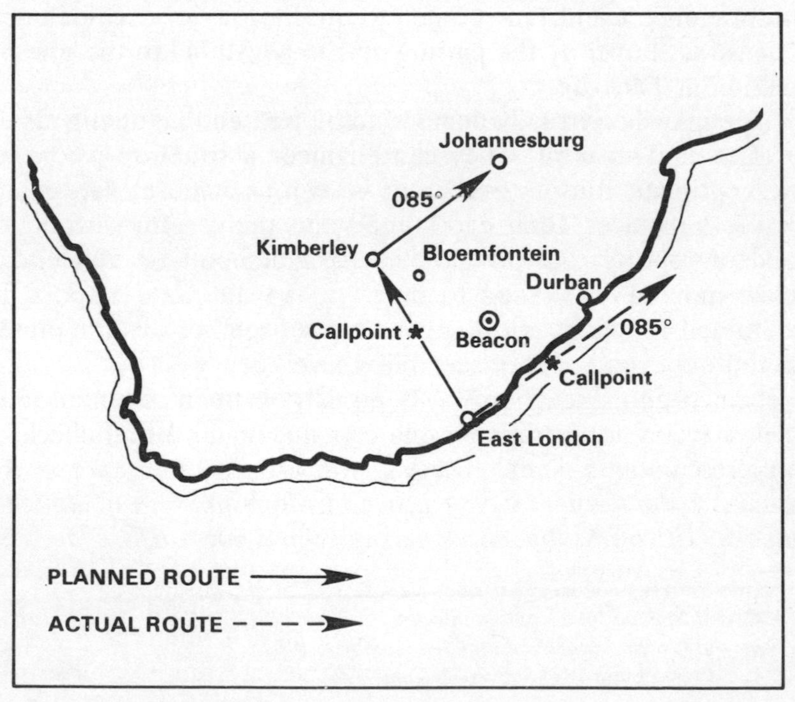

Kimberley would not answer them because it was too far: they were already well over the Indian Ocean, close to Durban, flying parallel to their would-be second leg. If nobody had realized the mistake, they could have flown on until they ran out of fuel and crashed into the sea. It could have become one of the great mysteries of the air. But it's no mystery why the captain lost his command.'

Such seemingly exaggerated examples led to special psychological studies of TAG (Transcockpit Authority Gradient) which reflects the relative strength and forcefulness of the personalities involved. Though anything that actually attains a name (let alone an acronym!) tends be acknowledged as something real, TAG reality still receives insufficient attention and so remains a cause of accidents.

The *Muser principle** defines the risk: 'Modern aircraft and operational techniques have become dangerously safe. Therefore safety needs a healthy dose of disagreement in the cockpit. Be suspicious of everything and everybody's actions at all times. It prevents the duplication of mistakes.' Even if the captain is not an overbearing, macho figure, a relatively junior pilot would need encouragement and reassurance to question a very senior one's actions, as shown by the pairing on the KLM 747 in the ground collision at Tenerife.†

Captains who virtually demand to be treated like demigods do so at their own peril. They can engineer a situation where an unexceptional, routine overshoot is seen as a moral defeat and total loss of face. Their ego is read into the careful wording of accident reports: the pilot 'failed to overshoot' or 'descended below minima' or 'failed to divert to an alternate airport' or 'continued the flight into known adverse weather' or 'attempted operation beyond experience/ability level'.‡

Their co-pilots' reaction may be equally obstinate, even suicidal. They may lay little traps for the captain; or do a fuel check or some calculations, shut the book with a bang, and stare at the captain: *I dare you – if you just nod, there goes your almighty authority, if you double-check me, and I'm right, you lose face . . .*

* Captain Hugo Muser, a former air safety chief of Swissair.

† CASB *Special Investigation into the Risk of Collisions Involving Aircraft on or Near the Ground at Canadian Civil Airports,* 1987.

‡ ICAO Accident Prevention Manual, 1984.

Such games of pyschological one up-manship may be counter-balanced by the now diminishing presence of a third man in the cockpit. On the other hand, the tensions of three people may have the effect, as a Swedish aircrash detective put it so succinctly, 'instead of one, now two people will say: "Let the bugger screw himself." '

Swissair learned its lesson in the days when there were four people, including a navigator and a radio operator, in their cockpits. On a flight from Zurich to Paris, a clash developed, and sparks were flying about soon after take-off. The captain told the radio operator in no uncertain terms to shut up: 'In future please keep your opinions to yourself unless I ask for them.' The flight deck went quiet, and, from then on, the radio operator thumped out his messages angrily in morse all the way to Paris. The touch-down seemed normal – until the blushing captain realized that he had landed at the wrong airport. The first officer and the navigator said nothing when the captain snapped at the radio operator: 'Tell Zurich we've landed at Orly and we'll be a bit late getting to Le Bourget.' To which the answer was 'I already sent that message to Zurich while we were on finals . . . sir.'

It is claimed that, on the current and future generations of aircraft, the presence of a third man in the cockpit is superfluous. People who feel uneasy about the disappearance of the flight engineer are deemed to be conservative or outright oldfashioned, but numerous incidents help to underline their reservations. For example, when, fairly soon after take-off, a British Airways 747 was beginning to cruise peacefully as if sleighing on the white expanse of clouds, senior Engineering Officer Bob Bradshaw carried out a routine check for a fuel configuration change, and discovered that a crossfeed valve had failed closed. He mentioned it in passing to the captain. The crew did not seem to be unduly concerned though they knew it meant that, if nothing was done about it, one engine would be starved to death by the time they approached Tokyo. Not a serious problem but a hindrance if anything else went wrong. They also remembered that on a PanAm Jumbo, also Tokyo-bound, a valve once failed similarly in the closed position, and while that tank remained full of fuel, three engines ran dry only because the three-man crew, all pilots, could not figure out how to circumvent the problem and transfer the fuel to empty tanks.

Had there been no engineer on that BA flight, Captain Torquil George would have had to ignore the problem for a while because he was too busy with various duties and watching out for heavy traffic over Europe. But Bradshaw could begin some quiet trouble-shooting, and quite casually, called the BA base: 'Thought you might like to know . . . ' He studied the schematic diagram of the fuel system, which looked like some advanced intelligence test. Eventually, he found a way to get around the failed valve: without actually opening the *jettison system*, he would utilize the *standby system* with which fuel could be jettisoned in an emergency landing. He told the captain, and a brief conference ended with the words: 'All things being equal, there shouldn't be any problem.' Despite the calm atmosphere, one sensed that the crew were earning their keep. And Bradshaw was still unhappy: the jettison system was not a neat enough answer, and might possibly cause complications later with out-of-balance wing loading.

At long last, without shouting *eureka*, the engineer hit on an ingenious solution: through loops and gates and valves in the maze of three systems, he could make the fuel transfer – and still keep the aircraft in perfect fuel trim – without using the jettison system. 'Just one of those things,' he told me. 'On the new 747-400 series, of course, there won't be an engineer, but they'll have an automated back-up system for – hopefully – any eventuality.'

The relief crew took over, and were fully briefed before the team would retire to their bunks. When they returned for duty, it was remarkable that the captain asked no questions about the valve. He just knew that Bradshaw still had his teeth into the problem.

It was an uneventful flight. Unaware of any problems, the passengers had only one serious decision to make: over the North Pole, they were torn between two appetizing sights – their meal and the stunning, shimmering vista of the *aurora borealis*.

Standards of cockpit resource management continue to vary vastly from airline to airline and from country to country.

At the British Caledonian Flight Training centre, where pilots of numerous nationalities turn up to get prepared for DC10 type-conversion, Captain Harrison and Captain Shirwell are used to shocking revelations. In some countries, for instance, pilots get no holding pattern training simply because at their home bases the weather is usually good and traffic is light – no wonder they create collision hazards by straying from the stacks over congested airports like New York's. Elsewhere, pilots may learn 'holding' on auto-pilot only, and never practise the art manually. 'The problem gets accentuated when one third world airline teaches conversion to another,' said the two training specialists. 'When their people come to us, we often discover that they had not flown their aircraft manually for six months or more even though the need, due to something like hydraulic or electrical failures, might arise any time to land manually – and on some abnormal approaches, a DC10 would land considerably faster than a Concorde. It's not the people but the system to be blamed: an African state, for instance, started from scratch but improved its training to stand comparison with anyone.' Harrison and Shirwell find that even the American licensing system is not up to the very high British standards. Jack Shirwell is also critical of the multiple choice exams (favoured by both the FAA and the CAA) because the pilot learns to *choose* rather than *think* of the correct answers.

Tacit acquiescence to some form of flying a *flag of convenience* is an all-too-real possibility. ICAO is committed to respect all signatories' right to choose: if a pilot is licensed according to one member's regulations, other members have no right to question their standards.

A MATTER OF ATTITUDE

Whispers of criticism can be heard about the training of Soviet and East European pilots. Only diplomacy and professional solidarity (essential when cooperation must be sought) prevent the whispers from growing into thunders of protest. Despite so-called *glasnost*, Soviet crashes are kept secret unless highly visible: outside their boundaries or at airports where Westerners witness them or when foreigners are among the victims. So it is only estimated that the general safety level of the huge Soviet air traffic is much lower than that of the Western world. Russian passengers cheer themselves up saying that 'the losses of Aeroflot are quite trivial – at least if compared with losses in the air during World War Two.' While much of the outside world's view must remain based on a combination of impressions, deduction and guesswork, an article*

* Development of Better Cockpit Training Programmes Under Way, *ICAO Bulletin*.

by N. F. Nikulin, chief of the Korovograd Superior Flying School, was most informative when you put together two of his seemingly unrelated sentences: 'A considerable proportion of the errors committed by crew members in flight may be due to inadequate training . . . Elements of the optimized pilot training system have now (*1987*) been introduced into the programme of the flying schools in the U.S.S.R.'

'Beyond training, the mistakes are often due to Soviet pilots' attitude to flying,' said a British specialist who was involved, in 1987, with the much disputed investigation of the TU134 that crashed into South African territory while 'looking for Maputo'. The location, and the fact that the President of Mozambique was among the victims, added heavy political loading to the case. 'Without the CVR, nobody would have believed how the crew behaved. The Russian checklist requirements were as good as any Western airline's – except that, at least in the last thirty minutes, none was complied with. They were operating in that area for eighteen months, and they must have gone *bush*. With not one check of radio aid frequency, they just flew the aircraft straight down. They used words like *check*, but it meant different things to them and the controller. It was ludicrous. They had a Russian five-man cockpit crew, but they said things like *check, no ILS . . . yes*. It made no sense. Reports about boozing in flight were nonsense. They crashed because of their lack of discipline.'

Gorbachev initiated a national campaign to fight heavy drinking, for many Russians find vodka irresistible. Aeroflot's house magazine exhorted against the evils of alcohol. Droves of staff had to be dismissed for driving airport vehicles 'under the influence'. A Soviet investigator told a British colleague that 'flightcrews are submitted to eyeball inspection for booze before take-off'. In December, 1987, *Pravda* revealed many of Aeroflot's management problems. Apart from heavy losses, half the safety incidents were due to 'low professional levels among staff'. In just nine months, 'major disciplinary transgressions rose by twenty per cent' and cases of drunkenness increased by sixty per cent. A transport journal admitted that Aeroflot, with its geriatric fleet and notoriously poor service, 'cannot achieve the level of even an average international airline'. In December, 1988, *Literaturnaya Gazeta* reported the following case: the last man aboard an Antonov-12 military transport aircraft failed to close the door properly, and when, with altitude, the oxygen level dropped, the six-man crew began to pass out one by one. The plane made random turns and went into the a spin. The bleary voice of the co-pilot came through to ATC: 'Where are we? Where are we flying? I feel terrible.' Then the captain woke up and tried to grab the controls as the co-pilot was being talked down by controllers at Ufa, in the Urals. It transpired, eventually, that the plane had been delayed at Chelyabinsk for three days, and, as reported in the *Japan Times*, the crew had used the whole period for a massive, non-stop drinking bout. The unmentioned fact that nobody tried to prevent them from boarding and taking off was an even more revealing aspect of their escapade.

Most Aeroflot pilots come out of the military, and carry old bad habits, a German investigator said. 'There're some hair-raising stories around. We've seen an East European BAC1-11 approach illegally with a sixteen-knot tailwind. He landed so hard on the nosegear that one wheel broke off with the axle, yet he didn't bother to seek costly repairs because that would be payable in hard currency and make him unpopular at home. So he just took off with one nosewheel on one axle. It would be totally unacceptable to us, but he made it. He was a good pilot. I just wouldn't like to fly with him.'

The first domestic Soviet air accident to be reported publicly was rather startling. On October 20, 1986, a TU-134 crashed at Kuybyshev. The report did not say what exactly the 'large number' of fatalities was, but the cause was another clear case of indiscipline: the captain, in a fit of machismo, shuttered the windscreen for a *simulated* blind landing. Such an exercise exceeded the capabilities of the aircraft and its systems, and was therefore forbidden by the flight manual. (To make things worse, when the aircraft broke up on the runway and burst into flames, the first rescue vehicle on the scene was four minutes late, and it had only half the required extinguishants. Other vehicles arrived 'much later', firemen did not carry enough breathing apparatus, medical staff had no equipment for resuscitation.) The report said that seventy-seven per cent of the Aeroflot crashes were caused by human error. Several high-ranking government officials were sacked to face disciplinary charges. The captain was sent to jail for fifteen years – the maximum term for such crimes according to the legal code. (Some signs of increased openness appeared in

October, 1988, when Moscow revealed three fatal crashes in a ten-day period.)

Soviet and East European carriers will now try to claw a bigger share out of international air traffic, offering cheap charter flights. The capacity is there – Aeroflot alone has 3,000 aircraft and 95,000 pilots, and carries just domestically 120 million passengers plus three million tons of freight a year – but the quality and safety of the service and the hardware are questionable.

Though – after numerous Chinese, Vietnamese, Cuban, Hungarian and Polish accidents and incidents – maintenance and the elderly Russian-built aircraft were blamed, inadequate operating principles were also often implicated.

Oxygen masks, standard on all airliners in the west, drop out automatically in case of sudden decompression. Aeroflot and the Bulgarian Balkan Airlines are, however, happy to fly without them: they claim to find it satisfactory that they have oxygen bottles on board, and the stewardesses will rush them to passengers who need them. They do not seem to consider at all how two hundred or more people would share two or three bottles.

On a London-Budapest Malév flight in May, 1988, the passengers received no briefing in any language.

If an aircraft flies over water beyond a limit of fifty miles from the shore, ICAO standards require that life jackets for everyone should be carried, but Aeroflot admitted to the *Sunday Times* on April 2, 1989 that they do not carry such lifesaving equipment for each passenger.

In many accidents, crew errors seemed the obvious causes, but that could not be documented because the quality of the investigations was unknown, and the final reports were not published.

When for instance, a passenger door of a TU154 flew open above Stockholm in 1987, two stewardesses fought a losing battle to close it. Hand-luggage and even a shoe of a stewardess were sucked through the gap. People aboard the Romanian Tarom charter flight felt the outward pull, held on to their seats, and feared the worst. A Romanian passenger, former world champion wrestler who now lives in Sweden, rushed down the aisle, wedged his body across the opening, and heaved for three minutes to shut the door at least partially. Fortunately, the aircraft was only 1,800 feet up at the time, and managed to descend quickly to avert a tragedy. 'If you have travelled on Romanian aircraft before,' said the athlete, 'you're quite prepared for something like this.' The investigation did not blame the old aircraft, but revealed severe negligence on the part of the pilot and co-pilot who failed to notice the *door unsecured* warning light. Åke Roed, who headed the Swedish inquiry, said the aircraft 'should have never left the ground' because, among other shortcomings, some seats were without seatbelts, and many life jackets were missing.

It is not very often that a country gets any clearcut evidence of the inferior training standard in another, but an opportunity arose in 1987. British holiday charter companies ran short of seats to the sun, and *wet leased* aircraft (i.e. complete with cockpit crews) from Tarom. For a while the pilots operated with a temporary British licence, but under the control of Romanian authorities. At some point, however, the airlines wanted to make the crews and aircraft British registered. To obtain the full licence, the pilots had to take the much more stringent British tests – and four out of five failed right away. Later a compromise was found: they passed a test which was identical to those used in Romania, therefore less rigorous than Britain's, but sufficient under international agreements to enter British airspace or fly holidaymakers from Britain. 'The episode lifted a veil, but what we saw was not much more than a drop of dirty water in an ocean,' a government official exclaimed. 'All that became obvious was that training in Romania was totally antiquated by our standards. Some people argue that when British passengers book seats on British charter flights, they ought to be told who and what aircraft will fly them.'

It defies logic, and it is a mockery of hard-fought-for international agreements that ICAO standards and recommendations are open to 'interpretation' or a watering down process. Within the European Community, there is still resistance to plans that any pilot, licensed by any member country, should be able to take a job with any airline within the group. If the proposal is accepted, piloting standards will need a tremendous array of extra safeguards. In November, 1989, Italian authorities started to investigate the entire licensing system when it was discovered that six of their pilots had false qualifications. If that can happen in the heart of Europe, if licensing may be a matter of free and nonchalant interpretation of recommendations, what irregularities or outright corruption could be expected elsewhere?

Fatigue and *stress* due to company pressures to fly to the outer limits of duty times with maximum allowable equipment deficiencies could also be merely a matter of attitudes to safety.

Crew fatigue has rarely been named as a proven causal factor, let alone the probable cause, of accidents. That is not because fatigue did not play a part, but because the condition cannot be measured with any precision retrospectively. When accidents were reviewed to see if fatigue was a significant causal factor, it was found that, over seventeen years, some Canadian regions had never recorded any data regarding fatigue. 'We found some good examples,' said Dr David Elcombe, CASB director of Safety Medicine, 'but the prevalence of the condition was probably understated. The human element is never in isolation. Before commenting on the difficult area of aircrew performance, we not only need to know if duty times have been exceeded but whether, for example, rest was of adequate quality, and whether the crew slept properly: hotels are not homes. Was the room noisy or hot? More importantly, what was the physical and mental status of the crew? Were they preoccupied or worried or suffering from insomnia? What was their personal time clock set to? Were there any early or subtle bodily conditions such as some of the more common forms of liver disease that could also give rise to fatigue? All of these areas and many more can act synergistically.'

That fatigue and stress are serious hazards is well-known. Air traffic controllers have long been complaining about excessive permissible duty hours, and now, at last, tighter limits may soon be introduced in several countries, including Britain. Both CHIRP and NASA receive dozens of reports from pilots admitting confidentially that they fell asleep in the cockpit during flights. One captain woke up 300 feet from the runway during his landing approach. Fatigue prevention is governed by complex sets of scales and regulations covering hours of duty including time spent on standby, mandatory rest periods, cumulative effects of flying time over months, stress exposure due to number of landings, the crossing of time zones that can interfere with the body clock, and other intangible factors. The rules vary from country to country, and as Sydney Lane of BALPA pointed out, 'the regulations set out the limits, the maximum allowable duty times, but some smaller, independent airlines *schedule* for that maximum without leaving the pilot any legal safety margin to exceed it if need be.'

In tightly scheduled short-haul charter operations, particularly in Europe, there is hardly any room to manoeuvre. The pressure to fly to the limit is explained by the fact that only maximum aircraft and crew utilization will ensure profitability during the key holiday periods. If strikes, congestion of airspace and airport, not to mention bad weather, cause long delays, 'captain's discretion' comes into play. If he is British, he may legally spend nine hours on standby and ten hours flying, and can still put in another three hours if he feels fit. But fatigue-ridden pilots, like those affected by anoxia, are the last people to judge their own fitness.

'Take the pilot who keeps getting his rest at the wrong time because of alternating night and day flights,' said Roger Green. 'He takes a plane to the Canary Islands, just a quick round trip, a milk run to the winter sun, they say, but there's a delay. Some minor technical hitch. By the time it's fixed, he knows the crew will be tired and may even exceed duty time on arrival back home. If he was in New York, he would not even hesitate to call it a day. But down there, it isn't easy to say "no" and postpone the flight. There's no relief crew: 150 people are waiting impatiently. Putting them in a hotel would cost a bomb to the airline, so they'd have to spend another twelve excruciating hours at the terminal. On top of all that, back at Gatwick another outgoing flight would be delayed for just as long, and the entire company schedule may be screwed up for days. So he says: "Let's go." He'll do his most difficult bit of flying in the most congested areas, maybe in snow storms, and he may have to divert or just hold, and hold above the airport, when he is at his lowest level of alertness. Admittedly, the landing process gets the adrenalin flowing, but if he then makes a mistake, will they call it pilot error?

'The CAA must approve every airline's operating manual. Most companies write it according to trade agreements that are more restrictive than the law. But if the others schedule their pilots right up to the brim of legal maximum, the CAA can't object to it because that would imply that the legislative "maximum" is seen as something unsafe and meaningless.* Regulations can only set acceptable minimum safety standards – not what true

* In 1989, the CAA argued that scheduling to the legal maximum negates the *spirit* of the law.

professionals would regard as 'good practice'. So it shouldn't be surprising that a company I know of has offered its pilots a thousand-pound bonus for pushing themselves and flying their aircraft to the legal limits throughout the profitable holiday seasons. The captains thought it wasn't very clever, but then shrugged their shoulders because a bonus is a bonus is a bonus.'*

On long-haul and super-long-haul non-stop flights, the big jets carry 'heavy crews' so that everybody gets some rest in bunks behind the cockpit. On those routes, upsets in the pilots' circadian rhythm, and the monotony of the job, are seen as a graver hazard than the occasional catnap in the driving seat while cruising at 35,000 feet. And if fatigue is hard to calibrate, how can anyone hope to measure these intangibles?

MACHINE V. MAN

Some airlines, like British Airways, use costly, sophisticated performance monitoring devices that help both flight crews and maintenance engineers to pinpoint developing bad habits, training loopholes and mechanical faults in their embryonic stage. But we are years, possibly decades, away from world-wide standardization of such advanced techniques.

Automation, fly by wire and glass cockpits are the rage of the age. New-technology display on the latest types has been hugely improved. The cathode ray tube offers striking forms of data presentation and warnings in multicolour. Crews can monitor progress, every automated manoeuvre, and demand even an analysis of the performance of the entire machine by calling up status information on television-type screens. Scientists are now working on aids like take-off performance monitors which could one day reduce the need of a pilot's gamble with split-second decisions when 'something seems to be wrong' as the speeding aircraft is about to lift off the runway. Pilot-proof Microwave Landing Systems, hailed as an enormous break-through, are still open to some doubts. With modern aircraft, smaller and less

* The Social Department of the European Commission is most anxious to tighten flight time limitations, but does not seem to be fully in touch with reality. Its proposals, if accepted, could increase European air fares by ten per cent, and could exacerbate the already serious manpower shortage by necessitating the employment of up to fifty per cent more pilots.

experienced crews, some new hazards may develop. The latest advanced simulators could be used to train pilots who could qualify with 'zero flight time' – no time spent actually in the air.

Revolutionary training techniques with the use of modern simulators (pioneered by Northwest Aerospace Training together with British Caledonian and the French Aeroformation), have opened the way to satisfy the huge demand for pilot conversion to new types of aircraft more realistically and cost-effectively. As a result, 'the FAA is moving towards more formal approval of [*conversion*] training courses tailored to individual needs, rather than requiring a set period of instruction regardless of previous experience.'* But as there is still some disagreement even between the FAA and the CAA, no one can foretell what will happen when less sophisticated authorities set their own conversion requirements, and choose to license their own, purely simulator-trained pilots to make their first real trip on a new type with a couple of hundred passengers behind them. No wonder there is a clamour for the enforcement of world-wide simulation standards, too.

Those who seem to place more faith in cockpit automation than in pilots' skill may find that their solution is to air safety what a new underpass or the addition of an extra lane on a five-mile stretch is to a busy motorway – it only shifts a perennial traffic jam black spot a little further down the road. Instead of looking for potential pilot errors, the investigators of future accidents may have to search for human causal factors among the people who design, programme and build the computers that design and programme the glass cockpits and control mechanisms for the flying robots. There is no doubt about it: computers make fewer errors than people. Though robots, too, may suffer from stress and fatigue due to sudden overloads, at least they are free from emotional upsets, and their decisions are not tinted by a surge of adrenalin.

Chuck Miller, the American investigation specialist, said: 'They say pilots will be safer because they'll have to do less work. Don't you believe it. Less work can be more dangerous. The boring task of monitoring instruments is something people are least suited for.' Fatigue and monotony combined to produce the aerobatics

* Harry Hopkins: Northwest Training Back on Course, *Flight International*, 22.11.1989.

performed by the Taiwanese Jumbo over California. And aviation psychologists also see a special risk in the false security of 'flying just dots on a small green screen as opposed to flying an aircraft through space you can see and feel all around you'.

At the 42nd Flight Safety Foundation seminar in Athens, 1989, Dr Agnes Huff described the new minefield opened up by automation. Her list of negative effects was headed by the erosion of flying skills, and the perceived loss of control. She denied that automation decreased the pilot's workload because, in proportion to the reduced manual duties, monitoring tasks increased and that only shifted the risk of human errors to the area of the man/device interaction.

Air transport modernization is faster than ever before – in some parts of the world. It could be disastrous to entrust the latest *toys* to immature hands when the industry and the international governing bodies cannot enforce universally even their minimal standards of training and attitudes, when massive commuter and charter airlines, not to mention entire countries and continents, are left behind in the stone age of the human factor concept, and when they would have to make a massive quantum leap from geriatric aircraft to press-button flying wonders. Perhaps that is why some engineers recommend that new aircraft should carry just one pilot and a dog in the cockpit: the pilot to feed the dog – and the dog to ensure that the pilot keeps his hands off the controls.

– 16 –

MURPHY'S LAW NUMBER TWO

Climbing effortlessly through 29,000 feet, the 737 approached its cruising height. Dinner service was in progress. A few passengers were already tucking into their pork salad, when loud bangs and sudden, mysterious vibration shattered the peaceful mood on the London-Belfast shuttle. To the accompaniment of wild rattling in the galley, smoke crept into the cabin. Terrifying blue sparks and fire around the left hand engine were noticed . . .

No, sadly, this was not fiction. It happened five minutes past eight o'clock on January 8, 1989.

A baby and 117 adult passengers with a crew of eight were aboard the British Midland jet, an almost brand new Boeing 737-400. Boosted by extra power from its two gleaming CFM 56-3C high bypass turbofan engines, the aircraft was the latest, modernized version of a well-tried workhorse – some 2,000 of the original model were still flying with 141 airlines all over the world.

The increasingly violent vibration baffled the pilots. They could smell burning, something like hot metal. An engine fire? They had no such warning from their instruments. Crews are trained to react fast to risk, but, at that altitude, they would have plenty of time to sort out what was wrong. The captain took command right away, and disengaged the auto-pilot. Now they could find out which of the two engines was the troublemaker. Like most pilots, they would proceed by trial and error. The crew picked the right hand engine to test first. When, just twenty seconds after the onset of the problem, the autothrottle was disengaged and the throttle was eased back, the heavy buffeting stopped, the burning smell and the vibration began to subside – so that was it, right first time.

They told Air Traffic Control that they had an emergency situation because of a probable fire in the right engine, and that they would divert to Castle Donington. There was no panic: their altitude gave them a fine cushion, and even just one engine would provide more than enough power for a safe landing. Passengers were told to fasten their seatbelts because there would be a diversion to the East Midlands Airport. The left engine was throttled back for the descent, and that caused yet another decrease in the vibration level. 127 seconds after the initial adversity, the right engine was shut down completely. The indications were that the problem had been eliminated.

In the cabin, the attendants were busy. Some extremely agitated passengers and a screaming woman needed reassurance; trays with untouched or half-finished meals had to be collected; preparations had to be made for a potential emergency landing. The flight service manager went to the cockpit to report that the passengers had been worried by the smoke and the vibration. The captain therefore broadcast an announcement that the right-hand engine, the source of their inconvenience, had been shut down, and people could expect to land in about ten minutes. Right-hand engine? Weren't the sparks and fire seen on the *left*-hand side? Some people were surprised and confused. Was there trouble with *both* engines? Was the engine on the right in even worse shape? Others, together with some cabin attendants, had not heard the details of the announcement. People trust the pilot. Nobody wants to be a busybody and make a fool of himself. Even cabin attendants think twice before asking silly questions, particularly when the captain must have plenty to do. He must know what he is doing.

The pilots would have agreed. They thought they knew what they were doing. The left-hand engine, on which they were flying, was still vibrating a little, but seemed to respond reasonably well. It would not be long now.

At about that time, an AAIB Principal Inspector of Accidents was enjoying a happy evening with his guests. It was the first occasion that his daughter's fiancé and his parents came to have dinner with the Trimbles.

After an uneventful quarter of an hour, the flight was cleared to approach Runway 27 at Castle Donington, just beyond a busy motorway. Descending through 2,000 feet, the landing gear was lowered. Fifteen-degree flaps were selected. Power was increased to keep on the glide slope and raise the nose so the aircraft would touch down in the proper attitude, wheels first.

A minute later, at 20.23 and fifty seconds, when the 737 was 900 feet above the ground and barely two miles from the runway, the left engine died without any warning.

The excessive vibration restarted with venom. Seventeen seconds later, the shrill of the left engine fire alarm filled the cockpit. The pilots tried to restart the right engine. Their desperate attempts to relight it were doomed. Time was running out. What could have been done comfortably thousands of feet up had now become an impossibility. With just a whiff of luck and with a few seconds more power from the left engine, they might have cleared the wide carriageways of the M1 and the embankment beyond. They did not. After an initial glancing impact in a field east of the motorway near Kegworth in Leicestershire, the aircraft crashed into the wooded western embankment and broke up.

Fifty seconds after the last heartbeat of the engine, thirty-nine passengers were dead. (Eight more died later, in hospital.) Almost everybody else aboard, including the baby and the crew, was seriously injured.

At 21.15, Eddie Trimble was just enjoying a glass of port and a good cigar, when a phone call from his chief put an end to all that. A plane was chartered (a rather unusual gesture of grandeur for the AAIB), and nine investigators were flown up north. Some of them, like Chris Pollard, Bob Carter and Steve Moss, were just back from a most arduous stint at Lockerbie, but nobody could be spared – such are the resources of the AAIB if hit by a spate of disasters. They were at the crash site at forty minutes past midnight, when rescue operations were still incomplete. Trimble spent the night setting up the investigation structure, appointing colleagues to be in charge of various specialist tasks, waiting impatiently until 04.20 for the removal of the last survivor from the wreckage (an injured young man, for instance, had been trapped with his dead wife for hours), so that another ten minutes later Pete Shepherd of the AAIB could get into the tail section to recover at long last the flight data and cockpit voice recorders.

They worked till six in the morning, then, knowing that a heavy day was ahead of them, they decided to have a well-earned rest – and grant themselves a whole hour's sleep.

The crash site became a hunting ground for rumour-mongers and know-alls. But the investigators' basic questions were clear: What went wrong with the engine/s? Whatever it was, did it make the crash inevitable? If not, what went wrong in the cockpit? And finally, could there have been more survivors?

The captain was very badly injured, but the CVR and flight recorder were intact, and there was an abundance of crucial evidence including the communications with ATC, ground witness accounts of fire in the left engine of the descending aircraft, the absence of any telltale sign of fire in the right engine, and a major discovery some two miles from the crash site: a thorough search of the area below the flight path recovered some fragments from the fan blades of the left-hand engine.

Fairly soon it became obvious that the pilots must have shut down the wrong engine. Why? Something must have misled them to make and persist with their first choice guess. It transpired as a vague possibility that the engine instruments might have been wired the wrong way round. Investigators hate to jump to unfounded conclusions, but that possibility seemed daunting enough to demand an immediate preliminary recommendation to check all new 737s for potential cross-wiring. The urgent process cleared the crashed aircraft and the fleet, though some mistakes of that kind were found on other types, like the 757.

The investigation continued on several parallel lines. With the help of the engine manufacturers it was established that the fan blades had failed. The initial causal theory – the ingestion of some object – would soon be discarded when signs of progressive *fatigue* explained the failure. But that should not have caused the disaster when there was another, perfectly functioning engine available to land the aircraft.

Attention had to focus on the pilots' actions. Shutting down the wrong engine, as we have seen, is far from being a unique error. It is often associated with the irresistible urge to do something, anything, to solve a problem. Correspondingly, making the decision to do nothing in an emergency at least for a while is difficult for a pilot who is trained to be positive. Captain Moody, the reader may recall, chose to *sit on his hands preventing them from doing things* when the volcanic eruption hit his aircraft, when contradictory information from his instruments flooded his eyes, ears and brain, and when he lost power from all four of his engines in quick succession. In the punchy language Americans can develop for any occasion, it is *a time for winding your stop-watch*. Buzz Aldrin, the space pilot, once revealed that before the flight to the moon, he had made a decision: if something unexpected and totally unforeseen happened, he would do nothing as long as possible in order to avoid some action that might only exacerbate the situation.

The pilots of the 737 were quick to hit back. Had their action failed to alleviate their problems, had it not appeared to cure the symptoms, they would have tried something else, no doubt. But the responses they received misled them through a series of coincidences, and dug a deeper and deeper hole for them to fall into. (The autothrottle, for instance, monitors the speed of the engine fan and, when that slows down, it reacts with dishing out more fuel. When they disconnected the autothrottle, the bad engine could settle to a lower level of vibration with fewer surges, so its performance felt much better. At the same time, the smoke and smells began to clear due to now reduced compression blade seal contact on the left engine. This case had, however, a most unusual element: thc crcw had some twenty minutes to take a second look at their initial decision. A couple of months after the accident, the captain issued a statement from his sickbed: 'Both the first officer and I agreed that the right engine was causing us the problem. Until just before the fire alarm rang, we had no indication of problems with the left engine.'

It sounded as if there had been nothing in that modern, sophisticated cockpit to alert them to their error and to the need to review their actions. But the investigators did find something that might have warned them: the vibration instruments.

The aircraft had an airborne vibration monitor which fed vibration signals from the engines to the flight recorder as well as a pair of dials in the cockpit. After the accident, both the monitor and the gauges were found to be serviceable. At some stage during the emergency, one dial would have swung to show the maximum measurable vibration in the left engine, and only a small amount – less than one notch on a scale of five – in the right engine.

Whether the pilots did look at the vibration gauges or not is for them to say. But it is another, entirely different consideration whether they would have reacted at all to those gauges if they had looked at them, and if the readings had been correct.

Rightly or wrongly, whatever the flight manuals say, the pilot community – not quite excluding even their training instructors – tends to learn a lot from experience, and develop habits, phobias and prejudices. In the early 1970s, pilots had a prejudice against the *stickshaker* which often gave spurious warnings of a stall, and which was one at least of the reasons why the pilot of a Trident reacted in *shut-up-gringo*-style by overriding the 'unreliable' stick-shaker – and crashed at Staines in 1972. In the 1980s, many, possibly the majority of pilots had such a prejudice against vibration indicators. 'They have been living with whole generations of gas-turbine driven aircraft on which the vibration alert system is pretty poor,' said Trimble. 'I'm not surprised that they don't trust it. Most pilots pay little or no attention to it, and perhaps the training routines do not lay sufficient emphasis on its use either. That may well be because the vibration instrumentation was created basically for engine trend monitoring. it was not meant to be a primary indicator of a serious engine problem. Only on the latest models is the system accepted as something to give reliable clues about the health of the engine. In the 737 case, those gauges alone could have prevented the accident.'

The position of those dials might have helped to make them seem superfluous on the crashed 737. On the new 757s and 767s, vibration gauges are a key part of the engine *alert system*. They can be seen on a computerized display of *primary* indicators that calls attention to any problem. When the old 737 model was modernized, and the 737-400 was born with extended power from better engines, the instrumentation was presented in a new glass cockpit form, but the manufacturers tried to make life easier for the pilots by keeping the new presentation close to that of the old mechanical indicators. Vibration gauges remained, therefore, on a *secondary* panel where they became only the third of five vertically placed pairs of small dials, and where nothing was deemed to deserve a red light of exceedence warning. (Some specialists believe that when such changes/improvements are made, below average line crews rather than professional test pilots should try them out because the latter are more used to watching everything and making rational decisions about the unexpected, whereas the former might fly for years without ever being called upon to evaluate new problems against old habits.)

Less than two months after the accident, the AAIB felt strongly enough about the problems to break its own habits: it issued 'not an accident report – either final or interim', but 'a special bulletin'. Apart from precautionary advice concerning the engines and their *modus operandi*, it contained recommendations to make pilots aware of the potential occurrence of a vibration/smoke/smell combination, to amend all aircraft manuals with suggestions for appropriate counter-measures, and to urge the CAA to 'review the current attitude of pilots to the engine vibration indicators . . .'

On June 9, 1989, a DanAir flight was 25,000 feet above the Channel when its pilots experienced sudden violent vibration and smoke. Unknown to them, the CFM 56-3C engine on the left had suffered a fan blade failure. The crew zeroed in automatically on the vibration gauges, shut down the faulty engine, and landed on the good one without any complications. Only two days later, another Sunday, another British Midland 737-400, also at cruising height on the London–Belfast route, suffered another fan blade failure accompanied by the worrying symptoms of vibration, but again, having been told where to look for answers, the crew made a troublefree landing. It confirmed that, on this occasion, the art of the state had genuinely improved the state of the art.*

If only similarly quick precautionary measures followed every good recommendation every time, there might have been no slaughter on the embankment of the M1.

The 'River Orrin' (*Juliet Lima*) tragedy, the burn-out of that older 737 at Manchester, occurred on August 22, 1985. The pilots were very much in the dark about the nature of the fire that raged behind them. In the Safety Recommendations of the accident report (published only in 1989, but known to the authorities much earlier), the following occupied the No. 2 position in a logical sequence: 'Research should be undertaken into the methods of providing the flight crew with an external view of the aircraft,

* How come that all three cases involved British users when many others also fly the 737-400? The answer is that not all airlines fitted GE/Snecma engines, and only some have specified the engine that can be operated at a higher thrust rating, the probable cause of fan blade fatigue.

enabling them to assess the nature and extent of external damage and fires'. *No aircraft is so equipped to this day*. Some people say a side-view mirror, common to every car, could do the trick. A video camera would certainly be more efficient. The cost would be minimal. Japanese aircraft on domestic routes have one in the nose just for fun: passengers enjoy the live TV views of the take-off and landing. So here is one more near-certainty: what specialists now call the 'error chain', which runs from the drawing board to flying, could have been broken by a single glance at a video monitor which would have told the pilot which engine was sparking and burning. (At the time of writing, BA is about to fit three small 'rear-view' video cameras on a jet for a twelve-month trial. This is the result of research by the CAA and the airline. If the test is successful, the CAA *may* instruct all British operators to install such cameras on the entire fleet. A marvel of tombstone engineering?)

The investigation of the survival aspects of the 737-400 accident reopened yet another bag of worms regarding seats. It is known that grave injuries tend to be suffered when passenger seats break free of their attachments. At Kegworth, however, even rearward facing seats *might not* have helped significantly. The aircraft was furnished with the new-type, doubly strong seats (withstanding a 16g force), but their potential was not fully realized because, particularly in the forward cabin, it was the floor itself that collapsed, and the seats, still firmly anchored, went with it. The investigators produced a first-ever computer simulation relating the widely varied pattern of injuries to seat/floor behaviour in an accident. They worked for some six months on it but could not continue beyond a certain point. After all, the AAIB is no research organization. An institute like Cranfield would be much better suited for extended research in such areas – but once again, it is a matter of funds which are hard to come by even if improved, safer floor design could be the payoff.

1989 was a terrible year for air safety. Yet, when airport loudspeakers spring to life to blare out the final call for passengers to proceed to the gates and board their flights, those who may have a tingle of unease in their collective stomach must remember that hundreds of millions travelled the airways in that year with nothing more dramatic than annoyance by delays and misrouted luggage.

There are no new causes of accidents: 1989 reproduced much of the range: human errors, old aircraft, adverse weather, sabotage, navigation, press-on-itis, fire and icing.

The statistics of safety are misleading only because they are meaningless to those who make up the few fatal exceptions. Considering their relatively small number, death by air accident is bad luck, indeed. But deaths by *foreseeable* and *avoidable* air accidents could be seen as murders by the negligence which renders hard-learned lessons so readily forgettable. Without greater accountability of governments and aviation authorities, and without passengers themselves voicing a strong interest in their own safety, many of those lessons and subsequent recommendations may continue to get buried in a maze of self-protecting bureaucracy and short-sighted profiteering.

If to err is only human, to open the gates to disastrous replays is bad government. If Murphy's Law warns us that anything that can go wrong will go wrong sooner or later, perhaps it is time to coin and enact a Murphy's Law Number Two as a reminder that whatever mistake is *allowed* to be repeated, *will* be repeated.

SELECTED BIBLIOGRAPHY

In additon to accident reports, statistics, regulations and special studies published in numerous countries, extensive use has been made of ICAO, IATA, ISASI and FSF seminars and publications; investigation course papers (University of Southern California, Cranfield, Stockholm); U.S. Congressional hearings, U.K. Select Committees; NASA *Callback* and CHIRP *Feedback*; and aviation journals and periodicals, particularly Aerospace, Air & Cosmos, Air Clues (RAF), Air & Space Magazine, Air Safety Week, Air Transport World, Aviation Daily, Aviation Week and Space Technology, Flight International, Flight Safety Digest, Flug Revue, Jane's Airport Review. Though many sources have been footnoted, some that may be useful to researchers have been included in this short bibliography.

AGARD Conference proceedings 1980–1989

ALPA: *Guides For Airport Standards*

– *Guides to Accident Survival Factors*

– *Compilation of Accidents/Incidents Involving Runway Overruns/Undershoots/Veeroffs* 1987

Ashford, R.: *Putting a Price on Safety*, Flight International, 19.4.86

Awford, I.: *Civil Liability Concerning Unlawful Interference with Civil Aviation*, Aviation Security Conference paper, Leiden University, 1987

Balfour, A.J.C.: *Aviation Pathology*, RAF Lecture, London, 1988

Birch, N.: *Passenger Protection Technology in Aircraft Accident Fires*, Gower Technical Press, Aldershot, 1988

Bruggink, G.M.: *The Safety Role of Regulatory, Corporate and Personal Initiatives in the 1980s*, paper at the U.S. General Accounting Office Conference on Transportation Issues, Washington, 1979

– *Catastrophic Inflight Occurrences* (two parts), ISASI forum, 1986, 1987

Brunetti, A.: *Safety Versus Economics*, FSF Flight Safety Digest, October, 1986

Burnett, J.: numerous lectures and submissions (referred to in the text)

Chalk, A.: *Air Travel: Safety Through the Market*, ISASI forum, November, 1987

Chaplin, J.: *The Role and Aims of Safety Regulation*, paper

Dillman, J.S.: *The Federal Tort Claims Act – Three Years After VARIG*, Remarks at 16th Aviation Law Seminar, Seattle-King County Bar Association

Dorey, F.C.: *Aviation Security*, Granada, London, 1983

Dussa–Fiala–Wagner–Zenzes: *Full Scale Study of a Cabin Fire*, Deutsche Forschungs- und Versuchsanstalt für Luft- und Raumfahrt e.V. Institut für Antriebstechnik, Köln, Germany, 1988

Eddy, P., Potter, E. & Page, B.: *Destination Disaster*, Hart–Davis–MacGibbon, London 1976

Ellis, G.: *Air Crash Investigation*, Glenndale Book, Greybull, Wyoming, 1984

FAA Annual Reports on the Effect of the Airline Deregulation Act on the Level of Air Safety

FAA Airport Certification; Revision and Reorganization; Final Rule, 18.11.87

Flugunfalluntersuchungsstelle, Luftfahrt-Bundesamt: Annual flight safety reviews

Fox, T.M.: *Aircraft Maintenance Management*, U.S. Air Force lecture, 1987

Fritsch,O.: *Accident Investigation Manual and Related Matters and Accident Prevention Data Banks*, presented at Accident Prevention and Investigation Course, Institute of Aviation Safety, Stockholm

Fujita, T.T.: *Microburst Wind Shear at New Orleans International Airport, Kenner, Louisiana, on July 9, 1982*, SMRP Research Paper 199, University of Chicago, 1983

Godson, J.: *The Rise and Fall of the DC10*, NEL, London, 1975

Golbey, S.B.: *Say Again?*, ISASI forum, April, 1988

Green, R. & Skinner, R.: *Chirp and Fatigue*, The Log, October 1987

Green, R.: *Finger Trouble*, Air Clues, August 1979

Hawkins, F.: *Human Factors in Flight*, Gower Technical Press, 1989

Hewes, V.: Paper at 7th World Airports Conference – Accident Survival, 1983

Higgins, E.A. & Vant, J.H.B.: *Operation Workload – a study of passenger energy expenditure during an emergency evacuation*, Office of Aviation Medicine, Washington D.C., 1989

Hill, I.R.: *The Air India Jumbo Jet Disaster: Kanishka – Injury Analysis*.

Hill, R.G.: *Investigation and Characteristics of Major Fire-Related Accidents in Civil Air Transports Over the Past Ten Years*, AGARD Conference, 1989

Hopkins, H.: *Accident Investigation*, Flight International, 28.9.85

Hurst, M.: *ATC at the Crossroads*, Aerogram, May, 1987

ICAO: *Accident Prevention Manual*, Montreal, 1984

ICAO: *Manual of Aircraft Accident Investigation*, Montreal, 1986

Jiwa, S.: *The Death of Air India Flight 182*, Star Books, London, 1986

Jones, F.: *Air Crash*, Robert Hale, London, 1985

Lautman, L.G. & Gallimore, P.L.: *Control of the Crew Caused Accident*, Airliner, Apr-Jun, 1987

Laynor, W.G.: *Summary of Windshear Accidents and Views About Prevention*, Aerospace Technology Conference paper, Long Beach, Calif., October, 1986

Lederer, J.: *Deregulation and Safety*, Chapter 10 in *Airline Deregulation*, ed. by Brenner et al, ENO Foundation, Westport, Ct., 1985

– *Economics and Air Safety*, FSF Flight Safety Digest, April & May, 1987

Leonard, D.R.: *Alpha-Numeric Call Signs for Aircraft*, paper presented to U.K. Flight Safety Committee, July, 1987

Lodge, J.: *Airport Emergency Procedures Involving Fire and Rescue Services*, lecture at Airport Operations course, Loughborough University, 1989

Martin, P. & Balfour, J.: *Carriage by Air – Limited or Unlimited Liability*, Business Law Review, July, 1983

Mason, J.K. & Reals, W.J.: *Aerospace Pathology*, College of American Pathologists Foundation, Chicago, Ill., 1973

Master–Balfour–Tettmar: *Alcohol, Microbes and Forensic Pathology*, The Society for General Microbiology Quarterly, February, 1988

McArtor, T.A.: speech at National Aviation Club, January, 1988

Miller, C.O.: *Human Factors in Accident Investigation*, Human Factors Symposium paper, The Hague, 1979

– *Management Factor Investigation Following Civil Aviation Mishaps*, ISASI forum, 1989

Muir, H.C. & Marrison, C.: *Human Factors in Cabin Safety*, Aerospace, April, 1989

Napfel, H.F.: *Data Compression in Flight Recorders*, Technical presentation to EUROCAE Working Group, 1988

Newton, E.: *Investigating Explosive Sabotage in Aircraft*, The International Journal of Aviation Safety, March, 1985

Northwestern University symposium on airline safety, Evanston, Ill., 1987

O'Brien, J.E.: *Deregulation and Safety*, ALPA's Perspective, Airline Pilot, August, 1987

Perrow, C.: *Normal Accidents*, Basic Books Inc., New York, 1984

Ramsden, J.M.: *Burning Questions*, Flight International, 13.6.87
– *Affordable Safety*, Flight International, 25.1.86

Reist, M. (SWISSAIR): *Communications, refresher course*
– *Safety by Stress Management*, training programme

Sarkos, C.P.: *Development of Improved Fire Safety Standards Adopted by the FAA*, AGARD Conference, 1989

SCI-SAFE Submission to Parliamentary Transport Committee, 1989

Speiser, S.M.: *Lawsuit*, Horizon Press, New York, 1980

Sutter, J.F.: *Changing Scene in the U.S. Air Transportation System*, 23rd Wings Club General Harris 'Sight' Lecture, 1986

Stevens, P.J.: *Fatal Civil Aircraft Accidents, Their Medical and Pathological Investigation*, John Wright & Sons, Bristol, 1970

Taylor, A.F.: *Investigator Training*, ISASI Forum, September 1984
– *Aircraft Fires – A study of transport accidents from 1975 to the present*, AGARD Conference, 1989

Taylor, L.: *Air Travel – How Safe Is It?*, BSP, Oxford, 1988

Taylor, R.W.: *Extended Range Operations of Twin-Engine Commercial Airplanes*, a Boeing paper, April, 1987

Tench, W.: *Safety Is No Accident*, Collins, London, 1985

Tucker, W.T.: *Learning from Past Accidents*, AGARD Conference, 1989

Vant, J.H.B.: *Smokehoods Donned Quickly – The Impact of Donning Smokehoods on Evacuation Times*, AGARD Conference, 1989

Weston and Hurst: *Zagreb One-Four: Cleared to Collide*, Granada, London, 1979

Wiegers, F. & Rosman, E.: *A Safety Profile of Widebody Commercial Aircraft*, a Flight Safety Foundation 39th Seminar paper, Vancouver, 1986

INDEX

The names of airlines and air-crews have not been used as index headings. Accidents and incidents are grouped according to aircraft types (location, date, operator, cause), and for cross reference, under the GEOGRAPHICAL LOCATION of each occurrence. The frequent appearance of certain models should not be taken as a comment on their reliability: mishaps have been chosen for discussion only to illustrate certain aspects of flight safety.